To Mom & Dad
June 1997
Love, Paul

EXPERIENCING NATURE

THE UNIVERSITY OF WESTERN ONTARIO SERIES IN PHILOSOPHY OF SCIENCE

A SERIES OF BOOKS
IN PHILOSOPHY OF SCIENCE, METHODOLOGY, EPISTEMOLOGY,
LOGIC, HISTORY OF SCIENCE, AND RELATED FIELDS

VOLUME 58

"To a great experience one thing is essential,
an experiencing nature."
Walter Bagehot

Allen G. Debus
(Courtesy of Allen G. Debus)

EXPERIENCING NATURE

Proceedings of a Conference in Honor of Allen G. Debus

Edited by

PAUL H. THEERMAN

Smithsonian Institution,
Washington, D.C., U.S.A.

and

KAREN HUNGER PARSHALL

University of Virginia,
Charlottesville, VA, U.S.A.

KLUWER ACADEMIC PUBLISHERS

DORDRECHT / BOSTON / LONDON

A C.I.P. Catalogue record for this book is available from the Library of Congress.

ISBN 0-7923-4477-4

Published by Kluwer Academic Publishers,
P.O. Box 17, 3300 AA Dordrecht, The Netherlands.

Kluwer Academic Publishers incorporates
the publishing programmes of
D. Reidel, Martinus Nijhoff, Dr W. Junk and MTP Press.

Sold and distributed in the U.S.A. and Canada
by Kluwer Academic Publishers,
101 Philip Drive, Norwell, MA 02061, U.S.A.

In all other countries, sold and distributed
by Kluwer Academic Publishers Group,
P.O. Box 322, 3300 AH Dordrecht, The Netherlands.

Printed on acid-free paper

Printed in the Netherlands

TABLE OF CONTENTS

ESTABLISHING AN HISTORIOGRAPHICAL TRADITION

INTRODUCTION

Paul H. Theerman and Karen Hunger Parshall

These papers, by Allen G. Debus, his students, and his adviser, I. Bernard Cohen, represent the proceedings of a conference held at the University of Chicago on 30 October 1991 to celebrate a triple anniversary—Allen Debus's sixty-fifth birthday, the thirtieth anniversary of his coming to the University of Chicago, and the twenty-fifth anniversary (plus one) of the publication of his celebrated work, *The English Paracelsians.* Coincidentally, it also marked the publication of his then most recent major study, *The French Paracelsians.* These books serve as benchmarks to an ongoing career, which this conference and these proceedings honor and acclaim.

In his scholarship, Allen Debus has consistently challenged traditional historiographical interpretations, a fact which shows through not only in his works on the Paracelsians in seventeenth-century England and in eighteenth-century France, respectively, but also in his numerous other publications. He has shown vividly and convincingly that the facile interpretation of an inevitable spread of the "Age of Reason," motivated and sustained by the empirical sciences, simply does not hold. He has enlarged this older interpretation through his demonstration of the persistent and important influence of Renaissance ideas—especially in the realm of religion and nature—well into the modern age. Two key historiographical lessons thus emerge from his work: simple typologies do not suffice for complex processes; and no substitute exists for the careful determination of the discrete and concrete ways in which ideas grow, change, and persist through time. Part of the intellectual delight that Allen Debus's work provides, lies in the rich historical analyses arising from his masterful application of these tenets.

Indeed, these two historiographical guides thematically unite the essays in the present volume. Far-removed from their original historical contexts, through agents who may not even be associated with science, ideas of science—"experiences of nature"—may exert profound and unexpected influences. Recent studies have tended to illuminate this process in two complementary ways: by focusing on changing intellectual contexts and by examining differing social milieus. To be sure, the more traditional, discipline-oriented approach to the history of science, in which theoretical ideas are traced and linked, sheds much light on the internal development of specific scientific fields. For example, viewed strictly as a work in the "new astronomy," Johannes Kepler's *Harmonices mundi* of 1619 yields a wealth of insights into its author's mathematical and astronomical thought when subjected to close textual and technical analysis.

Yet, viewed as a product of Kepler's larger intellectual milieu—one encompassing aspects of neo-Platonic and neo-Pythagorean thought—the text reveals underlying motivations and styles of argumentation that more purely mathematical and astronomical examinations may obscure or overlook. When ideas are borrowed from other disciplines or areas of thought and made to impinge on unaccustomed domains, potentially fruitful intellectual cross-fertilization can occur. Perspectives may be meaningfully narrowed or productively broadened with fresh insights as the result.

Bernard Cohen, Debus's major professor at Harvard, explores precisely this sort of cross-fertilization of ideas in his essay, given as the conference's keynote address. Broadly speaking, he examines the potency of concepts in the natural sciences—predominantly physics and physiology—as explanatory ideas in the social sciences. More specifically, Cohen investigates the body politic as modeled along physicalist lines by writers as diverse as William Harvey and Sigmund Freud. He focuses on the way in which different parts of theories cross disciplinary boundaries. In so doing, he tracks not only individual metaphors and structural analogies, but also the belief that the laws of society are, in fact, the very laws of physics and biology writ large. With great breadth of learning, Cohen illustrates the rule that ideas may have influence and consequence far beyond their initial points of departure.

Similarly, Michael and Phyllis Walton examine how ideas from one realm penetrate another and thereby acquire new and deeper meanings. They take, as their case in point, Kaballah, a method of close textual interpretation integral to the Jewish mystical tradition, which entered the intellectual mainstream during the Renaissance. The Waltons demonstrate how the notion of Kaballah—as a potent symbol of the mathematical unities of nature—motivated the ideas of two disparate thinkers: the Elizabethan mathematical magus, John Dee, and the idiosyncratic astronomer, Johannes Kepler. They underscore the interplay between purely mystical and more modern applications of mathematics.

Although different in time and topic, Robert Richards's study probes yet another aspect of the role of intellectual context in the development of scientific thought, namely, its effects on the first formulation of an individual scientist's theories. Standard accounts invoke Charles Darwin as the destroyer of natural theology, that is, the belief that the natural world demonstrates the wisdom and purposes of the creator. While Darwin's later thought posits an imperfect and contingent world antithetical to natural theology, Richards shows that Darwin hit upon natural selection using key assumptions of that older belief system. Indeed, natural selection took on certain attributes of God; Darwin made the doctrine, in a naturalistic way of course, the author of human morality, an altruistic morality quite different from the simple utilitarianism that dogged his reputation.

Karen Parshall also considers the fruitful conceptual cross-fertilization between scientific disciplines—specifically between invariant theory in mathematics and the theory of molecular structure in chemistry—in the research of the nineteenth-century mathematician, James Joseph Sylvester. Although chemists did not find useful the mathematical structures he developed, the perceived interrelations reflected, to Sylvester's mind at least, the fundamental unity of the mathematical and the natural sciences. Parshall isolates in Sylvester's positivistic philosophy the motivating force behind this novel and unexpected reciprocal application of mathematics and chemistry. She further explores the ideological uses Sylvester made of and envisioned for his unification of these two disparate areas.

As the papers of Cohen, the Waltons, Richards, and Parshall demonstrate, examination of the broader intellectual context in which scientific ideas emerge, mature, and interact reveals the richness and complexity of our attempts fully to experience and understand nature. The interaction of ideas, however, implies the interaction of people within their various social milieus. Just as an idea from one discipline may unexpectedly spark a discovery in a seemingly unrelated area, so the perspective of members of a particular social class or group may shed new light on our understanding of the pervasiveness of science and scientific thought. And the popular perception and comprehension of science represents but one aspect of the complex interrelations between science and society. Social mores, prevalent social, political, and cultural ideologies, these and other features of society may impact and be impacted by science.

In his essay, Terence Murphy gains insight into the world of French middle-class professionals of the sixteenth and seventeenth centuries by adopting just this sort of social perspective. He concentrates on the political and judicial instability occasioned by the religious civil wars in France as well as on some of its concomitant intellectual consequences: the metaphysical status of the Eucharistic host, the rising belief in sectarian uses of witchcraft, and the wavering opinion on the efficacy of the "royal touch" in curing scrofula. Murphy investigates the way in which political and legal problems acquired a medical dimension. He then shows how the restoration of political authority in France paralleled both the restabilization of the medical and physical basis of these practices and the ascendancy of medical authority in the corporate institutions of the physicians.

Other aspects of the impact of the social setting on medicine—the interplay between varied constituencies in an attempt to understand disease—define the contours of Martha Baldwin's contribution to the present volume. Tarantism, a malady resulting from the tarantula's bite, received much attention throughout Europe during the early modern period, despite the fact that it only manifested itself in a very limited geographical region and that only a handful of physicians had or would ever see an actual case of it. Baldwin documents numerous accounts of the disease reported by peasants, Catholic clergymen, learned physicians, and natural philosophers, in order to appreciate contemporary understanding of the affliction and its cure. These accounts then mark the point of departure for an analysis of the role of social class in the establishment of an "expert witness" as well as of experimental verification in seventeenth-century medicine. She concludes by showing how physicians used tarantism, a disease in the abstract for virtually all the commentators, as a platform for the espousal of their own particular natural philosophical views.

The Bureau d'Adresse, a highly eclectic intellectual clearinghouse sponsored in the mid-1600s by the growing bureaucracy of the Ancien Régime, reported on tarantism as well as many other widely ranging topics. In her essay, Kathleen Wellman examines this institution as a mirror reflective of the extent and depth of the diffusion of ideas of nature associated with the Scientific Revolution. The constituency of the Bureau d'Adresse, middle-class professionals who would increasingly come to exert power in the modern state, readily embraced an eclectic mix of Aristotelian, animistic, and mechanical notions. While not rigidly systematic, these ideas were easily adapted to support the authority of the emergent absolutist state. Wellman thus suggests rethinking the supposed connections between Cartesianism and absolutism.

Paul Theerman investigates the curious life of scientific ideas in still another venue by focusing on the phenomenal American career of the British popularizer and physics lecturer, Dionysius Lardner. Lardner's success in the 1840s belies the notion that Americans were indifferent or antagonistic to science. Instead, when presented with sufficient theatricality, when tied to public morality and progress, when backed by influential, reform-minded supporters such as Horace Greeley, scientific ideas reached a substantial proportion of the American population through lectures and their publication. Theerman's study points to the importance of such cultural concepts as "the sublime" in explaining the engagement of the popular imagination with scientific ideas.

This book closes with a perspective on the history of science different from those of the papers just described. In his essay "From the Sciences to History: A Personal and Intellectual Journey," Allen Debus provides an overview of his career in the field. He describes not only his development from a practicing chemist to a scholar of Renaissance science but also the concurrent development of the program in the history of science and medicine at the University of Chicago. Through his teaching and example, Debus has inspired a generation of students. Like him, these scholars have seen little inevitable in the way that modern science and its concepts have developed. They have taken up the challenge to follow the ideas wherever they lead. The corpus of Debus's work may thus be found not only in the bibliography which concludes the present volume but also in the historical research of his students, past, present, and future.

Acknowledgments

For assistance in the production of the conference and its proceedings, the authors and editors acknowledge the gracious aid provided by Michael and Phyllis Walton; the Morris Fishbein Center for the History of Science and Medicine at The University of Chicago and its director (from 1980 to 1992), George Stocking; the Joseph Henry Papers Project and the Smithsonian Institution Archives; the University of Virginia; and, especially, Elizabeth Bitoy of the Department of History and the Fishbein Center at the University of Chicago.

Three other former students also made presentations at the conference: Regis Cabral, "The Brazilian Way of Perceiving Nature: The 1925 Relativity Debate"; Ronald Calinger, "Leonard Euler: The First St. Petersburg Years"; and Sherrie Lyons, "Sea Monsters, Charles Lyell, and the History of Life on Earth." Their papers, however, were unavailable for publication here.

The Body Politic

before and after the Scientific Revolution

I. Bernard Cohen

Ever since the great revolution which produced modern science, there has been a hope that a science of society would be created on a par with the sciences of nature.[1] This new science of society, it has been supposed, would make use of the concepts, principles, theories, and methods of the physical and biological sciences. In this presentation[2] I shall explore some attempts to achieve this goal in relation to the body politic and, more generally, the organismic conception of the state.

The body politic is a name given to a comparison between the political system or the system of the state or of society and a living organism. After many centuries of continual use, this line of thought has left its mark in such common expressions as "head" of state, "nerves" and sinews" of government, a "healthy" economy, "ills" of society and their "cure," social "pathology," and many others.[3] My own concern for this topic is part of a general research program, of which the focus is a historical and critical analysis of the different ways in which the social sciences have interacted with the natural sciences.[4]

The central role of scientific discovery in the concept of the state arose early in the Scientific Revolution, when William Harvey in 1628 formally announced his radical reconstruction of animal and human physiology. Harvey differed from Descartes and Galileo in that he conceived his major discovery to have a direct paradigmatic value in the domain of social affairs. In introducing his great work *De Motu cordis,* in which he displayed his discovery of the circulation of the blood, Harvey used his new science of the body to transform the old notion of the body politic. This dramatic example of the use of the new science in a socio-political context occurs at the very beginning of the book, in the long and flowery dedication to the reigning monarch, Harvey's patron and patient, Charles I. The following passage expresses Harvey's view unambiguously:

> The Heart of all creatures is the foundation of their life, the Prince of all their parts, the Sun of their microcosm, that on which all growth depends and from whence all strength and vigour flows. In like manner the King is the foundation of his kingdom, the sun of his microcosm, the heart of his commonwealth, from whom all power flows and all mercy proceeds.[5]

P. H. Theerman and K. Hunger Parshall (eds.), Experiencing Nature, 1–41.

Harvey accepted as "fact" that "almost all our concepts of humanity are modeled on our knowledge of man himself, and several of our concepts of royalty on our knowledge of the heart." Accordingly, he wrote, "the knowledge of his own heart" must be of service to the king as "a divine example of his own actions," that is, a kind of portrayal, if on a smaller scale, of his functions, in the way that "men have ever been wont to compare great things with small."[6] Since King Charles was placed "at the summit of human affairs," he would be able, Harvey told him, "to contemplate simultaneously both the central organ of man's body and the likeness of [his] own royal power."[7]

The ancient organismic analogy of the "body politic," in which the state is compared to an animal or person, had long considered the sovereign as the head ruling the body. A few writers on the body politic prior to Harvey had given some importance to the heart, but in a framework of Aristotelian or Galenic thought. Thus, in 1565, John Halle had written that we might "call the heart of a man a king"; he added that the liver would then be one of "the chief governors under him,"[8] a reference to the Galenic principle that the liver is constantly generating new blood from digested food and sending it through the body to the heart. But in Harvey's system the ruling power of the heart became absolute because the blood is made to circulate as a result of the pumping action of the heart, rather than being constantly generated by the liver.[9]

Harvey thus transformed the traditional organismic analogy of the state (the "body politic") in the context of his own discoveries, creating a new socio-political expression of scientific discovery. Since he did so in the book disclosing his discoveries, he effectively gave sanction to further explorations of political systems based on the new human science. The inaugurator of modern physiology introduced his founding treatise with a bold declaration that true or experimentally based science may serve to better our understanding of the functioning of society, of the organization of the state. I know of no similar statement by any other founder of the new science. Such a sentiment would perhaps come more naturally to a Harvey than to a Galileo or a Kepler because the fabric of the human body shows the same kind of organization of complex and varied interacting parts that is found in socially organized humanity.

Lest it be thought that Harvey introduced the body politic only in the dedication, let me hasten to add that this theme appears again in the text itself, in the very important chapter seventeen, in which Harvey confirms the "motion and circulation of blood" by reference to the phenomena of the heart and the evidence of anatomy. The heart is the first organ to appear in a complete form in the embryo, Harvey wrote, and it "hath blood, life, sense, and motion before the brain or liver were made" and is "a sort of internal animal."[10] The heart, furthermore, "is as it were a Prince in the Commonwealth, in whose person is the first and highest government every where" and from which, "as from the original and foundation, all power in the animal is derived, and doth depend."[11]

We may discern the influence of Harvey in the writings of Thomas Hobbes, who was actually a legatee of Harvey's.[12] Hobbes's science of politics or of society had the dual aim of incorporating two new sciences—the new science of motion and the new physiology.[13] Hobbes was magnificently vain about this achievement. He deserved, he wrote, "the reputation of having been the first to lay the grounds of two new sciences": one "of *Optiques,* the most curious, and that other of *Natural Justice.*"[14] He boasted of his accomplishment in another declaration, where he began with a listing of the founders of the new natural science, beginning with "Galileus [who] in our time . . . was the first that opened . . . the gate of natural philosophy universal, which is the knowledge of the nature of motion." He ended with "the science of man's body," which he held to be "the most profitable part of natural science . . . first discovered with admirable sagacity by our countryman Doctor Harvey."[15] "Natural philosophy is therefore but young," he concluded; yet "civil philosophy is yet much younger, as being no older . . . than my own book *de Cive.*"

Fully convinced that a science of politics or of human society must be similar to a natural science, Hobbes based his social philosophy on the two primary concepts of the "mechanical philosophy": movement and matter or substance.[16] In transferring the science of motion from the inorganic to the organic world, Hobbes drew heavily on the discoveries of William Harvey, which must have had a special significance for him insofar as they were based on mathematics, i.e., on quantitative considerations. The circulation of the blood, the "vital motion" for Hobbes, became the very principle of life, so that the "original of life" was said by him to be "in the heart," which he described as being like a great "piece of machinery in which . . . one wheel gives motion to another."[17]

In a very real sense Hobbes's political system announces the demise of the traditional organismic metaphor in which the state had been considered to be like an animate being rather than a machine. Harvey's physiology, reinforced by the Cartesian mechanical philosophy, enabled Hobbes to envision the animate functions of a living being in terms of the operation of a complex mechanical device. He thereby transformed the old concept of the body politic from a purely animate status to that of a great animal machine, acting like an animal but composed of mechanical parts.[18]

Drawing directly on Harvey's comparison of the heart to a pump and of the circulatory system to a hydraulic network of pipes or conduits, Hobbes set forth—on the very first page of the Introduction to *Leviathan*—the analogy between a machine and an animal or human body. "The *Heart,*" he declares, is nothing "but a *Spring;* and the *Nerves,* but so many *Strings;* and the *Joynts,* but so many *Wheels,* giving motion to the whole Body." He then compares the state or commonwealth, which is "but an Artificiall Man," to a "Naturall" or biological man.

Like all other writers on the body politic, Hobbes sought for specific analogies. Thus in society, corporations are the muscles, public ministers are the organs or nerves, and the problems of the state are the diseases.[19] Yet we are not to make too much of this type of metaphor, since Hobbes warns us that the would-be scientific political scientist must abandon "the use of metaphors, tropes, and other rhetorical figures, instead of words proper."[20] His own Leviathan, or the commonwealth, is not a real natural being, but rather "an Artificiall Man," the creation of the human mind in what has been called an "imitation of God's art of creation." And so the "body politic" is nothing more than a "fictitious body" and its faculties and will are correspondingly "fictitious."

Whereas Hobbes attempted to introduce some aspects of the life sciences into a system of political thought based primarily on the physical science of motion, his somewhat junior contemporary John Harrington took a quite different tack. Consciously rejecting Hobbes's methodology, he based a socio-political system squarely on the new Harveyan biology without any dependence on the new science of motion of Galileo and Descartes.[21] Harrington's work is all the more significant in that he was "a political thinker of quite unusual power and independence," said to have been "the only observer of the Puritan Revolution who had any grasp of the social causes behind it."[22] Furthermore, Harrington was ultimately more influential in the sphere of practical politics than Hobbes since his doctrines were implemented in the following century, notably in the form of government adopted in the American Constitution, including such features as a bicameral legislature, elected officials, indirect election of the executive, and much else.[23] During the years of the American Revolution and the Constitutional Conventions, many American statesmen were aware that the concept of a "balance" in a socio-political context should be credited to John Harrington.

Harrington's social and political ideas are set forth primarily in *The Common-wealth of Oceana* (London, 1656), in which the island of

Oceana is a thinly disguised Britain. This work has been described as a "constitutional blueprint."[24] It is at once obvious to any reader of *Oceana* that Harrington was a great admirer of William Harvey, since he declared that his own work was a "political anatomy," an analogy of "the Harveyan anatomy of the animal body."[25] He firmly believed that his dissection of the problems of his age, together with his remedy in proposing new political institutions, constituted more than the traditional sort of historico-political analysis, but was an exact counterpart to the physiological anatomy of William Harvey. The construction of "a Model of Government," he wrote, must "embrace all those Muscles, Nerves, Arterys and Bones which are necessary to any direction of a well-order'd Commonwealth" and is to be likened to "the admirable structure and great fabric of the parts of a Man's Body" as revealed by "the Discoveries of Anatomists."[26] For Harrington this position implied that the political anatomist, like the physiological anatomist, must base his subject on the principles of nature, and not merely on one or two examples. William Harvey, he wrote, did not found his discovery of the circulation of the blood on "the *Anatomy* of this or that body" but rather on "the *Principles* of Nature."[27]

Harrington's appreciation of the Harveyan physiology was not limited to generalities, but invoked specific features of the new biological science. For example, in discussing the two chambers of his proposed legislature, Harrington drew directly on Harvey's *De Motu cordis,* arguing that "the parliament is the heart," which acts like a suction pump, first sucking in and then pumping out the "vital blood of *Oceana* in a continuous process of circulation."[28] His discussion of the different functions of the two chambers of his proposed legislature, based on the biology of the circulation, indicates his appreciation of Harvey's central idea that the heart is a pump. He even followed Harvey in using the mechanistic language of pump technology, and his concept of a continual process of blood circulation is clearly Harveyan. Thus for Harrington "the parliament is the heart," with its "two ventricles, the one greater and replenished with a grosser matter, the other less and full of a purer [matter]," which "sucks in and spouts forth the blood of *Oceana* by a perpetual circulation."[29] It was in terms of the anatomy and physiology of the heart that Harrington set forth in analogy his concept of a two-chambered legislature. He even called attention to the differences in quality between the blood emitted by the two ventricles as an analogy of the difference in function of the elected chamber of the legislature and the appointed senate.

Harrington's "comparison of the study of politics with anatomy," to quote Charles Blitzer, "was not simply a casual simile," but rather "represented a reasoned belief in the basic likeness of the two disciplines."[30] Both the body politic and the human body are composed of similarly interlocking machine-like structures which function in a coordinated manner. Harvey had achieved great success in studying the human body by the method of experience and reason; surely the political anatomist might hope for results of significance from the application of a similar method. In reply to a criticism leveled by Matthew Wren, Harrington reminded his critic that "anatomy is an art; but he who demonstrates by this art, demonstrates by nature."[31] So, is it "in the politicks," he wrote, "which are not to be erected upon fancy, but upon the known course of nature," just as anatomy "is not to be contradicted by fancy, but by demonstration of nature."[32] It is "no otherwise in the politicks," he concluded, than in anatomy.[33] In short, the study of politics and the study of anatomy are alike because they both seek principles of nature by reason and experience.

Harrington's knowledge of Harvey's science was not limited to the discoveries concerning the circulation of the blood, reported in Harvey's *De Motu cordis* of 1628. Harrington had also studied the work reporting the research activity which occupied Harvey for the greater part of his life, the study of embryology. Harvey's philosophy of animal generation was encapsulated in the allegorical frontispiece drawn for his work, *De Generatione animalium* (1651), showing Zeus opening a pyxis or egg-shaped box from which a variety of animals spring forth. On the egg there is inscribed boldly *Ex ovo omnia!*[34] These words do not occur as such in Harvey's text but they encapsulate his philosophy of generation, that "All things come from an egg." In a day when scientists tended to believe in some variety of preformation,[35] Harvey championed epigenesis and spent most of his research on an attempt to understand mammalian generation in terms of a fertilized ovum.

Not only was Harrington acquainted with the main findings of *De Generatione;* as a true student of Harvey, he made good use of this aspect of Harvey's science in his political thought. In a posthumously published work entitled "A System of Politics," Harrington wrote: "Those naturalists that have best written of generation, do observe that all things proceed from an eg."[36] Here is a direct English translation of the Latin motto in the frontispiece to Harvey's book. In this essay Harrington showed that he had more than just a general notion of epigenesis; he described aspects of the development of the fetus in the egg of a chick in some detail, following the line of Harvey's discoveries, and he used these facts of embryology as the basis of a political analogy.

* * *

Harrington began his presentation of embryology with a discussion of the "punctum saliens" or primordial heart of the chick:

> 1. Those naturalists that have best written of generation, do observe that all things proceed from an eg, and that there is in every eg a punctum saliens, or a part first mov'd, as the purple speck observ'd in those of hens; from the working whereof the other organs or fit members are delineated, distinguished, and wrought into one organical body.[37]

The "punctum saliens" had been known to embryologists long before Harvey and was considered the starting point of life, the embryonic heart. Aristotle found–as Harvey recorded–that the "punctum saliens" moved or "leaped" when physically stimulated.[38] Aristotle and later embryologists believed that the heart was the first organ to be formed in the development of the chick embryo and that the blood was formed later, after the appearance of the liver. It was a feature of the reigning Galenic physiology in Harvey's day that the blood is manufactured by the liver and so could not exist antecedent to the liver. But Harvey demonstrated by careful experiment that in the chick's egg the blood begins its existence before any organ such as the heart or liver takes form. Harvey's studies showed that in the early stages of development of the hen's egg there appears a little reddish purple point "which is yet so exceedingly small that in its dilation it flashes like the smallest spark of fire, and presently upon its contraction quite escapes the eye and disappears." This red palpitating (or salient) point, the "punctum saliens," was seen, he took note, to divide into two parts, pulsating in a reciprocating rhythm, so "that whilst one is contracting, the other is swollen with blood," the "which being presently contracted discharges itself of the blood that was in it" and so "in a continual reciprocating motion."[39] The "punctum saliens," Harvey reported, could be seen to develop into the heart after the blood had begun to form. Harvey's conclusions have been summed up as follows. The "blood exists before the pulse" and is "the first part of the embryo which may be said to live"; from the blood "the body of the embryo is made," that is, from it "are formed the blood vessels and the heart, and in due time the liver and the brain."[40] Harrington's paragraph number one summarizes Harvey's embryological findings concerning the "punctum saliens" and the way in which the organs develop from it.[41]

In the eighteenth century and early nineteenth centuries, there arose a considerable literature in which the new physiology was applied to concepts of society or to a social or political science. Jean-Jacques Rousseau wrote extensively on the body politic, a concept which he transformed in a number of different ways, for example, by stressing the notion of the death of the body politic.[42] As Judith Shklar has noted, unlike Hobbes's *Leviathan*, with its "'artificial eternity' which only unruly subjects and conquest could end, Rousseau's second body politic was born to die," not because "the people disappears, but because the general will is subverted by the magistrates, and justice and equality are destroyed." The body politic, she finds, "was of all metaphors of personification the one Rousseau used most frequently." Sometimes, as in his "Discourse on Political Economy," published in 1755 in volume five of Diderot's *Encyclopédie*, Rousseau went "into full anatomical detail." In this instance, while introducing the concept of the body politic, he said he was making use of "a common comparison, inaccurate in many respects, but suited to making myself better understood."[43] Concluding that the body politic is "a moral being which has a will," Rousseau set forth his analogy as follows:

> The body politic . . . can be considered as an organized, living body and similar to that of man. The sovereign power represents the head; the laws and customs are the brain, the center of the nervous system and seat of the understanding, the will and the senses, of which the judges and magistrates are the organs; commerce, industry, and agriculture are the mouth and stomach which prepare the common subsistence; public finances are the blood that a wise *economy*, performing the functions of the heart, sends back to distribute nourishment and life throughout the body; the citizens are the body and members which make the machine move, live, and work and which cannot be injured in any way without a painful sensation being transmitted right to the brain, if the animal is in a state of good health.[44]

It is to be observed that in the midst of a series of organismic analogies, Rousseau suddenly introduces the concept of society as a "machine."

Possibly the best known of all eighteenth-century applications of physiology to the social sciences is the analogy of the circulation of the blood that appears in the economic thought of François Quesnay, the founder of the school of *physiocrates*. Quesnay, said by Adam Smith to be "one of the worthiest men in France and one of the best Physicians that is to be met with in any country,"[45] first attracted public attention through a publication on blood-letting, which was based on his medical knowledge of the physiology of the circulatory system.[46]

An extreme example of a simple transfer of the physiology of the circulatory system to economics appears in the writings of Nicolas-François Canard. His system (1801) has been summarized by Claude Ménard as follows:

> In the giant organism of society, merchants occupy the center of the circulatory system, where the shop and the cash desk form the two ventricles. Labor and its products are represented by the arterial system, whilst monetary circulation is associated with the venous system. Equilibrium is simply the state of a mature social body; the circulation of the blood pulses in the active rhythm of procreation; the mother country literally engenders the colonies.[47]

Joseph Schumpeter believed that Canard's book would have partaken "of the blessings of deserved oblivion, had not a misfortune befallen it." This "misfortune consisted of its being 'crowned' by the same French Academy that later on failed to extend any recognition to Cournot and Walras." Those "Olympians who felt their neglect the more bitterly on account of the honor done to Canard visited him with a scathing contempt that bestowed upon him an unenviable immortality," making Canard "forever sure of a place" in "the history of scientific bodies." Schumpeter adds that Canard's book is, however, "far from being the worst that was ever written."[48]

* * *

Before proceeding to some later advances in the biological sciences and their influence on the body politic, and on organismic models of society, let me first introduce a distinction concerning levels of discourse. In my research I have found it useful to distinguish four separate such levels. The extremes are identity and metaphor, with analogue and homologue occupying intermediate positions. I shall first introduce analogue and homologue, reserving for later the problems of identity and metaphor.

The word "analogy" is generally used today for many kinds of similarity, but in the natural sciences, analogy is used to denote an equivalence of likeness of functions or of relations or of properties. Thus David Brewster (1843) wrote about waves or undulations as "a property of sound which has its analogy also in light."[49] In 1860, John Tyndall referred to "[t]he analogy between a river and a glacier moving through a sinuous valley."[50] In similar fashion John Lubbock observed that there "seem to be three principal types" of ants, thus "offering a curious analogy to the three great phases" in "the history of human development": "the hunting, pastoral, and agricultural."[51]

This particular sense of analogy is of special significance in writings on natural history: for instance, to express a similarity in function between organs "which are *essentially* different in different species." An example is the wing of a bird and the wing of a bat. Both types of wings enable their possessor to fly. Although they perform the same function, one is covered with feathers and the other is a stretched skin membrane.

In the language of the life sciences, the term "homology" (along with homologue and homologous) has a specific meaning which is quite distinct from that of analogy: to denote similarity in form as distinguished from similarity of function. The distinction becomes apparent once attention is focussed on bone structure (anatomical construction). An anatomical comparison shows that structurally the wings of a bat resemble the wings of a bird and are structurally similar to the forelegs of quadrupeds and the arms of humans. Accordingly, the wings of a bat and a bird are homologues of the arms of humans, flippers of seals, and the forelegs of quadrupeds.[52]

I use the terms analogy and homology to denote, respectively, the sense of similarity in function and the sense of similarity in structure. This distinction helps to specify the ways in which the social sciences have used the natural sciences, and equally the ways in which the natural sciences have used the social sciences. The same features may be seen in the ways in which the different natural sciences have made use of one another. These two concepts will not be taken to be mutually exclusive. That is, I shall use the term homologue for structurally equivalent concepts or structurally equivalent laws which themselves may have analogous functions.

The use of homologues appears as a prominent feature in nineteenth-century and even twentieth-century organismic sociology. Let me give two examples from very different cultures. The first in Thomas Carlyle's analysis of the problems of society is *Sartor Resartus* (1836). Here Carlyle introduces the social analogy of the skin. Note, however, that he does not merely present a general function of the epidermis as social analogue, but gives a detailed homology:

> For if Government is, so to speak, the outward SKIN of the Body Politic, holding the whole together and protecting it; and all your Craft-Guilds, and Associations for Industry, of hand or of head, are the . . . muscular and osseous Tissues (lying under such SKIN), where-by Society stands and works;—then is Religion the inmost Pericardial and Nervous Tissue, which ministers Life and War, providing Circulation to the whole. Without which Pericardial Tissue the Bones and Muscles (of Industry) were inert, or animated only by a Galvanic vitality; the SKIN would become a shrivelled pelt, or fast-rotting rawhide; and Society itself a dead carcass,—deserving to be buried.[53]

Carlyle appears to have been obsessed with such organismic analogies drawn from the realms of anatomy and medicine. For him, England was "in sick discontent," writhing "powerless on its fever bed," and the evils of his contemporary world were for him a kind of "Social Gangrene."[54]

Another nineteenth-century social thinker who made extensive use of extravagant organismic analogies was Johann Caspar Bluntschli (1808–1881), a Swiss-German jurist who spent a number of years as a Heidelberg professor. Author of many books on the state and on society, his major work on *The Theory of the State* (1851–1852) was in such demand that six editions were called for in the next three decades and an English translation was published by the Oxford University Press in 1895.[55] Convinced that both the state and the church are organisms similar to human beings, Bluntschli quite logically concluded that they both must have all the primary human attributes, including sexual characteristics, the state representing "the male, the church the female element." This attribution of sex led him to a theory of history, based on social-sexual development, in which the historical "evolution" of society and the state have followed the pattern of "evolution" of single individuals. Tracing the sexual history of church and state from childhood (the ancient Asiatic empires) through adolescence (the Jews of Biblical times) to early maturity (classical Greece), he found that in Greece, "the ecclesiastical organization" matured earlier "than the political institution," just as "the girl ripens earlier than the boy."[56] So extreme is Bluntschli's mismatched homology that it is difficult to imagine that he was developing a social analogue of Greek religion when he went on: "The sexual organs of the girl are sooner developed than those of the boy. The youthful breasts begin to swell; and the unfolding virgin turns into a beauty. Beauty was the soul of the cult of the Hellene . . . ." Bluntschli's sexism even led him to assert that the papal desire to subordinate the state to the church is as "unnatural" as "the subordination of a husband to his wife in a household."[57]

A similar extravagance occurs in the organismic conception of society proposed by the Russian sociologist, Paul von Lilienfeld (1829–1903). Lilienfeld was a Russian civil servant and social theorist who wrote extensively in Russian, German, and French. At the end of his life, Lilienfeld moved to Paris where he became (in 1897–1898) the President of the Institut International de Sociologie. In a late work on *Social Pathology* (1896), Lilienfeld proposed an analogy between the intellectual and moral state of a hysterical woman and a condition of society.[58] Quoting Dr. Edmond Dupouy, Lilienfeld described the condition of women suffering from hysteria.[59] They are, he noted, "mobile in their sentiments," and "they pass very easily from tears to laughter, from excessive joy to sadness, from passionate tenderness to haughty rage, from chastity to wanton purposes and lewd ideas." Additionally such women "love publicity, and to get themselves talked about they employ every means: denunciation, simulation of infirmities, or sicknesses, and the revolver." They find joy in pretending to be "victims of anything; they say they have been violated." In order to "achieve their goals they deceive everyone: husband, family, confessor, examining magistrate, and their doctor."[60]

The reader who is uninitiated in the literature of organismic sociology may wonder what social manifestations could possibly present analogous symptoms. Lilienfeld develops the comparison by a series of homologues. He begins by asking rhetorically whether the symptomatic behavior of women suffering from hysteria is not "perfectly analogous to the manner in which the population of a large city behaves during a financial crisis or on the occasion of civil disturbances." He finds in the behavior of such women "a faithful picture of the agitation of parties during elections." And when we consider the past, he asks, do we not find the same confused and disordered pattern of behavior, "caused by convulsive and contradictory reflexes of the social nervous system," as a feature of "all the religious, economic, and political revolutions with which humanity has been assailed?"[61] This complex nesting of mismatched homologies needs no comment.

In each of these examples, the author was so overcome by the power of his analogy that he developed absurd homologies. There is not a perfect fit. Such extravagances in homology suggest the general name of "fallacy of mis-matched homology" to anyone who is familiar with Alfred North Whitehead's concept of the fallacy of misplaced concreteness.[62] Such mismatched homology is a characteristic feature of many of the attempts to produce organismic analogues of society and may be observed in the writings of such diverse authors as Thomas Carlyle, Johann Caspar Bluntschli, Paul von Lilienfeld, Albert E. Schäffle, René Worms, Herbert Spencer, Walter B. Cannon, A. Lawrence Lowell (president of Harvard University) and Theodore Roosevelt (president of the United States).

Herbert Spencer indulged himself in analogies and homologies. An extreme example, which even his sympathetic biographer must admit is a case of "dubious biology . . . added to pedestrian sociology," is Spencer's likening of "the coalescence of the Anglo-Saxon kingdoms into England" and the formation of crustaceans. Here he was introducing his own rather odd notion that crustaceans, like insects, are "composite animals," in which the segments are independent life-units joined together.[63] Although organic analogies permeate Herbert Spencer's writings on sociology, he also drew heavily at times on analogies from engineering and the physical sciences. Some examples of extreme homologues are his comparison of "the undifferentiated and fragmented structures of Bushmen" and "the protozoa"; and "the ruling class, the trading or distributive classes, and the masses" and "the mucous, vascular and serous systems of the liver-fluke."[64] Perhaps the limit is reached when he refers to the two great national schools of France— the École Normale and the École Polytechnique—as "a double gland" intended "to secrete engineering faculty for public use."[65]

Walter B. Cannon differs from Herbert Spencer in that Spencer was a self-educated philosopher and social thinker making use of biology, whereas Cannon was one of the leading physiologists of his day, nominated several times for a Nobel Prize in Medicine. Cannon's first essay in biological sociology (1932) was titled "Biological and Social Homeostasis," an exploration of whether analogues of the "stabilizing processes" in animal organisms can be found in "other forms of organization—industrial, domestic or social."[66] In a manner reminiscent of Herbert Spencer and other nineteenth-century organicists, Cannon compared the circumstances of small groups of human beings living in "primitive conditions" to the "life of isolated single cells," and the grouping of "human beings . . . in large aggregations" to cells "grouped to form organisms."[67] Only in highly developed organisms, he reported, do the "automatic processes of stabilization" work "promptly and effectively." The comparison seemed to him to show that our present social system resembles organisms low on the evolutionary scale or organisms that have nor fully developed, in both of which "the physiological devices which preserve homeostasis are at first not fully developed."

Cannon's goal was to find an "equivalent" in "a state or nation" of the "fluid matrix of animal organisms." And it is here, in the suggestion of a homology that Cannon reveals the naive quality of his social thought. In the social body, he wrote, the equivalent ("in a functional sense") of the fluid matrix for maintaining homeostasis in the living body, is

> the system of distribution in all its aspects–canals, rivers, road and railroads, with boats, trucks and trains, serving, like the blood and lymph, as common carriers . . . [on which] the products of farm and factory, of mine and forest are borne to and fro.[68]

A decade later, Cannon returned to this topic, choosing it as the subject of his presidential address to the American Association for the Advancement of Science, delivered in December 1941.[69] In preparing the new version, Cannon sought help and advice from a young Harvard colleague, the sociologist Robert K. Merton, who sent him a list of books and articles on the subject of society as an organism. Cannon now withdrew his earlier analogies between cells and human members of society, and he declared that comparisons of "the body physiologic and the body politic" have been discredited in the past because they mistakenly had concentrated on "minutiae of structure."[70] We are "not illuminated," he said, "by a likening of manual laborers to muscle cells, manufacturers to gland cells, bankers to fat cells, and policemen to white corpuscles."[71] He, nevertheless, persisted in displaying extreme examples of mismatched homology.

Cannon's essay into sociology illustrates the danger of using analogies. On the level of general analogy, his suggestion that society might resemble an organism could be original and instructive, at least insofar as suggesting that the stability of a society is caused by certain self-regulating mechanisms. We may agree with Robert Merton, however, that Cannon made the mistake of introducing "substantive analogies and homologies between biological organisms and social systems." Merton went so far as to describe Cannon's result an "unexcelled . . . example of the fruitless extremes to which even a distinguished mind is driven." This comment is all the more significant in that it occurs in a methodological essay (on "Manifest and Latent Functions"),[72] in which Merton finds "Cannon's logic of procedure in physiology" to be a model for the sociological investigator, recommending to his readers the text proper of Cannon's book on the *Wisdom of the Body,* while warning them about "the unhappy epilogue on social homeostasis."

* * *

In considering these examples of mis-matched homology, our evaluations may be sharpened by attention to the reasons why these examples seem so outré to a critical reader of the end of the twentieth century. Why do we smile and assume a condescending air when we read the organicist sociologists like Bluntschli, Lilienfeld, Schäffle, Gini, and Spencer, but not when we encounter physical models such as William Stanley Jevons's lever of economics or Walras's economic machine or the numerous attempts to find in the realms of social sciences an analogue of the Newtonian universe?[73] Why are we not taken aback by Léon Walras's mechanics? Why do we give serious scholarly attention to the numerous attempts to find in the realms of social sciences an analogue of the Newtonian universe? The reason is not simply that one set are biologically based while the others come from physics. Henry Carey's attempt to produce a sociology based on electricity, a later rival to his astro-sociology, provokes smiles and giggles just as easily as the system of the organicists.[74]

I believe that the difference between the biological and the physical sources of social analogies lies in the fact that the biological analogue is usually a real object, an actual living being, endowed with all the forces of life and subject to all of life's problems: disease, aging, anxieties. By contrast, the physical analogues tend to be not concrete but abstract and theoretical. Jevons's economic analogue of the lever is a mathematical lever and accordingly does not have any physical properties such as color, hardness, weight, and so on. Even analogues made with a gravitational universe make use of abstract concepts—just as Newton did in Book One of the *Principia.* That is, in Book One there are no real planets with physical sizes, shapes, and similar physical properties but only mass points, whose only properties are position and instantaneous velocity in a mathematical space, mass, and the power to give rise to, and to be acted on by, a gravitating force.[75] Thus, unlike the earthy biological sources of analogy, those from physics are abstract and serve primarily as sources of equations and as symbols of their origin in the prestigious subject of mathematical physics.[76]

Whereas Bluntschli, Lilienfeld, Spencer, Schäffle (to some degree), Gini, and Worms (in his earlier writings) argue that society is itself an organism or is very much like an organism, the "mechanical economists" actually do no more than merely declare that there is an isomorphism between the equations of economics and those that originate in classical mechanics. The problem with the organicists' conception of society is, therefore, not that they found their analogies in living systems, but that they erred in not placing their considerations on the same plane of abstraction as did those who drew on physical analogies. They erred in their goal of creating a specific set of homologies, rather than a general analogy or metaphor.[77] It is in this sense that their procedure is very much like what Whitehead described as "the accidental error of mistaking the abstract for the concrete," mistaking the abstractions of social theory for the concreteness of an actual biological organism.

It is certainly not an error to make use of organicist metaphors or analogies in discussing society at large, the political system, or the economic system. As has been noted, we constantly use expressions deriving from the organic notion of the body politic, such as head of state, nerves of government, healthy state of society or of the economy, consumption, arteries, and many others. Even so severe a critic of organicist theories of society as Werner Stark finds himself using expressions based on animate nature. So extreme is Stark's hostility to organicism that he cannot help but pillory theories like Lilienfeld's as "ravings" and "nonsense." Even so, Stark was bound to admit that in writing about certain aspects of society, one is "constantly tempted" to make use of "organismic similes." Phrases like "one sector limps behind" or "one sector is out of joint with the rest" tend to "form themselves, as of their own volition, in one's mind, and try to push themselves into and flow out of one's pen." He concluded, "This alone shows that organicism has a deep root, and that its basic metaphor is not absurd, even if its votaries make is so."[78]

* * *

Before advancing to metaphor and identity, a word should be said about the general use of analogy in the natural and the social sciences. Analogy is not only a significant tool for concept formation and theory construction in the social sciences; analogy has also served in a number of different ways as a major tool in the advance of the biological and the physical sciences. Analogies, furthermore, constitute a primary instrument of interaction between the natural sciences and the social sciences. These interactions are very much like the interactions between one branch of the natural sciences and another. They arise from a recognition that an idea, concept, law, theory, system of equations, method of investigation, mathematical tool, or any other element of one subject is either similar in form to some element in another or has properties that enable it to be introduced usefully into that other subject. Analogy has always had the major function of being a tool of discovery, reducing a problem to another that has already been solved, or introducing some element or elements that have proved their worth in a quite different area of knowledge. Jeremy Bentham once said that hints from analogies constituted one of the most important tools available for scientific discovery.[79]

A traditional use of analogy is to be a source of justification for a novel or radical method or theory. An example would be to defend the introduction of higher mathematics (e.g., the calculus) into economics on the analogy that the calculus had been used successfully in rational mechanics. Yet another use of analogy is to make a difficult or strange idea seem reasonable, and hence acceptable to the scientific community, or to help in explaining an abstruse concept, as in all general presentations of relativity theory.

An analogy may even help to embolden a discoverer to announce a new concept or principle that may seem at first implausible. An example is seen in Sigmund Freud's hesitancy in presenting in full one of his radical and difficult concepts (introduced only as a "suspicion" in 1900 in his *Interpretation of Dreams*) that human beings have two different memory systems, one of which "receives perceptions but retains no perception of them," while the other preserves "permanent traces of the excitations" in "'mnenomic systems' lying behind the perceptual system."[80] In 1926, emboldened by his encounter with a mechanical device called the "Mystic Writing Pad," that seemed to simulate some main features of his concept, Freud described his ideas about human memory in full, suggesting that this writing pad could be considered an analogue of his "hypothetical structure of our perceptual apparatus."[81]

Analogy, we may note, was always an important part of Freud's thinking. In the *Standard Edition* of Freud's *Collected Works,* there is even a separate index of analogies. Best known of Freud's analogues are those he drew from literature, notably Greek tragedies, in formulating and describing (and even naming) concepts. Freud was consciously aware that in his cultural and anthropological studies—e.g., *Totem and Taboo* and *Moses and Monotheism*—"we are dealing only with analogies," and he fully recognized how dangerous it is, "not only with men but also with concepts, to tear them from the sphere in which they have originated and have been evolved." It was by invoking an analogy that Freud "likened religion to a collective obsessional neurosis, or allowed that Hamlet suffered unduly from an Oedipus concept."[82]

* * *

Thus far we have been considering analogies and homologies, but we have not yet addressed the general problem of metaphor. Traditionally, a metaphor is a rhetorical figure of speech, a literary device. For Aristotle, a metaphor gives something a name that belongs to something else, that does not properly belong to it.[83] Metaphor has long been used as a rhetorical device to give an enhancement to oral and written communication, an embellishment of speech and writing that increases the effectiveness with which a message is delivered. Rhetoric has usually implied eloquence on the Ciceronian model. During the Scientific Revolution of the seventeenth century, such rhetoric fell into disfavor. The advocates and practitioners of the "new philosophy" held that science should be presented in plain and descriptive terms of experiment and observation, followed by strict inductions and deductions, in which each step was to be plain and clearly understood—without any rhetorical flourishes to distract the reader from the evidence and the logic. This was one of the reasons for the great esteem given to mathematics, which is perhaps the most rhetoric-free discourse imaginable.[84]

A classic example of metaphor—in the sense of assigning a property or descriptive term to some object to which it is not strictly applicable—is the Scriptural comparison of life to a pilgrimage. Perhaps the most famous such metaphor is Shakespeare's comparison of life to a stage. Common metaphors are: heart of flint, sharp mind, head of state, and eye of the law. A striking metaphor was used by James I soon after gaining the crown of England. "I am the husband," he told Parliament, "and the whole isle is my lawful wife; I am the head and it is my body."[85]

In considering metaphors in our present context, a helpful distinction can be made between a general kind of metaphor and one that specifically invokes some element of natural sciences.[86] The difference between the two may be seen clearly by considering "body" in the metaphor of the body politic. An example of a general metaphor of the body, one that does not invoke the natural sciences, is found in 2 Corinthians, where St. Paul sets forth a crude hierarchy of organs and parts of the body—from head and heart to limbs and belly—without any reference to medicine and physiology. It is the same for the oft-repeated Aesopian fable of legs and the belly, in which the legs revolt because they believe they do all the work while the belly merely lies at ease above them doing nothing.[87] These examples may be contrasted with a statement in which James I likened the expanding metropolis of London to the spleen, "whose increase wastes the body."[88] For here he was basing his metaphor on a physician's acquaintance with the function of the spleen. That is, he was invoking an analogy of the operations of a city and the functions of a specific organ of the human body. In this context, we must remember that in those days the king performed a number of medical roles: his "royal touch," for example, could allegedly induce a cure for scrofula.[89]

* * *

Let me now say a few words about identity. "What is society?" asked Herbert Spencer; his reply was simple and unequivocal, "An organism."[90] Two others who were of the "identity" persuasion were Johann Bluntschli (who, we have seen, endowed society and its institutions with sex) and Paul von Lilienfeld (who declared explicitly that he considered society to be a real organism). Those whose belief was only on the level of metaphor include the sociologists Émile Durkheim (notably in his *The Division of Labor in Society*), René Worms (in his later works), and others.[91]

The case of René Worms may particularly attract our attention in this context because of his influence as Secrétaire Général of the Paris-based Institut International de Sociologie. In his book on *Organism and Society,*[92] a detailed development of a system of organismic sociology, Worms introduced extensive biological tutorials and drew heavily on recent work in histology, cellular morphology and physiology, and pathology. A major source was Spencer's *Principles of Sociology* (1876), a work frequently cited in terms of the highest respect. Proceeding in the manner of a biological analyst, Worms began his treatise with a discussion of the anatomy of societies, then their physiology, and lastly their pathology. In his later *Philosophy of the Social Sciences* (1903), Worms confessed that several years after publishing his *Organism and Society,* he had been led, "by personal reflection and by discussion," to moderate the "intransigence of his earlier conclusions." Primarily, he admitted to having underestimated the "true value of the individual by making him a simple cell in the social body" and by believing him to be "chained by physio-biological laws."[93] He had accordingly neglected the power of "free will" and the degree to which man is regulated by "the laws which he has given to himself" and the "contracts he has made."[94]

Those who held that it is only on the level of metaphor that society can be considered an organism nevertheless usually assumed that there is some kind of analogy between functions of the animal and the human body; most often this was expressed as a homology between social institutions and animal organs. The particular choice of the organ or biological entity to be used in a given homology, is not, however, always immediate and obvious. For example, both Paul von Lilienfeld and Albert E. Schäffle were convinced that the cell theory is fundamental to biology and must accordingly find its analogue or counterpart in the social organism. But whereas Lilienfeld believed that the human individual is the social homologue of the biological cell, Schäffle assigned this role to the family.[95]

In their writings both Lilienfeld and Schäffle did not present their respective sociological counterparts of the biological cell as mere homologues or analogues. Pursuing their respective themes of identity, they wrote as if the individual or the family were actually the cell of the social organism. It is manifest, however, that a single human individual or a family is not identical in every respect to a biological cell that is part of an organism. For instance, if a single cell is separated from the parent organism, it cannot survive, whereas there are many conditions in which an isolated human individual or an isolated family can survive. Schäffle went to the extreme of finding "homologues of such protective tissues of the animal body as hair, nails, and horny skin," whose social counterparts (Schäffle himself says they occur "analogically in the body social") are "roofs, coverings, wrappings, fences, walls, clothes, even picture-frames and book-covers . . . ."[96] How could he help but be aware that in every aspect of form or structure the listed entities from society differ in obvious and significant ways from any biological counterpart in a living organism? This example indicates that great care must be exercised in any critical analysis of the identity school of organismic social thought, especially in distinguishing levels of actual identity from levels of analogy and homology and even metaphor.

* * *

Metaphor embraces both analogy and homology, but it goes further to include the whole gamut of concepts, laws, theories, techniques, models, standards, and even values of the natural sciences (and mathematics) that social scientists have sought to borrow, emulate, imitate, or use in any way. And so our attention is directed to a very important aspect of the interactions between the natural and the social sciences, the transfer of values. We may see such a transfer in a notable example from economics.[97] In his *Theory of Political Economy,* Jevons took note of the "objections made to the general character of the [differential] equations" he had employed, defending his position by making an analogy between economics and physics, declaring that economics is similar to physics insofar as "the equations employed do not differ in general character from those which are really treated in many branches of physical science." The example he chose to develop was the use of the principle of virtual velocities (or virtual displacements) applied to the lever, where there is a homology of equations; that is, the equations for the case of the lever "have exactly the form of the equations of exchange [in economics]."[98] He even introduced a diagram in order to "put this analogy of the theories of exchange and of the lever in the clearest possible light."

Jevons penned this defense for the Preface to the second edition of his *Theory of Economics,* where he was responding to certain critics of the first edition. As I read what Jevons had to say on this subject, it seems obvious that he is making several points at once. He is declaring that the experience of rational mechanics shows that the use of the calculus is appropriate in "science" and that this form of mathematics may be expected to be just as beneficial to other sciences, including economic science, as it has been for the science of rational mechanics. So much for analogy. Jevons is also taking pride in showing that the form of equations is the same in both domains, introducing what I would call a factor of homology. But Jevons is also making another point: that the reason why economics can be treated mathematically in the same way as rational mechanics is that economics is "like" rational mechanics, which is the reason why there is a parallelism of concepts and a homology of equations. Since rational mechanics was then esteemed as the paradigmatic exact science, Jevons was consciously or unconsciously expressing his belief that economics could be a science on the same high level.

Léon Walras expressed a similar thesis, that mathematical economics is like rational mechanics, displaying a similar structure, including a parallelism or identity of equations. In his *Elements of Pure Economics,* he even went so far as to declare that "the pure science of economics is a science which resembles the physico-mathematical sciences in every respect."[99] This point of view suffuses Walras's preface to the fourth edition (1900) of his *Elements of Pure Economics,* where he asserted that "the whole theory is mathematical," in the sense that "the proof" of economic theory "must be given mathematically," even though the theory "may be described in ordinary language."[100] He also believed that "only with the aid of mathematics can we understand what is meant by the condition of maximum utility" and by other concepts. "Nor," he added, "can we understand without mathematics why or how current equilibrium prices are arrived at." He then asserted that "economics, like astronomy and mechanics is both an empirical and a rational science." He foresaw that in the twentieth century "mathematical economics will rank with the mathematical sciences of astronomy and mechanics."[101] Here—like Jevons—he was making a statement about values, on the level of metaphor, rather than putting forth methodological precepts.

Walras was rather explicit on this feature of economics in an essay on the analogy, "Economique et Mécanique," published in 1909,[102] in which he boasted that identical differential equations appear in his analysis of economics and in two examples from mathematical physics: the equilibrium of a lever and the motion of planets according to gravitational celestial mechanics. Claude Ménard has described Walras's text on "Economics and Mechanics" as a term-by-term comparison of "the principle of the proportionality of the *raretés* to values, revealing the parallels of the theorem of maximum satisfaction with the equation of maximal energy of rational mechanics."[103] Ménard and others have pointed out that Walras introduced features of rational mechanics in analogy with his economics for the most part only after he had already formulated his economics.[104] Apparently Walras was invoking the analogy in justification of his economic theory, rather than using the analogy as a tool in the discovery process. He was drawing on the metaphor of rational mechanics in order to transfer to his economics the values that had been traditionally assigned to this paradigmatic exact science.

Very often in the social sciences, there is declared a kind of parallelism that goes beyond any mere analogy (similarity in function) or homology (similarity in form of concept or equations) between the concepts, principles, laws, or equations of that discipline and one of the natural sciences. The examples of Jevons and Walras are particularly noteworthy, however, because there is implied a transfer to economics of the values of rational mechanics, the science established by Isaac Newton and which obtained its name from the Preface written by Newton for his great *Principia* (published in 1687). Such a transfer of values that goes beyond the formal transfer of concepts, equations, laws, and principles is well described as metaphor. That is, the founders of neoclassical economics not only introduced into their subject the specific analogues and homologues that Philip Mirowski and others have delineated for us in critical perspective; they also adopted the metaphor of rational mechanics or mathematical physics—attributing to their own endeavors, and to their subject at large, the values associated with the exact sciences that they were attempting to emulate.[105]

The extreme of this position is perhaps to be found in the writings of Vilfredo Pareto and Irving Fisher. Pareto went so far as to publish a table giving in parallel columns the primary concepts and principles of rational mechanics and economics.[106] Irving Fisher later produced a somewhat similar table, even specifying which pairs are vector quantities and which are scalars.[107]

The value-laden aspect of metaphor shows itself clearly in shifts of the specific part of the natural sciences that the social sciences are supposed to emulate. In this connection, the example of Jeremy Bentham is very illuminating. At different times in his life he considered that the art and science of society should be modelled on medicine, a subject with which he was well acquainted, even writing that "the art of legislation is but the art of healing practiced on a large scale," which—he added—was not a "mere fanciful" image. But at other times, he chose as his paradigm the new chemistry, conceiving that he would be its Lavoisier.[108]

We may see this feature of metaphor even more clearly in Engels's two eulogies of Karl Marx. At Marx's graveside, Engel's laudation took the form of a comparison with Darwin, indicating both the greatness of the effect of Darwin and Marx on contemporary thought and the revolutionary character of their ideas.[109] Later, by the time he was editing Marx's Nachlass to produce volume two of *Das Kapital,* Engels changed his mind about Marx's place in history. Now, as he wrote in his introduction, he found that Marx's counterpart in producing a revolution in the natural sciences was Lavoisier, the chief author (as Engels spent several paragraphs to prove) of the Chemical Revolution.[110] While both Darwin and Lavoisier were symbols of scientific greatness, each stood for a quite different kind of science or implied a metaphor comprising different sets of values and achievements.

This example may remind us that any choice of a particular analogy in the interactions of the natural and social sciences carries with it the metaphor, the system of values that may even be more important than the apparent closeness of the fit between concepts or principles or the isomorphism of systems of equations. This is especially the case when analogies serve to legitimize a research program, to justify the use of a concept, a principle, or a theory taken from the natural sciences, as we have seen to have been the case with regard to neoclassical economics and rational mechanics.

* * *

The "organismic" sociologists of the nineteenth century made continual and overt references to their sources of analogies and homologies in current biology and medicine. They seem to have gloried in displaying the bio-medical roots of their social thought and in showing how up-to-date was their biological knowledge. They even went so far as to insert into their sociological treatises some extensive biological tutorials on the latest developments in the natural sciences. In their writings we may see the joy and satisfaction of using the latest findings in biology, even the historical recognition that the cell theory had been an organizing principle that had brought the life sciences to a state of maturity—a conclusion which led them to hope that the use of an analogue of the cell theory would produce a similar effect in sociology. One can trace in these sociological works the successive ideas of Karl Ernst von Baer on embryological development (and the increase of complexity as a part of development), the doctrine of Henri Milne-Edwards and others on division of labor in relation to cell function and structure, Rudolf Virchow's concept of cellular pathology, the new ideas of germ theory of disease, and Claude Bernard's concept of the *milieu intérieure.*[111]

Why was there in the late nineteenth century, so vigorous a school of social thought based on a close analogy with the life sciences? In order to understand why this occurred, we must take into account the great succession of achievements of the biological sciences in the nineteenth century, together with the extraordinary successes of medicine. Not only were there significant theoretical advances such as those mentioned in the previous paragraph and the Darwinian theory of evolution; at long last, after many centuries of ignorance, the cause of many diseases had definitely been found in the activity of certain identifiable micro-organisms. Nor did success stop with mere identification since knowledge of the cause of disease and of curing them should occur. Although there were still important diseases to be understood and conquered, the means seemed at hand to make still further great discoveries. Such a joyful and triumphant mood may be contrasted with the more dismal state of mind of many physicists at the end of the nineteenth century, expressed in the prediction that the assignment for the future was to measure the constants of nature to yet one more place of decimals.[112]

There is no difficulty, therefore, in reconstructing the outlook of the many social scientists who believed that a great age of biology was taking the place of the older great age of physics. The economist Alfred Marshall was probably speaking for many when he said that at "the beginning of the nineteenth century the mathematico-physical group of sciences was in the ascendent," but now, at last, "the speculations of biology [have] made a great stride forwards." The discoveries in biology, he continued, now attract "the attention of all men [just] as those of physics had done in earlier years." The result was that the "moral and historical sciences of the day have in consequence changed their tone, and Economics has shared in the general movement."[113] In considering the importance of the biological sciences for the social sciences, furthermore, we should take note that many sociologists were convinced, as Comte had taught explicitly, that sociology must be a science very close, and even dependent on, human biology because its subject is human behavior and human organization.

* * *

In this presentation, I have not as yet taken note that some bio-medical scientists have used social analogies just as social scientists have drawn on analogies from the bio-medical sciences. An outstanding example is Rudolf Virchow, originator of the doctrine of "cellular pathology," introduced in 1858. Virchow's version of the cell theory provided a direct model for such social scientists as Lilienfeld and Schäffle, but he himself introduced his biological ideas in a social matrix.[114] Believing that all plants and animals are aggregates of cells as the fundamental life-units, Virchow concluded that all structural and functional properties of organisms are determined by relations among individual cells. This line of thought led him to believe, as he wrote in his *Treatise on Cellular Pathology,* that "the composition of a larger body, the so-called individual, always amounts to some kind of social institution."[115]

In referring to the cells as providing the "living organism" with a "multiplicity of vital foci," Virchow explained that every organism "is a free state of individuals with equal rights though not with equal endowments, which keeps together because the individuals are dependent upon one another."[116] This quotation clearly shows the validity of Owsei Temkin's judgment that "the metaphor of the cell state for Virchow was not a mere manner of speech, but an integral part of his biological theory."[117]

* * *

Today's literature exhibits an almost universal disdain for the older organismic sociology. Even the usually staid *Encyclopaedia of the Social Sciences* abandons its customary restraint when presenting Lilienfeld's ideas. In that work Howard Becker concludes that Lilienfeld's "influence was negative," that he "was a horrible example of the consequence of taking seriously his own dictum, *Sociologus nemo nisi biologus.*"[118] Such critics usually make no attempt to find out whether this school of thought, admittedly influential in its own time, may have left us a permanent legacy. Let me conclude, therefore, by referring to a major contribution of the organicists to social thinking, a transfer of some concepts and principles developed in medical sciences. Perhaps the primary mark of the organicists is the concept of normal and pathological conditions of society. Comte, Lilienfeld, Schäffle, and Worms (and other adherents of organicist sociology) introduced into social thought the medical concepts of normal and pathological states, which they transferred to the social domain of thought from medicine. They advocated the important medical principle that normal and pathological social states should not be conceived as wholly different types or classes of conditions, but rather extreme stages of a single type or class of condition.

Auguste Comte seems to have initiated the doctrine that social disturbances should be considered as pathological cases of the "social body," being, as he said, "in the social body, exactly analogous to diseases in the individual organism." He stated unequivocally that pathology, "the study of malady, is the way to understand the healthy state."[119] Primarily, it was the "principle of continuity" of the medical physiologist Broussais, which taught Comte and others "that the phenomena of the pathological state are a simple prolongation of the phenomena of the normal state, beyond the ordinary limits of variation."[120] Lilienfeld wrote at length on the pathology of society as an extreme of the normal condition. He gained this insight not only from Comte but also from Virchow, whose "cellular pathology" he esteemed as "one of the most striking conquests of modern science." From Virchow he had learned that "every pathological state of the human body derives from degeneration or an abnormal activity of the simple cell, as the elementary anatomical unit from which every organism is constructed."[121] Furthermore, he explained, Virchow taught that "there is no essential and absolute difference between the normal state and the pathological state of an organism." A "deviation from the normal state" occurs—Lilienfeld declared (on the authority of Virchow)—when "a cell or a group of cells manifests an activity outside the necessary time, outside the necessary place, or outside the limits of excitation prescribed by the normal state."[122]

The medical analogy of the normal and the pathological, introduced into social science by the organismic sociologists, has become a cornerstone of our social thinking, a striking example of the way in which the concept of the body politic has developed under the impact of the natural sciences.

* * *

One of Michel Foucault's many prescient though startling conclusions was that only "ignorance" has caused sociologists to seek their origins in Montesquieu and Comte, whereas they should have recognized that "sociological knowledge was formed in practices like those of the doctors."[123] This organicist theme of an analogy between sociology and medicine is a feature of older, historically oriented textbooks of sociology and the theme appears again explicitly in Louis Wirth's notion (in 1931) of "clinical sociology" and "sociological clinics."[124] In 1935 L. J. Henderson put forth the case that sociologists should adopt the analogy of clinical medicine (and even its techniques) and anticipated Foucault in conceiving "the practice of medicine as applied sociology."[125]

To organic sociologists it seemed an obvious analogical conclusion from medicine that social ills or diseases are caused by ailing individuals, just as Virchow taught that medical disorders should be reduced to a pathological condition in individual cells. This belief can be traced back even earlier, to the eighteenth century, when there was a strong current of thought linking individual health or well-being to the health of society. Utopians such as Condorcet drew an analogy between the eventual achievement of perfect condition of individual health and the creation of perfect society. There were predictions of a time when people would become so healthy and long lived that death would become a "curious accident."[126]

Such unbridled optimism suffered a deadly blow when the studies of Malthus on population showed that an analogy between the health of the individual members of society and the health of society as a whole might be too facile. Malthus demonstrated with grim examples that health and natural vigor in procreating could be a cause of social ills and diseases, producing a "power of population" restrained only by misery and disease, and the exercise of "moral restraint."[127] The effect of human health and the healthy, natural "desire and power of generation" (as Hume described this human drive[128]) would naturally lead to poverty, hunger, and misery, Malthus argued, because any possible increase in food supply (limited to an arithmetic ratio) could never keep up with the increase of population (in a geometric or exponential ratio).[129]

The conclusions of Malthus did not acquire a significant place in the thought of the late nineteenth-century organicists, perhaps because there were still vast underpopulated regions of the earth that could absorb the overflow. Additionally, advances in scientific agriculture (such as control of plant and animal diseases, improved rotation and management of crops, new fertilizers, and advances in plant and animal breeding) could have given hope that Malthus's law of an arithmetic increase in food production might no longer have the same force as in a condition of nature without human–i.e., scientific–intervention. Later discussions of the social implications of population studies, however, give proof that biological considerations are an essential part of social thought. This conclusion may give us a measure of the importance and originality of the nineteenth-century organicist sociologists, and their seventeenth- and eighteenth-century predecessors, even though they explored a different set of analogies.

Notes

1. The research on which this chapter is based has been generously supported by the Richard Lounsbery Foundation of New York City. In making this acknowledgement, I should like particularly to take cognizance of the sympathetic and enthusiastic personal support given to me by the Foundation's president, Mr. Alan McHenry.

2. This presentation draws heavily on two chapters (ch. 1, "An Analysis of Interactions between the Natural Sciences & the Social Sciences"; ch. 4, "The Scientific Revolution and the Social Sciences") from a book I have edited, *The Natural Sciences & the Social Sciences: Some Critical and Historical Perspectives,* published in 1993 in Boston Studies in the Philosophy of Science (Dordrecht/Boston/London: Kluwer Academic Publishers, 1993). My contributions to this volume have been published as *Interactions: Some Contacts between the Natural Sciences and the Social Sciences* (Cambridge, Mass./London: The MIT Press, 1994) and have appeared in translation as *Scienze della Natura e Scienze sociale* (Rome/Bari: Editori Laterza, 1993).

3. See, notably, David George Hale, *The Body Politic: A Political Metaphor in Renaissance English Literature* (The Hague/ Paris: Mouton, 1971); and Judith Schlanger, *Les Métaphores de l'Organisme* (Paris: Librairie Philosophique J. Vrin, 1971).

4. Some published results of this research, in addition to the work cited in note 2, include my forthcoming book on *Science and the Founding Fathers: Science in the Political Thought of Jefferson, Franklin, Adams, and Madison* (New York: W. W. Norton & Co., 1995); "Newton & The Social Sciences, with Special Reference to Economics, or, The Case of the Missing Paradigm," in *Natural Images in Economic Thought: "Markets Read in Tooth and Claw,"* ed. Philip Mirowski (Cambridge/New York: Cambridge University Press, 1994); and "Analogy, Homology, and Metaphor in the Interactions between the Natural Sciences & the Social Sciences, Especially Economics," *History of Political Economy* 25 (1993) supplement: 7–44, volume edited by Neil de Marchi.

5. William Harvey, *An Anatomical Disputation Concerning the Movement of the Heart and Blood in Living Creatures,* trans. Gweneth

Whitteridge (Oxford/London: Blackwell Scientific Publications, 1976), p. 3.

6. Ibid.

7. Ibid., from the translation by Kenneth J. Franklin (Oxford: Blackwell Scientific Publications, 1962).

8. From "The Prologue to the Reader," in John Halle, comp., *A Very Frutefull and Necessary Briefe Worke of Anatomie, or Dissection of the Body of Man . . . , with a Commodious Order of Notes, Leading the Chirurgien's Hande from all Offence and Error . . . Compiled in Three Treatises* (London: Thomas Marshe, 1565), published as part of *A Most Excellent and Learned Worke of Chirurgerie, Called Chirurgia parva Lanfranchi . . . ,* (London: Thomas Marshe, 1565). See Hale, *Body Politic.*

9. The difference between the status assigned to the heart by Harvey in his two major work–*De Generatione* and *De Motu cordis*–is set forth by Gweneth Whitteridge, *William Harvey and the Circulation of the Blood* (London: Macdonald; New York Elsevier, 1971).

10. *The Anatomical Exercises of Dr. William Harvey . . . : The First English Text of 1653,* ed. Geoffrey Keynes (London: The Nonesuch Press, [1928]) ch. 17, p. 115.

11. Ibid. In *De Motu cordis* Harvey was almost exclusively concerned with the function of the heart as the primary agent producing the circulation and not with the question of whether the heart comes into being in the embryo before the blood. In various works, and notably in the *De Generatione animalium,* Harvey made it plain that the blood appears in the development of the embryo before the heart or the liver or any other organ. On Harvey's views concerning the primacy of the heart, especially the difference between Harvey's and Aristotle's positions on this topic, see Whitteridge, *William Harvey,* pp. 227–235.

12. John Aubrey recorded that William Harvey, in his will, "left his old friend Thomas Hobbes 10 pounds as a token of his Love"; see Oliver Lawson Dick, ed., *Aubrey's Brief Lives, Edited from the Original Manuscripts* (London: Secker and Warburg, 1949; reprint ed., Ann Arbor: The University of Michigan Press, 1962), p. 133.

13. Many studies of Hobbes have taken cognizance of the fact that he based his system to some degree on the new physics of motion, but less attention has been paid to his use of Harveyan physiology. Hence, my presentation of Hobbes's use of the natural sciences stresses the biomedical basis of his political thought rather than his use of mathematics and the physical sciences.

There are many presentations of the thought of Hobbes, among them Leo Strauss, *The Political Philosophy of Hobbes: Its Birth and Its Genesis,* trans. from the German manuscript by Elsa M. Sinclair (Oxford: The Clarendon Press, 1936; reprint ed., Chicago: University of Chicago Press, 1966); Thomas A. Spragens, *The Politics of Motion: The World of Thomas Hobbes* (London: Croon Helm, 1973); and M. M. Goldsmith, *Hobbes's Science of Politics* (London/New York: Columbia University Press, 1966).

Also David Johnston, *The Rhetoric of Leviathan: Thomas Hobbes and the Politics of Cultural Transformations* (Princeton: Princeton University Press, 1986); Tom Sorell, *Hobbes* (London/New York: Routledge & Kegan Paul, 1986—"The Arguments of the Philosophers"); Richard Tuck, *Hobbes* (Oxford/New York: Oxford University Press, 1989—"Past Masters"); and Frithiof Brandt, *Thomas Hobbes' Mechanical Conception of Nature* (Copenhagen: Levin & Munksgaard, 1928).

There is much to be learned from two volumes by C. B. MacPherson, *The Political Theory of Possessive Individualism, Hobbes to Locke* (Oxford: Clarendon Press, 1962), *Democratic Theory: Essays in Retrieval* (Oxford: Clarendon Press, 1973). Especially important in the present context is an essay by J. W. N. Watkins, "Philosophy and Politics in Hobbes," *Philosophical Quarterly* 5 (1955): 125–146; expanded into the book, *Hobbes's System of Ideas: A Study in the Political Significance of Philosophical Theories* (London: Hutchinson & Co., 1965; 2d ed., 1973).

14. The writings of Hobbes have been collected in two sets —Sir William Molesworth, ed., *The English Works of Thomas Hobbes,* 11 vols. (London: John Bohn, 1839–1845; reprint ed., Aalen [Germany]: Scientia Verlag, 1966); Sir William Molesworth, ed., *Thomae Hobbes malmesburiensis Opera philosophica quae Latinae scripsit omnia,* 5 vols. (London: John Bohn, 1839- 1845; reprint ed., Aalen [Germany]: Scientia Verlag, 1961). There are also articles on Hobbes in the *Encyclopaedia of the Social Sciences,* 15 vols. (New York: The Macmillan Company, 1937), vol. 4, and the *International Encyclopedia of the Social Sciences* (1968). The quotation comes from *English Works,* 7:470–471.

15. *English Works,* 1:viii ("Epistle Dedicatory").

16. On the "mechanical philosophy," see R. S. Westfall, *Force in Newton's Physics: The Science of Dynamics in the Seventeenth Century* (London: Macdonald; New York: American Elzevier, 1971); also William R. Shea, *The Magic of Numbers and Motion: The Scientific Career of René Descartes* (Canton [Ma.]: Science History Publications, 1991).

17. Hobbes's *Leviathan,* his major work, is available in many editions and reprints; among them the Pelican Classics edition, ed. C. B. MacPherson (Harmondsworth: Penguin Books, 1968). The most recent edition, ed. Richard Tuck (Cambridge/New York: Cambridge University Press, 1991), has indexes of subjects and of names and places and a

concordance with earlier editions. The quotation is from the Tuck edition, p. 3.

18. For details, see "The Scientific Revolution and the Social Sciences" in Cohen, ed., *The Natural Sciences & Social Sciences.*

19. See *Leviathan,* Tuck ed., ch. 29, pp. 228–230.

20. *English Works,* 3:29.

21. Hobbes is a particular subject of attack in Harrington's major work, *Oceana,* cited in note 24 below.

22. R. Tawney, "Harrington's Interpretation of his Age," *Proceedings of the British Academy* 27 (1941): 199–223, especially p. 200.

23. H. F. Russell Smith, *Harrington and His Oceana* (Cambridge: Cambridge University Press, 1914), ch. 8; Theodore Dwight, "Harrington and his Influence upon American Political Institutions and Political Thought," *Political Science Quarterly* 2 (1887): 1–44. This subject is explored further in my *Science and the Founding Fathers.*

24. Charles Blitzer, *An Immortal Commonwealth* (New Haven: Yale University Press, 1960), pp. 160–169, and 210–279. This is the best informed and most authoritative work on Harrington; a convenient list of Harrington's publications is given on pp. 337– 339. A useful, briefer presentation is given in Michael Downs, *James Harrington* (Boston: Twayne Publishers, 1977).

Harrington's *Oceana* is available in a number of current editions: John Toland, ed., *Works: The Oceana and Other Works* (London: 1771; reprint ed., Aalen [Germany]: Scientia Verlag, 1980); S. B. Liljegren, ed., *James Harrington's Oceana, Edited with Notes* (Heidelberg: Carl Winters Universitätsbuchhandlung, 1924 —Skrifter ultgivna av Vetenskaps-Societeten i Lund, 4; reprint ed., Westport [Conn.]: Hyperion Press, 1979).

25. Toland, *Works,* p. 403; also Blitzer, *Immortal Commonwealth,* pp. 89–108.

26. Ibid.

27. *Oceana,* Liljegren ed., p. 13; Toland ed., p. 36. See Blitzer, *Immortal Commonwealth,* p. 99. On p. 110 of *Oceana* (Toland ed.), Harrington compares the Council of Trade to the Vena Porta.

28. *Oceana,* Liljegren ed., p. 149; Toland ed., p. 149.

29. Ibid.

30. Blitzer, *Immortal Commonwealth,* p. 99.

31. Toland ed., p. 560.

32. Ibid.

33. Ibid.

34. See I. B. Cohen, "A Note on Harvey's 'Egg' as Pandora's 'Box,'" in *Changing Perspectives in the History of Science: Essays in Honour of*

Joseph Needham, ed. Mikulas Teich & Robert Young (London: Heinemann, 1973), pp. 232–249.

35. See F. J. Cole, *Early Theories of Sexual Generation* (Oxford: Clarendon Press, 1930).

36. Toland ed., p. 470.

37. Ibid.

38. William Harvey, *Disputations Touching the Generation of Animals,* trans. Gweneth Whitteridge (Oxford/London: Blackwell Scientific Publications, 1981), ch. 47, pp. 214–215.

39. Ibid., ch. 51, p. 241; quoted from the 1653 translation, as given in Kenneth D. Keele, *William Harvey: The Man, the Physician, and the Scientist* (London: Nelson, 1965), p. 185; also William Harvey, *Anatomical Exercises in the Generation of Animals,* trans. Robert Willis, in *The Works of William Harvey* (London: The Sydenham Society, 1847; reprint ed., New York/London: Johnson Reprint Corporation, 1965), p. 235.

40. Keele, *William Harvey,* pp. 183–186. See Gweneth Whitteridge's summary in the introduction to her translation, p. liii.

41. For an account of Harrington's use of the "punctum saliens" in a political context, see my chapter 4, "The Scientific Revolution & the Social Sciences," in *The Natural Sciences & Social Sciences.*

42. See ch. 11 of Rousseau's *Social Contract.*

43. Translated in Allen Ritter & Julia Conaway Bondanella, eds., *Rousseau's Political Writings* (New York/London: W. W. Norton & Company, 1988), p. 61.

44. An alternative translation, made by G. D. H. Cole, is published in Charles M. Sherover, ed., *Annotated Edition: The Social Contract or Principles of Political Right by Jean Jacques Rousseau* (New York: New American Library, 1974).

45. Adam Smith to Lady Frances Scott, 15 October 1766, in Ernest Campbell Mossner & Ian Simpson Ross, eds., *The Correspondence of Adam Smith* (Oxford: Oxford University Press, 1977; reprint ed., Indianapolis: Liberty Classics, 1987—The Glasgow Edition of the Works and Correspondence of Adam Smith, vol. 6), p. 120.

46. On the significance of the circulation of the blood in the economics of Quesnay, see Vernon Foley, "An Origin of the Tableau Oeconomique," *History of Political Economy* 5 (1973): 121–150.

47. Nicolas-François Canard, *Principes d'Économie politique* (Paris: Biusson, 1801), summarized in Claude Ménard, "The Machine and the Heart: An Essay on Analogies in Economic Reasoning," *Social Concept* 5 (1988): 81–95, esp. p. 82.

48. Joseph A. Schumpeter, *History of Economic Analysis* (New York: Oxford University Press, 1954), p. 498.

49. David Brewster, *Letters on Natural Magic* (New York: Harper & Brothers, 1843), p. 181.

50. John Tyndall, *The Glaciers of the Alps* (Boston: Ticknor & Fields, 1861), p. 285.

51. John Lubbock [Lord Avery], *Scientific Lectures,* X vols. (London: Macmillan, 1879), 4:137.

52. See Ernst Mayr, *The Growth of Biological Thought: Diversity, Evolution, and Inheritance* (Cambridge: The Belknap Press of Harvard University Press, 1982), p. 45. According to Mayr, the "term 'homologous' existed already prior to 1859, but it acquired its currently accepted meaning only when Darwin established the theory of common descent. Under this theory the biologically most meaningful definition of 'homologous' is: 'A feature in two or more taxa is homologous when it is derived from the same (or a corresponding) feature of their common ancestor.'"

53. Thomas Carlyle, *Sartor Resartus* (London: James Munroe & Co, 1836); H. D. Traill, ed., *The Works of Thomas Carlyle,* 30 vols. (London: Chapman & Hall, 1896–1899; reprint ed., New York: AMS Press, 1969), 1:172. See Frederick W. Roe, *The Social Philosophy of Carlyle and Ruskin* (New York: Harcourt, Brace & Co., 1921); also Hale, *Body Politic,* pp. 134–135.

54. Hale, *Body Politic,* p. 135.

55. Johann Caspar Bluntschli, *Lehre vom modernen Staat,* 6th ed., 2 vols. (Stuttgart: J. G. Cotta, 1885–1886), the first volume of which was translated into English as *Theory of the State* (Oxford: Oxford University Press, 1892); *Psychologische Studien über Staat und Kirche* (Zurich/ Frauenfeld: C. Beyel, 1844). Bluntschli was also author of a widely used reference work, *Deutsches Staats-Wörterbuch* (Stuttgart/Leipzig: Expedition des Staats-Wörterbuchs, 1857–1870).

56. Selections from Bluntschli's *Psychologische Studien über Staat und Kirche* (pp. 54, 86–87) are quoted from the translation in Werner Stark, *The Fundamental Forms of Social Thought* (London: Routledge & Kegan Paul, 1962), pp. 61–62.

57. Ibid.

58. Paul von Lilienfeld, or Paul de Lilienfeld, or Pavel Fedorovich Lilienfeld-Toailles, or Pavel Fedorovich Lilienfel'd Toal' (1829–1903) published a book in Russian on the elements of political economy in 1860 under the pseudonym Lileyewa. This was followed by a first version, in Russian (under the initials P. L.), of *Thoughts on the Social Science of the Future* in 1872; a five-volume German version of this work, *Gedanken über die Socialwissenschaft der Zukunft* (vols. 1–4: Mitau: E. Behre's Verlag, 1873–1879; vol. 5: Hamburg: Gebr. Behre's

Verlag; Mitau: E. Behre's Verlag, 1881). Later works include *La Pathologie sociale* (Paris: V. Giard & E. Brière, 1896); and *Zur Vertheidigung der organischen Methode in der Sociologie* (Berlin: Druck und Verlag von Georg Reimer, 1898).

See, e.g., Otto Henne am Rhyn, *Paul von Lilienfeld* (Gdansk, Leipzig, Vienna: Carl Hinstorff's Verlagsbuchhandlung, [n.d.]; *Deutsche Denker und ihre Geistesschöpfungen,* ed. Adolf Hinrichsen, vol. 6). For further bibliography related to Lilienfeld, see Howard Becker, "Lilienfeld-Toailles, Pavel Fedorovich," *Encyclopaedia of the Social Sciences,* 9:474–474.

59. *La Pathologie sociale.*

60. Translated from Paul von Lilienfeld, *La Pathologie sociale,* p. 59.

61. Ibid., pp. 59–60.

62. Although "fallacy" is often used in a narrow technical sense to denote a flaw (or type of flaw) that "vitiates a syllogism," a primary meaning in every dictionary I have consulted (*Oxford English Dictionary, Oxford English Dictionary*—suppl., *Oxford English Dictionary*—2d. ed.; *Concise Oxford Dictionary*—6th ed.; *Webster's New International*—2d & 3d eds.) is a "misleading argument" or a delusion or error, or some unsoundness or delusiveness or disappointing character of an argument or belief. *The American Heritage Dictionary* gives as the first meaning: "An idea or opinion founded on mistaken logic or perception; a false notion"; other meanings include "the quality of being deceptive" and "incorrectness of reasoning or belief." The only example given is a "romantic fallacy, that Shakespeare was superhuman." This example displays features in common with two frequently encountered uses of "fallacy" today: John Ruskin's notion of the "pathetic fallacy" (in which inanimate objects are supposed to have human emotions) and W. K. Wimsatt and Monroe Beardsley's "intentional fallacy" (overstressing the author's intentions in assessing a literary work). These usages are somewhat similar to Alfred North Whitehead's "fallacy of misplaced concreteness."

63. J. D. Y. Peel, *Herbert Spencer: The Evolution of a Sociologist* (New York: Basic Books, 1971), p. 178; see Herbert Spencer, "The Social Organism," *Essays: Scientific, Political, and Speculative,* 3 vols. (New York: D. Appleton & Co., 1883), 1:287–289.

64. Herbert Spencer, "The Social Organism," *Essays,* 2:277– 279, 283–286.

65. Herbert Spencer, "Specialized Administration," *Essays: Scientific, Political, and Speculative,* 3 vols. (New York: D. Appleton & Co., 1896), 3:427–428. Theodore Porter suggests that Spencer may have had his tongue in cheek when writing about the engineering schools as a "dou-

ble gland," but a sense of humor was not a primary characteristic of Spencer's prose style.

For an understanding of Spencer's thought, including his drawing upon analogues and homologues from the biological sciences, see Robert J. Richards, *Darwin and the Emergence of Evolutionary Theories of Mind and Behavior* (Chicago/London: The University of Chicago Press, 1987), one of the few studies in which the author has attempted so to steep himself in the literature of the period that he can interpret Spencer's ideas (and those of others) in terms of the standards and currents of thought of those bygone days.

66. Walter Cannon, "Relations of Biological and Social Homeostasis," *The Wisdom of the Body* (New York: W. W. Norton & Company, 1932; reprint ed., 1939), pp. 305–324.

67. Ibid., p. 309.

68. Ibid., p. 314.

69. Walter B. Cannon, "The Body Physiologic and the Body Politic," *Science* 93 (1941): 1–10.

70. Ibid.

71. Ibid. Cannon had in mind the writings of certain extremist organicists, of whom Albert Schäffle is an outstanding example. See Stark, *Fundamental Forms of Social Thought,* and Cohen "An Analysis of Interactions between the Natural Sciences & the Social Sciences," in *The Natural Sciences & Social Sciences.*

72. Robert K. Merton, *Social Theory and Social Structure* (New York: The Free Press, 1968), ch. 3, pp. 101n, 102–103.

73. The analogies used in economics by Jevons and Walras are introduced below; for more information, see Philip Mirowski, *More Heat Than Light* (Cambridge/New York: Cambridge University Press, 1989); also my own paper, "Analogy, Homology, and Metaphor in the Natural Sciences & the Social Sciences."

74. See Stark, *Fundamental Forms of Social Thought,* pp. 156–160; also H. C. Carey, *The Unity of Law; As Exhibited in the Relations of Physical, Social, Mental, and Moral Science* (Philadelphia: Henry Carey Baird, 1872), pp. 116–127.

75. See my article "Newton and the Social Sciences" in *Natural Images,* ed. Mirowski.

76. On the dependence of neo-classical economists on rational mechanics (including energy physics), see Mirowski, *More Heat than Light.* See the valuable discussion of these topics in Claude Ménard, "The Machine and the Heart: An Essay on Analogies in Economics Reasoning," *Social Concept* 5(1) (December 1988): 81–95.

77. An example of this kind of error in the use of a biological analogue or metaphor (in the controversy between Armen Alchian and Edith Penrose) is discussed in my article, "Analogy, Homology, and Metaphor in the Natural Sciences & the Social Sciences."

78. See Stark, *Fundamental Forms of Social Thought,* pp. 73–74.

79. Mary P. Mack, *Jeremy Bentham: An Odyssey of Ideas, 1748–1792* (London: Heinemann, 1962), pp. 275–281.

80. See Sigmund Freud, "A Note on the 'Mystic Writing Pad,'" *The Standard Edition of the Complete Psychological Works of Sigmund Freud,* 24 vols., trans. under the general editorship of James Strachey, in collaboration with Anna Freud, assisted by Alix Strachey & Alan Tyson (London: The Hogarth Press and the Institute of Psycho-analysis, 1953–1974), 19:228.

Two decades later, in *Beyond the Pleasure Principle* (1920), Freud noted further that "the inexplicable phenomenon of consciousness arises in the perceptual system *instead of* the permanent traces." For references to this topic, see *Standard Edition,* 5:40 and 18:25. The editor (John Strachey) notes that this distinction had already been made by Josef Breuer.

81. See Freud, "A Note on the 'Mystic Writing Pad,'" *Standard Edition,* 19:228.

This mechanical analogue served two functions often found in the use of analogues: (1) to many to make his earlier hypothetical conjecture seem reasonable enough for him to set forth his ideas in full, and (2) to make his difficult concept of the structure of memory understandable and thus acceptable to the psychoanalytic community.

82. Donald H. Kaplan, "The Psychoanalysis of Art: Some Ends, Some Means," *Journal of the American Psychoanalytic Association* 36 (1988): 259–302, esp. p. 260.

83. *Poetics,* 1457b, 1459a, 1458a. For a brief but incisive history of the uses of metaphor from antiquity to the present, see Mark Johnson, ed., *Philosophical Perspectives on Metaphor* (Minneapolis: University of Minnesota Press, 1981), introductory chapter.

84. There is a large and ever-growing literature on metaphor and also on rhetorical aspects of science. On the age of the scientific revolution, see Peter Dear, "*Totius in verba:* Rhetoric and Authority in the Early Royal Society," *Isis* 76 (1985): 145–161; Steven Shapin, "Pump and Circumstance: Robert Boyle's Literary Technology," *Social Studies of Science* 14 (1984): 487–494. Also Steven Shapin & Simon Schaffer, *Leviathan and the Air-Pump: Hobbes, Boyle, and the Experimental Life* (Princeton: Princeton University Press, 1985); Alan C. Gross, *The Rhetoric of Science* (Cambridge/London: Harvard University Press,

1990). Also Marcello Pera, *Scienza e Retorica* (Rome/Bari: Gius. Laterza & Figli, 1991); M. Pera & William R. Shea, eds., *Persuading Science: The Art of Scientific Rhetoric* (Canton, [Ma.]: Science History Publications, USA, 1991); Peter Dear, ed., *The Literary Structure of Scientific Argument: Historical Studies* (Philadelphia: University of Pennsylvania Press, 1991). On metaphor in general see Max Black, *Models and Metaphors: Studies in Language and Philosophy* (Ithaca: Cornell University Press, 1962); Johnson, ed., *Philosophical Perspectives on Metaphor* (cited in the previous note); Arjo Klamer, ed., *Conversations with Economists* (Totowa, [N.J.]: Rowman & Lilienfeld, 1983); Donald N. McCloskey, *If You're So Smart: The Narrative of Economic Expertise* (Chicago/London: The University of Chicago Press, 1990); and Andrew Ortony, *Metaphor and Thought* (Cambridge/London/New York: Cambridge University Press, 1979). This topic also appears in discussions of economics, notably in Mirowski, *More Heat than Light.*

85. James I, "Speech of 1603," in *The Political Works of James I,* ed. Charles H. McIlwain (Cambridge: Harvard University Press, 1918), p. 272; see Hale, *Body Politic,* p. 111.

86. Metaphor was a lively topic of discussion at two recent conferences relating to the development of economics. The first (held at Duke University in April 1991) had special reference to Philip Mirowski's *More Heat Than Light;* a volume of proceedings edited by Neil de Marchi appeared in 1993 as a supplement to volume 25 of *History of Political Economy.* A conference, organized by Philip Mirowski at Notre Dame on "Natural Images in Economics," is published in *Natural Images,* ed. Mirowski.

87. See Hale, *Body Politic,* pp. 99–102.

88. "Speech in Star Chamber 1616," *Political Works,* p. 343; see Hale, *Body Politic,* p. 111, n. 19.

89. See, on this subject, Marc Bloch, *The Royal Touch: Sacred Monarchy and Scrofula in England and France,* trans. J. E. Anderson (London: Routledge & Kegan Paul, 1973).

90. Herbert Spencer, *The Principles of Sociology,* 3rd ed., 3 vols. (New York: D. Appleton and Company, 1897), vol. 1, part 2, § 1, "What Is a Society ?"; § 2, "A Society Is an Organism."

91. See, further, my "Analysis of Interactions between the Natural Sciences & the Social Sciences," in *The Natural Sciences & the Social Sciences.*

92. René Worms, *Organisme et Société* (Paris: V. Giard & E. Brière, 1896).

93. René Worms, *Philosophie des Sciences sociales,* 3 vols. (Paris: V. Giard & E. Brière, 1903), 1:12.

94. In discussing the causes of his having altered his point of view, Worms assigned primary importance to the ideas of Ernest Haeckel and the recent discovery of the relative discontinuities among cerebral cells, which he said is "the glory of Golgi and Ramón y Cajal." He also pointed to E. Metchnikoff's research on phagocytes, which had indicated new aspects of cells, even to the extent of suggesting that cells may exercise an art or even a science in defending themselves against their enemies.

In short, Worms said, his sociological critics had undermined his former overly simplistic view of the social organism, while the advances of the bio-medical sciences had radically altered the scientific base on which his earlier organismic sociology was founded, primarily by enlarging the knowledge and understanding of the life and functions of the cells of which living organisms are made. In the present context, the details of the differences between Worms's two treatises are of less interest than the very fact that in each of these works the social analogues drawn from the life sciences were direct reflections of the rapidly changing current state of biological and medical knowledge.

95. See my "Analysis of Interactions between the Natural Sciences and the Social Sciences," in *The Natural Sciences & the Social Sciences,* §1.6.

96. Albert Schäffle, *Bau und Leben des socialen Körpers: Encyclopädischer Entwurf einer realen Anatomie, Physiologie und Psychologie der menschlichen Gesellschaft mit besonderer Rücksicht auf die Volkswirthschaft als socialen Stoffwechsel,* 4 vols. (Tübingen: H. Laupp'sche Buchhandlung, 1881), 1:327, 329; translated in Stark, *Fundamental Forms of Social Thought,* p. 68.

97. The purpose of the Notre Dame conference (referred to in note 86) was a critical assessment of the thesis of Philip Mirowski's *More Heat Than Light,* primarily that neoclassical economics was founded on the "metaphor" of rational mechanics cum energy physics. See my article, "Analogy, Homology, and Metaphor in the Natural Sciences & the Social Sciences."

98. See W. Stanley Jevons, *The Theory of Political Economy,* 5th ed. (1911; reprint ed., New York: Augustus Kelley, 1965), p. 105.

99. Similar sentiments were expressed by F. Y. Edgeworth, *Mathematical Psychics: An Essay on The Application of Mathematics to The Moral Sciences* (London: C. Kegan & Co., 1881); for details see my "An Analysis of Interactions between the Natural Sciences & the Social Sciences," in *The Natural Sciences & the Social Sciences.*

100. Léon Walras, "Preface to the Fourth Edition," *Elements of Pure Economics or the Theory of Social Wealth,* trans. William Jaffé (Home-

wood [Il.]: Richard D. Irwin [for the American Economic Association & the Royal Economic Society], 1954), p. 43.

101. Ibid., pp. 47–48.

102. Léon Walras, "Economique et Mécanique," *Bulletin de la Société vaudoise de Sciences naturelles* 45 (1909): 313–325; translated, with a commentary, in Philip Mirowski & Pamela Cook, "Walras' 'Economics and Mechanics,'" *Economics as Discourse,* ed. Warren J. Samuels (Boston/Dordrecht/London: Kluwer, 1990).

103. Claude Ménard, "The Machine and the Heart: An Essay on Analogies in Economic Reasoning," *Social Concept* 4 (1988): 81–95, p. 84. Here, according to Ménard, Walras "also defines the properties of general equilibrium for traded goods, services and money." This is "the infamous 'Walras Law,' based upon the example of uniformly accelerated motion in celestial mechanics, invoking the equation of constraint between mass and acceleration."

104. In a scholarly study of Walras's economic ideas in relation to Mirowski's analysis, Albert Jolink denies "strenuously" that Walras's economics theory "slavishly imitates physics"; see Albert Jolink, "'Procrustean Beds and All That': The Irrelevance of Walras for a Mirowski-Thesis," which appeared in 1993 in *History of Political Economy,* as referred to in note 86 above. Jolink cites evidence that Walras had little or no understanding of energy physics before 1906, and he concludes that even after 1906 it is doubtful whether "Walras had any understanding at all concerning a proto-energetic metaphor."

105. On this topic see Mirowski's *More Heat Than Light* and also my article "Newton and the Social Sciences" in *Natural Images,* ed. Mirowski.

106. This table is available in Vilfredo Pareto, *Sociological Writings,* ed. S. E. Finer, trans. Derick Mirfin (Oxford: Basil Blackwell, 1966), pp. 106–107. It is reproduced in Bruna Ingrao, "Physics and Pareto's Economics," in *Natural Images,* ed. Mirowski; and also in my "Analogy, Homology, and Metaphor in the Natural Sciences & the Social Sciences." I wish to thank Bruna Ingrao for bringing Pareto's table to my attention.

See further Bruna Ingrao, "L'Analogia meccanica nel Pensiero di Pareto," in *Pareto oggi,* ed. G. Busino (Bologna: Il Mulino, 1991).

107. Irving Fisher, *Mathematical Investigations in the Theory of Value and Prices* (New Haven: Yale University Press, 1926).

108. Mack, *Jeremy Bentham,* p. 264.

109. See Giuliano Pancaldi, "The Technology of Nature: Marx's Thoughts on Darwin," in *The Natural Sciences & the Social Sciences.*

110. This episode and its significance are discussed in I. B. Cohen, *Revolution in Science* (Cambridge: The Belknap Press of Harvard University Press, 1985), suppl. § 14.1.

111. For details see my "Analysis of the Interactions between the Natural Sciences & the Social Sciences," in *The Natural Sciences & the Social Sciences,* and, especially, Camille Limoges's chapter 10 in the same work.

112. See Lawrence Badash, "The Completeness of Nineteenth-Century Science," *Isis* 63 (1972): 45–58. James Clerk Maxwell, in his inaugural address at Cambridge University, said "that in a few years . . . the only occupation which will be left to men of science will be to carry on these measurements [of the great physical constants] to another place of decimals"; A. A. Michelson published in 1898 his prediction "that most of the grand underlying principles [of physical science] have been firmly established."

113. Alfred Marshall, "The Present Position of Economics" (1885), an inaugural lecture given after his election to the professorship at Cam bridge, printed in A. C. Pigou, ed., *Memorials of Alfred Marshall* (London: Macmillan & Co, 1925), p. 154.

114. For Virchow, the concept of the "cell state" was particularly significant because there was always a close parallel between his "biological views and his liberal political opinions." See Irwin H. Ackerknecht, *Rudolph Virchow: Doctor, Statesman, Anthropologist* (Madison: The University of Wisconsin Press, 1953; reprint ed., New York: Arno Press, 1981).

115. Virchow, we may note, was not the only nineteenth-century biologist to use social analogies in scientific discourse. Thomas Henry Huxley made use of a social analogy in describing the sponge, which—he said—represents a kind of sub-aqueous city, "in which the people are arranged about the streets and roads, in such a manner, that each can easily appropriate his food from the water as it passes along." This is an example of the use of analogy to illustrate a scientific concept, making such a concept easier to visualize or to understand.

116. See Owsei Temkin, "Metaphors of Human Biology," in *Science and Civilization,* ed. Robert C. Stauffer (Madison: University of Wisconsin Press, 1949), pp. 167–194, especially p. 175.

117. Ibid., p. 175. On this point, see further Ernst Hirschfeld, "Virchow," *Kyklos* 2 (1929): 106–116.

118. Howard Becker, "Lilienfeld-Toailles," pp. 473–474.

119. Gertrude Lenzer, *Auguste Comte and Posititivism: The Essential Writings* (New York/Evanston: Harper & Row, 1975), p. 191, taken from *The Positive Philosophy of Auguste Comte,* trans. [and condensed

by] Harriet Martineau (London: Longmans, Green, 1853), book 5, ch. 6.

120. Ibid. Comte also wrote that a consequence of "the admirable axiom of Broussais" is that "it destroys the old absolute distinction between health and disease"; see Auguste Comte, *The Foundations of Sociology,* ed. Kenneth Thompson (New York: John Wiley & Sons, 1975), p. 141, based on Auguste Comte's *System of Positive Polity,* 4 vols. (London: Longmans, Green, 1848–1854), 2:367–382.

121. Translated from Lilienfeld's *La Pathologie sociale,* pp. 20–21.

122. Ibid., p. 21.

123. Michel Foucault, *Power/Knowledge: Selected Interviews and Other Writings, 1972–1977* (Brighton, [Sussex]: Harvester Press, 1980), p. 151.

See, on the subject of this, Bryan S. Turner, "Sociology and the Body," chapter 2 of his *The Body and Society: Explorations in Social Theory* (Oxford: Basil Blackwell, 1984)

124. Louis Wirth, "Clinical Sociology," *American Journal of Sociology* 37 (1931): 49–66.

125. Lawrence J. Henderson, "Physician and Patient as a Social System," *New England Journal of Medicine* 212 (1935): 819–823. This essay is reprinted in Bernard Barber, ed., *L. J. Henderson on the Social System: Selected Writings* (Chicago: University of Chicago Press, 1970), a work introduced by an informative and insightful editor's foreword.

126. See Keith M. Baker, *Condorcet, from Natural Philosophy to Social Mathematics* (Chicago: University of Chicago Press, 1975), pp. 348–349, 368–369.

127. On Malthus's views, see the informative introduction in Anthony Flew, ed., *An Essay on the Principle of Population and A Summary View of the Principle of Population* (Harmondsworth: Penguin Books, 1970); and also Anthony Flew, *Thinking about Social Thinking: The Philosophy of the Social Sciences* (Oxford: Basil Blackwell, 1985), ch. 4, §1.

128. David Hume, "Of the Populousness of Ancient Nations," *Essays Moral, Political, and Literary* 2 vols. (Edinburgh: R. Fleming and A. Alison for A. Kincaid, 1742), 1:376.

129. Catherine Gallagher has observed that Malthus effectively "occludes the possibility of using the healthy body to signify the healthy society"; see her essay on "The Body Versus the Social Body in Malthus and Mahew," chapter 3 in *The Making of the Modern Body: Sexuality and Society in the Nineteenth Century,* ed. Catherine Gallagher & Thomas Laqueur (Berkeley/Los Angeles: University of California Press, 1987), especially p. 85.

THE GEOMETRICAL KABBALAHS OF JOHN DEE AND JOHANNES KEPLER:

THE HEBREW TRADITION AND THE MATHEMATICAL STUDY OF NATURE

Michael T. Walton and Phyllis J. Walton

Renaissance Neoplatonism was marked by a preoccupation with restoring ancient wisdom as a foundation for present and future knowledge. Although its adherents often had divergent and idiosyncratic agendas, there was a consistent emphasis on developing a Christian world view grounded in knowledge of the harmony of God's creation.[1] The Renaissance Neoplatonic synthesis of Christianity and ancient philosophy was significantly reinforced by the discovery of the Kabbalah, the Jewish esoteric tradition. The Kabbalah taught a descending order of creation from the perfection of God to the imperfect material world. The letters of the Hebrew alphabet, which are also numbers, were the basic units of creation. The notions of creation by descent and numerical harmonies indicated similarities between the Kabbalah and the ideas of Pythagoras and of Plato. The correspondences were believed to demonstrate that both traditions shared a common origin, i.e., divine revelation to Adam, Abraham, or Moses. Renaissance Christians, like Pico della Mirandola, Johannes Reuchlin, and Franciscus Georgius, looked to the Jewish mystical tradition in their attempts to rediscover the wisdom behind the Hebrew scriptures. They also viewed the Kabbalah as the sacred original from which the pagan philosophers derived their knowledge.[2]

Numerous studies have established the importance of Neoplatonism and the Christian Neoplatonic Kabbalah to the intellectual life of the sixteenth century.[3] The lasting effects of this world view on European thought, however, are much less clearly understood. One area in which the Neoplatonic-Kabbalistic philosophy can be shown to have shaped future thought is in helping to establish the idea that mathematical abstraction is descriptive of nature.

How easily Neoplatonism could be connected to Hebrew Kabbalah is seen in the similarity between the ideas of the sixth-century philosopher Boethius and the fundamental Kabbalistic text *Sepher Yezirah.* Both were popular sources for sixteenth-century Neoplatonism. Boethius taught that "all things . . . do appeare to be formed by the reason of numbers. For this was the principall example or patterne in the minde of the Creator."[4] This belief was supported by *Sepher Yezirah,* or *Book of Formation.*[5] *Yezirah* declared that God created the world

P. H. Theerman and K. Hunger Parshall (eds.), Experiencing Nature, 43–59.

through three aspects of the verb "to number." He created His world [cosmos] through three numberings, "by writing [sefer], number [sofor], and telling [sippur]."[6] For Renaissance Neoplatonists the view that number underlies all creation was axiomatic.

The success of sixteenth-century mathematicians in showing correspondences between observable phenomena and mathematical abstractions, such as geometrical figures, was proof that number *is* the basis of the natural world. This served to establish mathematics as an indispensable tool for reading the "Book of Nature," God's second great scripture. Mathematics thus became an integral part of natural philosophy, not only for Renaissance Neoplatonists, but also for succeeding generations. How the Neoplatonic-Kabbalistic tradition helped to routinize mathematics into the study of nature is illustrated by an analysis of the sources for and the impact of the "geometrical Kabbalahs" of John Dee and Johannes Kepler.

John Dee

John Dee, English magus-mathematician, focused the general elements of Neoplatonism into his area of special interest—the relationship of mathematics to nature. He was enormously well-read and wrote on a wide variety of topics.[7] Two of his works, however, the *Monas hieroglyphica* and the "Mathematical Praeface" are crucial to understanding the role played by the Neoplatonic-Kabbalistic philosophy in his mathematics.

During his lifetime, the *Monas hieroglyphica* was, perhaps, Dee's most acclaimed work.[8] In the *Monas,* he combined geometry with Kabbalistic techniques to explore a geometrical symbol, the hieroglyphic monad, from which other symbols could be derived. Dee believed that the monad was revealed to him by "our IEOVA."[9] It is composed of the point, the circle, the cross, and the sign for Aries. The symbol was first mentioned in Dee's astronomical treatise, ΠΡΟΠΑΙΔΕΥΜΑΤΑ ΑΦΟΡΙΣΤΙΚΑ (*Propaedeumata aphoristica*).[10] Dee delved into Hebrew and the Kabbalah before he composed the treatise on the monad. The results of Dee's Kabbalistic researches was the unpublished and now lost *Cabbalae hebraicae compendiosa Tabella.*[11] In the *Monas,* Dee's studies of mathematics and Kabbalah were blended to reveal the true order of the Universe. The fact that Hebrew letters were also numbers and that they were constructed of points and lines, the bases of geometry and the primary components of the monad, led Dee to the conclusion that the mathematical aspects of the alphabet were the fittest objects for Kabbalistic analysis.[12]

Dee's Hebrew studies acquainted him with the Kabbalistic interpretive techniques, "Gematriam, Notariacon, & Tzyruph." *Gematria* is the numerical interpretation of scripture by adding the value of the letters of a word or phrase and comparing them to the sums of their words and phrases. *Notariacon* is the doctrine that letters represent words of concepts. *Tzyruph* is the derivation of meaning by changing the order of letters in words and phrases, e.g., evil = live. These techniques were the essence of geometrical analysis in his view. Dee felt that Kabbalah would reform geometry and that the "geometer . . . will begin to feel embarrassed, and the principles of his art will seem to him insufficiently established" when he sees them applied to the monad.[13] Likewise, "the Hebrew cabbalist who, when he will see that (the three principal keys to his art, called) Gematria, Notariacon, and Tzyruph, are used outside the confines of the language called holy . . . will call this art [geometrical Kabbalah] holy, too. . . ."[14] Dee believed not that his art was the equal of the Hebrew Kabbalah, but rather, that geometrical Kabbalah was the "real cabbala" which had been lost, leaving to man only the lesser, "vulgar" alphabetical Kabbalah.[15]

For Dee, the monad was not an abstract symbol divorced from reality. It was, rather, a projection of underlying mathematical truth, which, when Kabbalistically interpreted, was "the rebuilder and restorer of all astronomy and an astronomical messenger."[16] The monad was an astronomical reformer in that it contained the elements of astronomy, the circle and the point. From these, the orbits of the planets and the astronomical symbols, such as ♈ (Aries) and ♉ (Taurus), could be derived. Thus Dee described the movements of the Sun and its illumination of the moon in Theorem 15 of the *Monas:* "the Moon, while the Sun's splendour is in Aries, receives a new dignity of light in the next adjoining sign (namely Taurus) and is exalted above her powers."[17] In addition to enabling the astronomer to observe the planets' orbits exactly, the monad would also reveal to the musician celestial harmonies and allow the optician to shape a mirror capable of reducing stones and metals to impalpable powders.[18] Far from fostering vain speculations, Dee hoped that Kabbalistic manipulation of his geometrical symbol would lead to a knowledge of things "concrete and corporeal."[19] He wanted the geometrician, adept in the true Kabbalah, to gain an understanding of nature which conferred genuine power.

In addition to its natural philosophical consequences, he envisioned spiritual insights arising from mastery of the monad. A grasp of the monad would lead to a heightened awareness of the Unity, the source of all harmonies. Knowledge of the Unity gained through geometrical Kabbalah would compel Hebrew Kabbalists to acknowledge God's gifts to all nations and presumably facilitate the unification of Jews and Christians into one faith.[20] Dee thus shared the goal of a unified European religion with the Christian Kabbalists Pico, Reuchlin, and Giorgius, whose primary interests were theological.

Although Dee did not expressly acknowledge the source of his geometrical Kabbalah (other than divine revelation) in the *Monas,* the elements of the monad and his Kabbalistic techniques are closely related to Kabbalistic ideas in Heinrich Cornelius Agrippa's *De occulta Philosophia.* Agrippa, like Dee, quoted Boethius's passage that all was created from number.[21] Furthermore, he stressed the numerical value of Hebrew letters in words which paraphrase the *Sepher Yezirah.*[22] Agrippa wrote that all was formed by letters, their figure, their form, and their signification. He followed *Yezirah* in dividing the Hebrew alphabet into twelve simple, seven double, and three mother letters.[23] The divisions of the alphabet corresponded to the twelve zodiacal signs, the seven planets, and the three substantial elements, fire, water, and earth.[24] Agrippa taught the rudiments of Kabbalistic manipulation and its significance for geometrical figures. He further discussed the circle or "unity," the center and circumference of all things, and symbols, including the cross, which became the components of Dee's monad.

Dee may have learned of yeziratic Kabbalah from Guillaume Postel. During his continental travels, Dee met Postel, who was one of the most learned and idiosyncratic of the Christian Hebraists. Within two years after meeting Dee in France in 1550, Postel published a Latin translation of *Yezirah.*[25] It is difficult to believe that these two magi failed to discuss the Hebrew tradition, Kabbalistic manipulation, and universal language, all elements of the Christian Neoplatonic philosophy they shared.

Dee was also indebted to Pico and Reuchlin. In his "Mathematical Praeface" to Henry Billingsley's translation of the *Elements of Euclid,* Dee used ideas articulated by Pico della Mirandola. For example, the significance of mathematics, Dee taught, arises from its intermediate position between the natural and the supernatural worlds. Dee states that according to Pico's mathematical theses, mathematics is "a way . . . to the searchying out, and understandying of everything, hable to be known."[26] Reuchlin in *De Arte cabalistica* had also stressed the intermediate position of mathematics as a concept taught by Pythagoras in discussions structured to tie that philosopher to the Kabbalah. Dee made identical connections in various theorems of the *Monas.*[27]

The "Praeface" is essentially a plea for the application of mathematics to the searching out of all knowledge. If all is formed of number, then only through mathematics can the cosmos be understood. Ocular demonstrations of this Neoplatonic truth were to be found in architecture and perspective drawing which duplicate nature through numerical harmonies. Dee directed his reader to Vitruvius, Albrecht Dürer, and Agrippa for proof of the "Harmonius and Microcosmicall constitution."[28] Not only scholars, but also tradesmen, artisans, and warriors would benefit from a knowledge of mathematics. The study of mathematics, indeed, was primary to any proper educational reform.

The *Monas* and the "Mathematical Praeface" illustrate John Dee's method of unifying Kabbalah with traditional ideas of Renaissance Neoplatonism. He drew on both Pico and Agrippa for Kabbalistic knowledge. Although Dee does not quote Pico's numerous Kabbalistic theses, they were published with the mathematical theses which he does cite. Pico's infusion of Kabbalah into Renaissance Neoplatonism reinforced the connection between the two traditions. Indeed, the essential teachings of *Yezirah,* that the world was created from letters, numbers, and words, were available to Dee from several sources. The connection of *Yezirah* with Boethius's famous statement that numbers underlay nature was reasonable if not inevitable for Neoplatonists with an interest in the Hebrew tradition. For Dee, the completion of geometry, the highest mathematical art, was found in its coupling with Kabbalistic analysis. The "restored" mathematics was a key both to purify the spirit and to master nature, two goals which Dee shared with his younger contemporary, Johannes Kepler.

Johannes Kepler

Unlike John Dee, who attempted to circumnavigate the Neoplatonic world while emphasizing the mathematical study of nature, Johannes Kepler, after leaving his theological studies, occupied himself primarily, even obsessively, with applying mathematics to the discovery of nature's heavenly harmonies. Kepler was well-versed in the greater intellectual world, but his career, from its beginning as a mathematics teacher through his appointment as court astronomer, did not foster the composition of broader humanistic works. Nonetheless, in Kepler's astronomical works, one can glimpse his broad learning and theological disposition in his connection of Neoplatonism and Holy Writ in an effort to understand God's revelation to man's senses, the Book of Nature. Kepler's devotion to reading the Book of Nature was fired by a vision he had while teaching a mathematics class. During this religious experience, not unlike "IEOVA's" revelation of the monad to Dee, Kepler perceived that the intervals of the planetary orbits corresponded to the five regular geometrical solids. In his words, "the Allpowerful appeared in the summer, after tiring and relentless labor, and revealed to me an important discovery in astronomy: which I have set forth in a special tract."[29]

The *Mysterium cosmographicum* was the tract that grew out of Kepler's encounter with the Divine.[30] The *Mysterium* established Kepler both as a passionate supporter of Copernican astronomy and as an advocate of the Neoplatonic quest for mathematical harmonies. In the poem which greeted the reader of the *Mysterium,* Kepler presented his vision of a world created from number in which Copernicus was the restorer of Pythagorean truth, if not Pythagoras reborn.

> The nature of the Universe, God's motive and plan for creating it, God's source for the numbers, the law for such a great mass, the reason why there are six orbits, the spaces which fall between all spheres . . . here Pythagoras reveals all this to you by five figures. Clearly he has revealed by this example that we can be born again after two thousand years of error, until the appearance of Copernicus, in virtue of his name, a better explorer of the Universe. But hold back no longer from the fruits found within these rinds.[31]

Kepler considered the calling of astronomer to be as sacred as that of theologian. Both pursued the Divine vision. If the revelation underlying the *Mysterium* proved faulty as to its specifics, it was sufficient to motivate Kepler through the years of frustration and trial and error which produced the *Astronomia nova* in 1609.[32] The *Astronomia nova* recreated Copernicus's universe by proving that planets travel in elliptical orbits. However, the culmination of Kepler's efforts to find nature's fundamental mathematical truths was the *Harmonices mundi,* begun in 1599 and completed in 1618.[33]

The *Harmonices mundi* is analogous to Dee's *Monas,* at once theoretical and practical. More sophisticated to modern eyes than Dee's work because it uses a recognizable and rigorous mathematical analysis, it is above all the attempt of a sixteenth-century mathematical adept to comprehend the harmonies underlying nature. The discussion of music theory in Book 2 evinces Kepler's abiding belief that harmonies control the cosmos.[34] Like the *Monas,* it is a guide for one who wishes to discover mathematical relationships in music, astrology, architecture, poetry, and astronomy. As in Dee's great work, these mathematical harmonies are geometrical, not arithmetical, because, as Robert Westman has noted, Kepler was "guided by his 'a priori' assumption that geometrical statements are logically prior to physical propositions about the world."[35]

The *Harmonices* presented Kepler's geometrical Kabbalah, as he declared in a letter to the natural philosopher and alchemist Joachim Tanckius. Kepler wrote of the embryonic *Harmonices:*

> I also play with symbols; I have begun a little work, a *geometrical kabbalah,* which deals with the ideas of the thing of nature in geometry. But when I play I do not forget I am playing. For nothing is proved with symbols, nothing hidden is shown in natural philosophy through geometrical symbols except that known before, unless it is shown by certain reasons to be not only symbolic but also a description of the connection between things and causes.[36]

Kepler apparently did not see his "geometrical kabbalah" as an outgrowth of either the Christian Kabbalah or of traditional Kabbalistic methodology. He, like Dee, treated the manipulation of geometrical figures as the true Kabbalah, which revealed nature's truths, the forms in the mind of the Creator. As he stated in the *Harmonices:*

> Geometry existed before the Creation, is co-eternal with the mind of God, "is God himself" (what exists in God that is not God himself?); geometry provided God with a model for the creation and was emplanted into man, together with God's own likeness–and not merely conveyed to his mind through the eyes.[37]

The alphabetical Kabbalah was a fanciful counterfeit. Kepler took pointed exception to alphabetical, and even to numerical, Kabbalah in an appendix to the *Harmonices* directed at Robert Fludd's *Utriusque cosmici.*[38] Kepler lumped Fludd and his attempts to arrive at cosmic harmonies through Kabbalistic processes with what he viewed as the discredited speculative notions of chemists, Hermetics, and Paracelsians.[39] Fludd was not to be considered a genuine mathematician, because his ideas were mere phantasms divorced from the observable. Kepler distinguished himself from Fludd unequivocally: "One sees that Fludd takes great pleasure in incomprehensible picture puzzles," wrote

Kepler, "whereas I go forth from obscurity into the bright light of knowledge."[40]

Kepler's rejection of Fludd and the fanciful knowledge yielded by his Kabbalah places Kepler's own "geometrical kabbalah" in perspective. Kepler was uninterested in mere mathematical manipulation. The truths revealed on Sinai about the creation could only concern geometry because geometry was prior to creation in the mind of the creator. From Kepler's deep religious convictions and his belief in Pythagoreanism reborn, it was logical for him to assume that God had revealed the mathematical means to understand the geometrical underpinnings of the physical universe. The patriarch Abraham, the prophet Moses, and Plato received the true methodology, the Kabbalah, rooted in geometry and observations. Plato's *Timaeus* was a commentary on Genesis.[41] Kepler, like Dee, looked beyond the ramblings of "vulgar" Kabbalists to a tradition which was rigorous and mathematical.

Kepler's criticism of Fludd is instructive as to the rigor which he expected from the mathematical analysis of observable phenomena. Dee suggested in the "Mathematical Praeface" that mathematics would change the world if applied to physics, architecture, and even theater. Kepler extended that line of thought. From his first work, the *Mysterium,* through the *Harmonices,* Kepler lent no credence to speculative harmonies. Unlike other Neoplatonists, he required his theories to express the geometrical statement which accorded with the empirical data.[42] This goes far to explains the intense effort he put into deriving through trial and error his important third law of planetary motion: the squares of the periods of the orbits of any two planets are as the cube of their mean distances from the Sun. Unlike Fludd, Kepler worked diligently to fit his mathematics to observation and to bring the obscure into the bright light of knowledge. The result was a geometrical Kabbalah imbued with the power to describe and explain nature. Kepler's geometrical Kabbalah did in fact what Dee had hoped the manipulation of his monad would do. It reformed astronomy and revealed the orbits of the planets.

Conclusion

Renaissance Neoplatonism focused attention on harmony. It fostered the belief that the world was created from number and that mathematics, therefore, was the proper tool for discovering the natural harmonies. The discovery of the Jewish Kabbalah and its adaptation to Christian needs reinforced the notions that Neoplatonism was a tradition dating from the patriarchs and that the world was created from number, an aspect of the Hebrew alphabet. One of the proponents of the Christian Kabbalah, Heinrich Cornelius Agrippa, reproduced the basic Kabbalistic ideas of creation and techniques for Kabbalistic interpretation in his *De occulta Philosophia.* In part Agrippa's discussion paraphrased the basic Kabbalistic text *Sepher Yezirah,* which was rendered into Latin by Guillaume Postel. Agrippa and *Yezirah* were sources for those, like Dee, who saw in the Kabbalistic interpretation of numbers a way of discovering harmonies which had been revealed to Abraham and Moses and which had been explored by Pythagoras and Plato.

John Dee represented a branch of the Neoplatonic-Kabbalistic world view which took on a life of its own. He combined general notions of harmony with specific ideas of creation from number, geometry, and Kabbalistic interpretation in his *Monas hieroglyphica.* In the *Monas* and the "Mathematicall Praeface," Dee made a plea for the study of nature through mathematics. The true Kabbalah, geometrical Kabbalah, would reform man's view of the universe and produce changes in technology. Johannes Kepler went beyond Dee in the rigor of his geometrical Kabbalah. He rejected Kabbalistic speculations in favor of a careful, tedious process of fitting mathematics to what was observable in nature. Kepler believed he was called to study astronomy through revelation, but his fidelity to both mathematics and observation required him to perfect the ancient tool of geometry in order to reform astronomy. He was called to continue the task of Copernicus.

The geometrical Kabbalah helped to establish the assumption that mathematical abstractions properly fit with data are descriptive of nature. In time, Kepler's notions of harmony and the use of terms like "geometrical Kabbalah" fell out of vogue. His methodology, the mathematical description of reality, however, remained an indispensable part of the study of nature.

Notes

1. The development of Renaissance Neoplatonism (Neopythagoreanism) is the subject of numerous works. Frances Yates provides an excellent overview of the leitmotifs of the philosophy in *Giordano Bruno and Hermetic Tradition* (London: Routledge & Kegan Paul, Ltd., 1964). From this study, she focused on narrower areas in *The Rosicrucian Enlightenment* (London: Routledge & Kegan Paul, Ltd., 1972), and *The Occult Philosophy in the Elizabethan Age* (London: Routledge & Kegan Paul, Ltd., 1979). In all three works, John Dee is mentioned

prominently as is Robert Fludd. Yates's use of the term "Hermetic" to describe an aspect of Renaissance Neoplatonism was criticized in Robert S. Westman and J. E. McGuire, *Hermeticism and the Scientific Revolution* (Los Angeles: University of California Press, 1977). Charles B. Schmitt, however, produced a more general criticism of the emphasis placed on Neoplatonism by modern Renaissance scholars. He makes a strong case for the vitality of Aristotelianism in the period in *John Case and Aristotelianism in Renaissance England* (Montreal: McGill-Queens University Press, 1983). Specific studies of the Neoplatonic belief in an esoteric tradition, dating from the patriarchs and rediscovered in the Renaissance, are presented by Danton B. Sailor, "Moses and Atomism," *Journal of the History of Ideas* 25 (1964): 3–16, and D. P. Walker, "The *Prisca Theologia* in France," *Journal of the Warburg and Courtauld Institutes* 17 (1954): 204–259. The idea that Moses taught Hebrew wisdom to the Greeks was extremely important in linking the Kabbalah with Neoplatonism. The idea became a commonplace found in sources as divergent as Heinrich Cornelius Agrippa's *De Incertitudine et Vanitate scientiarum et artium* (Cologne, 1530) and Robert Boyle's *The Sceptical Chymist* (London, 1661).

2. Pico della Mirandola's interest in Hebrew was occasioned by his belief that all wisdom was revealed to Moses, a doctrine know as the *prisca theologica.* Pico believed that the study of the Jewish esoteric tradition would not only reveal wisdom but also prove to Jews the truth of Christianity. This doctrine was articulated by Johannes Reuchlin and the Franciscans Franciscus Georgius and Archangelus of Borgo Nuova, among others, who adopted the Kabbalistic view that Kabbalah was part of the oral tradition given to Moses. Impressed with the apparent parallels between Kabbalah and Pythagoreanism, Georgius also suggested that there was a correspondence between the Kabbalistic system and Aristotelianism. Reuchlin has one of the characters in his dialogue, *De Arte cabalistica,* express the opinion that not only Pythagoreanism but all philosophy was rooted in Hebrew sources. (*De Arte cabalistica* [Hagenau, 1516] has been issued in facsimile as *On the Art of the Kabbalah,* with a translation by Martin and Sarah Goodman [New York: Abaris Books, 1983], and reissued with an introduction by Moshe Idel [Lincoln: University of Nebraska Press, 1993]. The original introduction by G. Lloyd Jones confuses John Dee with Robert Fludd as the author of the *Moysiacal Philosophy* [pp. 130 and 131], while the Idel introduction does much to identify the sources of Reuchlin's Kabbalism.)

3. Among the best works dealing with Christian Hebraists and the Kabbalah are Joseph Leon Blau, *The Christian Interpretation of the*

Cabala in the Renaissance (New York: Columbia University Press, 1944); François Secret, *Les Kabbalistes chrétiens de la Renaissance* (Paris, Dunod, 1964); Jerome Friedman, *The Most Ancient Testimony* (Athens, Ohio: Ohio University Press, 1983); and the above-cited works of Frances Yates.

4. John Dee's translation from Boethius's *De Arithmetica* in the "Mathematicall Praeface" to *The Elements of Geometrie of Euclide,* trans. Henry Billingsley (London, 1570), p. 3. The "Praeface" has been reprinted with an introduction by Allen G. Debus as *John Dee, The Mathematicall Praeface* (New York: Science History Publications, 1975); all references to Dee's "Praeface" hereafter may be found in this reprint. Boethius was, of course, but one of a number of Neopythagorean thinkers. Robin Waterfield has translated an anonymous but important late classical Neopythagorean work, *The Theology of Arithmetic* (Grand Rapids, Michigan: Phanes Press, 1988), which discusses the underlying mathematical nature of the cosmos.

5. The *Sepher Yezirah* is probably the earliest extant Kabbalistic work. Its position in Jewish Kabbalah is discussed by Gershom Sholem in *Kabbalah* (New York: Keter, 1974) and by Joseph Dan in *Jewish Mysticism and Jewish Ethics* (Seattle: University of Washington Press, 1986). There are several modern translations: Knut Stenrig (London: Wm. Rider and Sons, Ltd, 1923); Irving Friedman (New York: Samuel Weiser, Inc., 1977); and Harris Lenowitz in *Origins,* ed. Jerome Rothenberg (New York: Anchor Books, 1976), pp. 7–78. Guillaume Postel translated *Yezirah* into Latin as *Abrahami patriarchae Liber Jezirah, sive Formationis mundi* (Paris, 1552). (For biographical studies of Postel, see W. J. Bouwsma, *Concordia mundi: The Career and Thought of Guillaume Postel* [Cambridge, Mass.: Harvard University Press, 1957], and Marion L. Kunz, *Guilluame Postel, Prophet of the Restitution of All Things: His Life and Thought* [The Hague: Martinus Nijhoff, 1981].) *Yezirah* exists in two Hebrew versions, one long and one short. Both describe the same basic alphabetic formation. Awareness of *Yezirah* seems to have increased in the seventeenth century. It is quoted often by Athanasius Kircher in *Obiliscus pamphilius* (Rome, 1650). In that work, Kircher also reproduced and interpreted Dee's monad: pp. 364–379.

6. Hebrew: ברא את עולמו בשלשה ספרים בספר וסופר וספור

7. John Dee had the largest private library in England; see Peter J. French, *John Dee: The World of an Elizabethan Magus* (London: Routledge & Kegan Paul, Ltd., 1972), pp. 40–61. The catalogue of Dee's books has been analyzed and published in Julian Roberts and Andrew G. Watson, *John Dee's Library Catalogue* (London: The Bibliographical

Society, 1990). The library contained not only Neoplatonic and general scientific works, but also histories and a large number of Hebrew books. G. Lloyd George in his *The Discovery of Hebrew in Tudor England: A Third Language* (Manchester: University of Manchester Press, 1983), Appendix 2, has listed Dee's Hebrew library.

8. Two translations of Dee's *Monas hieroglyphica* (Antwerp, 1564) have been made into English: J. W. Hamilton Jones (London: Watkins, 1947); and the acknowledged best translation, C. H. Josten, "A Translation of John Dee's *Monas hieroglyphica* (Antwerp, 1564), with an Introduction and Annotations," *Ambix* 12 (1964): 84–221. A discussion of the Kabbalistic elements of the *Monas* is found in Michael T. Walton, "John Dee's *Monas hieroglyphica:* Geometrical Cabala," *Ambix* 23 (1976): 116–123.

9. Josten, "Dee's *Monas,*" p. 123.

10. The monad appeared on the title page of the ΠΡΟΠΑΙΔΕΥΜΑΤΑ ΑΦΟΡΙΣΤΙΚΑ *Propaedeumata aphoristica* (London, 1558; reprinted London, 1568). In this work, dedicated to Gerard Mercator, the monad is called the "insignia" of the "inferior astronomy," that is, alchemy. The text, however, deals primarily with mathematics and astronomy. See Debus, *John Dee,* p. 6, and Josten, "Dee's *Monas,*" p. 86.

11. Just what the *Cabbalae* was is difficult to determine. It seems to have been a collection of Kabbalastic materials gleaned through Dee's studies of Latin and Hebrew books. Dee's knowledge of Hebrew is uncertain. He possessed several grammars and basic study texts, and in the *Monas* he uses some technical grammar and cryptic terms. *Voarchadumico, voarch beth adumoth,* and *edom* seem to refer to the person illuminated by primordial light who can see in the red stone carbuncle the two "red things," the reigns of fire and water. See Josten, "Dee's *Monas,*" pp. 126 and 137. Dee was acquainted with Reuchlin's *De Arte cabalistica* and the Kabbalah in Agrippa's *De occulta Philosophia.* He owned two copies of the Agrippa and noted it and Reuchlin's work in the margin of British Library, Sloane MS 3188 fol. 12v. See French, *John Dee,* p. 52 and 53. Dee met Guilluame Postel in France prior to Postel's publication of his translation of *Yezirah* and the publication of the *Monas,* and may have benefited from the Hebraist's knowledge. N. H. Clulee discusses Dee's connnection to Postel in *John Dee's Natural Philosophy: Between Science anad Religion* (London: Routledge, 1988), pp. 88 and 209. He ties the two together in a search for the universal language.

12. Josten, "Dee's *Monas,*" pp. 132–133. Dee pointed out in the *Monas* that alphabets are derived from the *sureq* or dot and the *yod* or

line of the Hebrew alphabet. The point and the line are also the source of geometrical figures. See ibid., p. 127.

13. Ibid., pp. 128–129.

14. Ibid., pp. 132–133. The term *tzyruph* is an unusual Kabbalistic term. The usual term is *temunah.* Reuchlin uses *temunah,* but also *tzyruph.* See *De Arte cabalistica,* lib. 3, cap. 16. Given the popularity of *De Arte,* it is probably the source of Dee's use of *tzyruph.*

15. Ibid., pp. 134–135.

16. Ibid., p. 123.

17. Ibid., p. 164.

18. Ibid., p. 131.

19. Ibid., p. 129. As the reference to the monad in the *Propaedeumata* as the insignia of the lesser astronomy indicates, the monad was not only an astronomical messenger, but also an alchemical messenger. The alchemical applications of the monad are touched on by Josten. N. H. Clulee has also illuminated this aspect of Dee's philosophy in "John Dee's Mathematics and the Grading of Compound Qualities," *Ambix* 18 (1971): 178–211.

20. Josten, "Dee's *Monas,*" p. 133. French discusses Dee's religious views and his desire to unite Christians and Jews in *John Dee,* p. 135.

21. Henry Cornelius Agrippa, *Of Occult Philosophy,* trans. J. F. [John French] (London, 1651), book 2:170.

22. Ibid., book 1:161. Agrippa discussed the techniques of Kabbalistic exegesis in book 2. Agrippa's and Dee's ideas were remarkably similar to those of the author of *The Theology of Arithmetic* (note 4, above). The *Theology* discusses the monad, the dyad, the triad, etc., and their meanings.

23. Agrippa, *Occult Philosophy,* book 1:161. The paraphrasing is apparent when compared to the Hebrew phrase and English translation cited in note 5, above.

24. Ibid., book 1:161. Frances Yates discusses Agrippa at length in her *Occult Philosophy in the Elizabethan Age,* pp. 37–60. For a more complete study of Agrippa, see Charles Nauret, *Agrippa and the Crisis of Renaissance Thought* (Urbana, Ill.: University of Illinois Press, 1965).

25. See note 5.

26. Dee, "Praeface," 1 recto. Dee owned a set of Pico's works. French, *John Dee,* p. 51.

27. *De Arte cabalistica,* lib. 2, cap. 43. Cusanus was a major popularizer of the Platonic idea that mathematics was intermediate between God and the creation. Dee also cited Cusanus: "Praeface," Aiii verso. Clulee discusses Cusanus and Dee in *John Dee's Natural Philosophy,* pp. 153–154. In an unpublished paper on the numerical game *rithmomati-*

cia, Anne E. Moyer illustrates the widespread acceptance of Neoplatonic mathematical ideas: "Mathematics, Philosophy, and Learned Leisure: The Philosopher's Game in Sixteenth-Century Europe."

28. Dee, "Praeface," ciiij recto. The importance of Neoplatonic mathematics to art in the sixteenth century is discussed by Charles Carmen in an unpublished paper, "Leonardo's *Vitruvian Man:* A Renaissance Microcosm." Professor Carmen argues that the two positions occupied by Leonardo's Vitruvian Man depart from Vitruvius, but are true to the geometrical harmonies which put man between the greater and the lesser worlds. Dürer's work on perspective, *Underweysung der Messung mit Zirkel und Richtscheyt* (Nuremberg, 1525), as well as Alberti's *Della Pittura* (1436; first printed Basel, 1540), emphasize art as a replication of nature and society.

29. Johannes Kepler, letter to Duke Frederick of Württemberg, February 17, 1596, in Johannes Kepler, *Gesammelte Werke,* ed. W. v. Dyck and Max Caspar (Munich: C. H. Beckische Verlag, 1938–), 13:218, translation ours. The original German reads: "Demnach der Allmechtig verschinen Sommer nach langwürringer ungesparter mühe und vleiss mir ein Haupt*inventum* in der *Astronomia* geoffenbaret: Wie solliches ich in ein besondern Tractätl aussgefuhrt, und allberaitt Zupublicirn in willens." Agrippa discussed the celestial powers and the regular solids in *Occult Philosophy,* book 2:253–254.

30. Kepler reprinted the *Mysterium,* originally published in Tübingen in 1596, twenty-five years later, Frankfurt in 1621, with commentary. This edition has been translated into English by A. M. Duncan in Johannes Kepler, *Mysterium cosmographicum* (New York: Abaris, 1981).

31. The poem exists in two version, one appended to Kepler's letter to Duke Frederick concerning his revelation (see note 27) and one in the *Mysterium.* The translation of the *Mysterium* text is by Duncan (see note 28). Both versions look to Copernicus as the restorer and improver of Pythagorean knowledge. The variations in the Latin text are interesting.

The letter version reads:

Quid mundus, quae causa Deo, ratioque creandi.
Unde Deo numeri; quae tantae regula moti;
Quod faciat sex circuitus; quo quaelibet orbe
Intervalla cadant; cur tanto Jupiter et Mars
Orbibus haud primis interstinguantur hiatu;
Hic te Pythagoras docet omnia quinque figuris.
Scillicet exemplo docuit, nos posse renasci.

Bis mille erratis dum fit Copernicus annis,
Hoc, meliar mundi specalator, nominis. At tu
Frogibus inventis, quibus est e glande voluptas,
Cum grege porcorum poseas ex lintre subulcos.

The poem is the same in the *Mysterium,* except for the last two lines, which are replaced by "Glandibus inventas noli postponere fruges." Kepler's printed version thus does not associate his readers with swine. We translate the letter version thus:

What is this Universe, its creation's need?
How from God did its being proceed?
Whence by Deity was it counted?
What great pattern laid, before its construction mounted?
How came the planets circuits six?
In which orb does the rhythm of their intervals mix?
Though not among the first, why are Jupiter and Mars
Divided by an expanse more suitable for stars?
All this does Pythagoras teach with five figures pure of form.
And shows by example that we can be reborn.
Two thousand years you wandered until Copernicus came,
A better cosmic student, a man renowned of name.
Now you, by his fruitful invention, know pleasure sublime
Yet still you feed the herdsmen from the same trough as the swine.

The poem is especially interesting in that it evinces Kepler's adherence to the Neoplatonic idea of Pythagoras, or Pythagoreanism, reborn. See Blau, *Christian Interpretation,* pp. 41–64, for a discussion of "Pythagoras Redivivus."

32. Johannes Kepler, *Astronomia nova* (Heidelberg, 1609). The liveliest discussion of Kepler and his discoveries is found in Arthur Koestler, *The Sleepwalkers* (New York: Macmillan Co., 1959). The classic and most complete biography of Kepler is Max Caspar, *Johannes Kepler* (Stuttgart: Kohlhammer, 1948), trans. ed. C. Doris Hellman (London: Abelard-Schuman, 1959).

33. Johannes Kepler, *Harmonices mundi* (Frankfurt, 1619), translated into German by Max Caspar, *Johannes Kepler, Welt Harmonik* (Munich: R. Oldenbourg Verlag, 1967).

34. J. V. Field in *Kepler's Geometrical Cosmology* (Chicago: University of Chicago Press, 1988) discusses book 2, Kepler's search for geometrical harmony, in great detail. Field stresses Kepler's Platonism and

its divergence from that of Neoplatonists like Robert Fludd. See especially pp. 171–190.

35. Westman, *Hermeticism,* p. 67.

36. Johannes Kepler, letter to Joachim Tanckius, May 12, 1608, *Gesammelte Werke,* 16:158, translation and emphasis ours. The letter was written in response to some speculative harmonies suggested by Reinhard. After attacking speculative harmonies, Kepler declared, "I believe geometrical proportions were the pattern of the Creator which introduced like ideas into the world."

> Hanc ego geometicam proportionem puto Ideam fuisse Creatori ad introducendam generationem similis ex simili:

He then went on to say that he also played games, like those of Reinhard, but unlike Reinhard, he knew them to be games. He then commented on symbols and natural philosophy:

> Ludo Quippe et ego Symbolis, et opusculum institui,
> Cabalam Geometricam, quae est de Ideis rerum Naturalium
> in Geometria: sed ita ludo, ut me ludere no obliuiscar.
> Nihil enim probatur symbolis, nihil abstrusi eruitur
> in Naturali philosophia, per Symbolas geometricas,
> tantum ante nota accomodantur: nisi certis rationibus
> euincatur, non tantum esse Symbolica sed esse
> descriptos connexionis rei utruisque modos et causes:

37. Kepler, *Harmonices,* in *Gesammelte Werke,* 6:223, translated by Koestler, *The Sleepwalkers,* p. 262.

38. Robert Fludd's *Utriusque cosmici* (Oppenheim, 1617). Robert Fludd is one of the most important figures in the late sixteenth- and early seventeenth-century debate over the subject matter and methodology of natural philosophy. Allen G. Debus has ably and exhaustively explored Fludd's role in natural philosophy in numerous articles and monographs. Perhaps his most succinct study is "The Chemical Debates of the Seventeenth Century: The Reaction to Robert Fludd and John Baptiste van Helmont," in *Reason, Experiment, and Mysticism,* ed. M. L. Righini Bonelli and William R. Shea (New York: Science History Publications, 1975), pp. 19–47. An excellent summary of Fludd's philosophy has been published by Jocelyn Godwin, *Robert Fludd: Hermetic Philosopher and Surveyor of Two Worlds* (Boulder, Colo.: Shambhala, 1979). William H. Huffman, *Robert Fludd and the End of the Renais-*

sance (London: Routledge, 1988), reworks the traditional materials and adds some insights from social history.

39. Kepler, *Harmonices,* in *Gesammelte Werke,* 6:374. The passage reads, "Illud quidem familiare est Chymicis, Hermeticis, Paracelsistis; hoc proprium habent Mathematici."

40. Ibid., translation ours. The passage reads, "Videas etiam, ipsum plurimum delectari rerum aenigmatibus tenebrosis, cum ego res ipsas obscuritate involutas in lucem intellectus proferre nitar." Caspar comments on the passage in *Johannes Kepler* (English trans.), p. 292, as does Westman, *Hermeticism,* pp. 59–67, and Field, *Cosmology,* pp. 179–190.

41. Kepler discusses Moses' Egyptian knowledge in *De Stella nova,* in *Gesammelte Werke,* 1:173 (originally published in Prague, 1606). He identifies the *Timaeus* as a commentary on Genesis in *Harmonices mundi,* in *Gesammelte Werke,* 6:221. Kepler clearly expresses the idea that Pythagoras and Plato were part of a tradition begun by the Hebrews and seen in its purity in Genesis. See also his *Astronomiae Pars optica,* in *Gesammelte Werke,* 2:198–199.

Kepler's use of the word Kaballah to describe the harmonies is best seen in light of the *prisca* tradition. Kabbalah commonly referred to mystical knowledge revealed at Sinai. John French, in his translation of Sendivogius's *New Light of Alchemy* (London, 1650), pp. 315–316, defined Kabbalah as "a most secret science, which is said to be delivered by Divine Inspiration, together with the Law of Moses."

42. Westman in *Hermeticism* expresses this crucial difference between Kepler and other Neoplatonists as the "important recognition, never appreciated by Hermetic natural philosophers, that a theory of physical causes is insufficient without the primacy of empirically and geometrically controlled statements about nature" (pp. 67–68).

THE THEOLOGICAL FOUNDATIONS OF DARWIN'S THEORY OF EVOLUTION

Robert J. Richards

Allen Debus has striven to demonstrate that religion, in the early modern period, was not a foreign agent that bent science away from its true course, nor even an extrinsic force that benignly pushed thinkers more quickly along a path they would have otherwise trod. Allen has tried to show that religion was enmeshed within the interstices of scientific thought, that only from our contemporary perspective can we pick out those strands that unrolled toward modern science and those that folded around the Bible. Under this conception it would be foolish to sift out religion from our understanding of early modern science. Among recent historians, some would concede Allen's historiographic axiom, but they would undoubtedly confine its use to the time when Paracelsus commanded attention. I think, however, that Allen's perspective can be useful in trying to understand even that science which was believed to replace religion during the last half of the nineteenth century, namely Darwin's theory of evolution.

Andrew Dickson White found in the confrontation of evolutionary theory with traditional orthodoxy the very archetype of the warfare that he chronicled in his book *History of the Warfare of Science with Theology in Christendom,* which appeared at the end of the nineteenth century. "Darwin's *Origin of Species,*" White declared, "had come into the theological world like a plow into an ant-hill. Everywhere those thus rudely awakened from their old comfort and repose had swarmed forth angry and confused."[1] Many churchmen were indeed angry and confused. The Anglican Bishop of Oxford, Samuel Wilberforce, was a divine in high dudgeon when he inquired of Thomas Henry Huxley, Darwin's champion, whether it was through his grandmother or through his grandfather that he claimed his descent from a monkey.[2] Cardinal Manning, the Catholic primate of England, certainly objected to what he thought "a brutal philosophy—to wit, there is no God and the Ape is our Adam."[3]

Most contemporary thinkers who have reflected on what Darwin wrought would probably agree with the judgment, though surely not the sentiment, of a correspondent at that time for the *Catholic World,* who wrote: "Mr. Darwin is, we have reason to believe, the mouth-piece or chief trumpeter of that infidel clique whose well known object is to do away with all idea of a God."[4] The "infidel clique," which the author decried, has grown ever larger as Darwinian theory has developed toward the present time. Certainly such scientists as Richard Lewontin,

P. H. Theerman and K. Hunger Parshall (eds.), Experiencing Nature, 61–79.

Steven Rose, and Leon Kamin would apparently wish to claim membership. In their book *Not in Our Genes,* they express both intellectual and historical orthodoxy when they avow: "Natural-selection theory and physiological reductionism were explosive and powerful enough statements of a research program to occasion the replacement of one ideology—of God—by another: a mechanical, materialist science."[5] According to the historian of science Susan Cannon—to add one more voice to this contemporary chorus—Darwin's research program drained nature of moral significance; he had shown nature to be, as Cannon put it, "morally meaningless."[6]

Upon closer inspection, however, the war declared by White and apparently concluded by the victory of evolutionary theory even before Darwin's death seems more like ritualized combat: as with bighorn sheep, much head-butting, but no real damage done. That is, if you consider the views of Darwin's colleagues and the next generation of his disciples, you quickly come to two conclusions: first, that evolutionary theory more readily supported metaphysical spiritualism than materialism; and second, that God was made to feel right at home in this evolving universe.[7] Charles Lyell, Darwin's close scientific friend, finally converted to evolutionism, but secured his new belief within a more general theism. Alfred Russel Wallace, co-founder of evolution by natural selection, broke with Darwin on the question of man; he maintained that human mental and moral development was under the aegis of higher spiritual powers. George Romanes, Darwin's anointed disciple, thought materialism to be destroyed by evolutionary theory. And William James employed natural selection to argue for an independent mind, human freedom, moral choice, and divine hegemony. If Darwin's research program required the rejection of God and his replacement by universal mechanism, few later evolutionists seem to have been aware of this.[8]

Maybe, though, they really did not notice what was actually there. Darwin's own inchoately expressed metaphysical views, after all, did seem materialistic—though benignly so, I believe; and he did profess agnosticism in his later years, even going so far as to condemn Christianity. In his *Autobiography* he described his religious trajectory in this way:

> disbelief crept over me at a very slow rate, but was at last complete. The rate was so slow that I felt no distress, and have never since doubted even for a single second that my conclusion was correct. I can indeed hardly see how anyone ought to wish Christianity to be true: for if so, the plain language of the text

> seems to show that men who do not believe, and this would include my Father, Brother and almost all of my best friends, will be everlastingly punished. And this is a damnable doctrine.[9]

Passages such as this appear to endorse the traditional view of Darwin's accomplishment, namely, that he rejected a universe ruled by divine agency, one in which all of nature works toward the good. Nonetheless, I think the traditional account misses the deeper layers of Darwin's theory, which, I believe, was molded in the forms of nineteenth-century theology. A more perspicuous rendering of Darwin's ideas would find, I think, that Darwin created natural selection in the image of God and that he understood its action to infuse moral values into nature, not to suck them from nature. Had Darwin not theologized natural selection, had he been of the naturalizing mind of, say, Thomas Henry Huxley, I believe his theory would have been constructed quite differently, or so I will argue below.

In what follows, I will first indicate how Darwin constructed his theory of evolution against the doctrine of special creation. In this construction, he cast the familiar God out of the immediate biological sphere, just as he had been rejected from the astrophysical sphere sometime earlier. Second, I will argue that Darwin's description of the process of evolution and the law he formulated to explain the process were yet designed according to theological models. The divine ghost could not be exorcised. Finally, I will try to show how Darwin specifically reconstructed both human nature and Nature writ large with a moral spine.

Darwin's Opposition to Creationism

While on the five-year voyage of the *Beagle,* which began in 1831, Darwin remained both biologically and theologically orthodox. He recalled that while on board, he was "heartily laughed at by several of the officers (though themselves orthodox) for quoting the Bible as an unanswerable authority on some point of morality."[10] His biology was just as bully.

Yet Darwin's biological faith must have swayed as a bending reed. After his return to England in October, 1836, he rapidly concluded that species were not stable, rather that they changed over time. From early spring of 1837 through 1839, he worked out his theory of evolution, and just as steadily the religious creed of his birth ebbed away. He recorded his intellectual development in a series of notebooks in which he pitted his new theory against the theological and yet scientifically respectable conception of special creation. In his first transmutation notebook, *Notebook B,* he argued that certain useless anatomical structures gave the lie to the belief that the Creator immediately designed each of his creatures. As Darwin put it in this notebook: "When one sees nipple on man's breast, one does not say some use, but sex not having been determined.—so with useless wings under elytra of beetles.—born from beetles with wings & modified—if simple creation, surely would have been born without them."[11] Male nipples seemed to Darwin only a vague biological memory of a time when hermaphrodites, something like Empedocles' round-men, wandered the earth.

In the notebooks and even when he wrote the *Origin of Species,* Darwin had not yet given up theism. But he had reduced God's activity to that assigned by Enlightenment Deism, and indeed, by advanced theology: God might have set the universe rolling as a primary cause, but then he left its particular destiny to the inexorable workings of the secondary causes described by natural law.[12]

Darwin had fair precedent for rejecting the idea that God immediately intervened in the world to set it aright. By the early nineteenth century, British natural theologians had begun to argue that it ascribed a greater intellectual power to the Creator to have formed the world through law than through dim brute force. Charles Babbage, author of the Ninth Bridgewater Treatise on the "Power Wisdom and Goodness of God as Manifested in his Creation," contriver of the analytical engine, and frequent dining companion of the young Darwin[13]—declared that "to have foreseen the changes to occur after the creation and to have provided, by one comprehensive law, for all that should ever occur, either to the races themselves, to the individuals of which they are composed, or to the globe which they inhabit, manifests a degree of power and of knowledge of a far higher order [than that required to intercede at each step]."[14] So when Darwin, in his early notebooks, suggested that it bespoke a grander idea of the Creator to believe that he worked through fixed laws, he merely endorsed the growing modern consensus of theologians of his day.

In his Essay of 1842, in which he brought together his evolutionary ideas for the first time in a coherent and continuous sketch, Darwin even considered some theological justification for making natural law the substitute for God's direct hand: it seemed to him beneath the dignity of God to have directly created every creature, especially the aesthetically unpleasing ones and those of disgusting habits, like sea slugs and maggots. On the other hand, if creation were through law, then God's dignity might be preserved, while the dirty work would be accomplished through lawful processes—or at least so Darwin suggested.[15]

In the *Origin of Species,* which was published in 1859, Darwin argued, of course, that we ought to regard the particular formation of animals to be the work of evolution by natural selection rather than the result of special creation. Yet in this work of twenty-years maturation, he continued to suggest that the laws of evolution, those secondary laws, ought best be conceived as God's commands. As he concluded in the last chapter of the *Origin:*

> To my mind it accords better with what we know of the laws impressed on matter by the Creator, that the production and extinction of the past and present inhabitants of the world should have been due to secondary causes, like those determining the birth and death of the individual. When I view all beings not as special creations, but as the lineal descendants of some few beings which lived long before the first bed of the Silurian system was deposited, they seem to me to be ennobled.[16]

Darwin's remarks about the role of the Creator in establishing laws appears in the *Origin*—though not, I think, in the earlier essays—to have been uttered as much in defense against the rigidly righteous who might quickly slam his book shut before carefully considering his argument—at least as much as that kind of rhetorical defense, as an expression of firm religious conviction. Yet it would be a mistake to think that the *Origin of Species* was composed, as Lewontin and such historians of science as Will Provine believe, in the dead night of atheism; though, perhaps, the twilight of agnosticism was beginning to spread over Darwin's vision. The pall of disbelief fairly exudes from his letter to Asa Gray, an American botanist and divine, in May of 1860:

> With respect to the theological view of the question. This is always painful to me. I am bewildered. I had no intention to write atheistically. But I own that I cannot see as plainly as others do, and as I should wish to do, evidence of design and beneficence on all sides of us. There seems to me too much misery in the world. I cannot persuade myself that a beneficent and omnipotent God would have designedly created the Ichneumonidae with the express intention of their feeding within the living bodies of Caterpillars, or that a cat should play with mice. Not believing this, I see no necessity in the belief that the eye

> was expressly designed. On the other hand, I cannot anyhow be contented to view this wonderful universe, and especially the nature of man, and to conclude that everything is the result of brute force. I am inclined to look at everything as resulting from designed laws, with the details, whether good or bad, left to the working out of what we may call change. Not that this notion *at all* satisfies me. I feel most deeply that the whole subject is too profound for the human intellect. A dog might as well speculate on the mind of Newton. Let each man hope and believe what he can.[17]

Later, however, even these sentiments gradually eroded, and Darwin came to think himself a miserably weak coward to have capitulated to religious opinion by including references to the Creator in the *Origin.* In the *Descent of Man,* though, he redeemed himself. In that publication of 1871, Darwin guardedly described the evolution of religious sentiment, suggesting that it arose naturally from man's fear and superstition. He even made bold to intimate that religious belief in invisible spirits had the same origins as the apparently similar belief his little dog must have harbored as it went barking after a wind-blown parasol that appeared to be carried aloft by an invisible hand.[18]

The Theological Structure of Darwin's Theory

So Darwin cast out the enemy, slowly but finally, or so it seems. Yet, a more careful examination of the way Darwin constructed his theory reveals, think, that the ghost of the Creator had come to invest the very clever and beneficent hand of natural selection itself. From the time he first began to work on his theory of evolution until he penned the *Origin,* Darwin believed in God's governance; and, as I wish to argue, he thought of natural selection as the mediating instrument of that governance. In the years after 1859, Darwin slipped the grasp of the traditional Creator, but the divine spirit lingered within the very mechanism that seems to represent the golem of modern thought, natural selection.

The device of natural selection, as it has been subsequently refined out of Darwin's theory, expresses his discovery that more creatures are produced within a species than the environment can support. Those born with any accidental modifications giving a slight advantage over others of their kind will stand a better chance of reaching reproductive age and of passing on their heritable modifications. Thus the species will be gradually altered. For contemporary biologists, "natural selection" is merely the term that expresses these systematic causal relationships obtaining among the environment, an individual, its traits, and its offspring.

This skeletal idea, however, was initially formulated by Darwin in a livelier way, fleshed out and animated by a different spirit. In its original guise, natural selection seemed to have properties that many later Darwinians would wish to deny, and even the elder Darwin reflected with some hesitation on his original conception. In the *Descent of Man,* he recalled how he had portrayed natural selection in the *Origin* as operating with too much omniscience and thus as producing creatures all of whose traits were beneficially adapted. He later wished to acknowledge that many traits displayed by creatures neither worked to their advantage nor to their disadvantage; they were benignly neutral. As he confessed in the *Descent of Man:* "I was not able to annul the influence of my former belief, then widely prevalent, that each species had been purposely created; and this led to my tacitly assuming that every detail of structure, excepting rudiments, was of some special, though unrecognized, service."[19] In this typical crisis of scientific conscience, Darwin thought he had invested natural selection with too much of the wisdom of Paley's God.

In his Essay of 1842, it is even clearer that Darwin theologized natural selection. The essay, really a set of rough notes, shows him struggling with those ideas that appear to have dropped so easily from the pen that composed the *Origin of Species.* When he begins to formulate for himself the process of natural selection, he pictures it, if not as identical to the Creator, then as a supernaturally wise being that has an eternity to work its will. At the end of the following drifting passage, however, Darwin does seem to unmask the being as God himself.

> If a being infinitely more sagacious than man (not an omniscient creator) during thousands and thousands of years were to select all the variations which tended towards certain ends [(or were to produce causes which tended to the same end)], for instance, if he foresaw a canine animal would be better off, owing to the country producing more hares, if he were longer legged and keener sight,—, greyhound produced. . . . Who, seeing how plants vary in garden, what blind foolish man has done in a few years, will deny an all seeing being in thousands of years could effect (if the Creator chooses to do so), either by his own direct foresight or by intermediate means,—which will represent the creator of this universe.[20]

By 1844, Darwin was very sure that his theory of evolution by natural selection was both revolutionary and, more importantly, powerful enough to raze the citadels of orthodoxy. He thought it behooved him, then, to take measures to preserve his theory and his own fame. He wrote out another essay at greater length and had a copyist produce a fair copy. Darwin then made out his will, leaving four hundred pounds for the publication of his essay. He did not wish his ideas to be interred with him.

In the 1844 essay, Darwin developed, over several pages, the model of an"imaginary Being," as he called it, who would carefully select advantageous traits.[21] And in the *Origin of Species,* this being reappears, infinitely sagacious and determined to produce the most good possible in creation. Darwin described its activities, now in the guise of natural selection, this way in the *Origin:* "It may be said that natural selection is daily and hourly scrutinising, throughout the world, every variation, even the slightest; rejecting that which is bad, preserving and adding up all that is good; silently and insensibly working, whenever and wherever opportunity offers, at the improvement of each organic being in relation to its organic and inorganic conditions of life."[22]

Now it might be thought that Darwin had merely clothed an entirely cold and bloodless mechanism in some picturesque metaphor in order to make more vivid for his readers the reality of natural selection. Darwin's strategy, of course, was at least this. But the image really did, I think, control the development of his entire theory of evolution, leading him to describe the process in a way differently than he otherwise might have. Let me point to six immediate consequences of his theologized conception of natural selection.

First, from the early notebooks through the essays to the *Origin of Species,* Darwin, while he rejected special creation, retained the notion of creation through law—and his divinized idea of natural selection formed the link joining God as primary cause with biological nature as product. In the Essay of 1842, he emphasized the difficulty of conceiving of law as having the requisite power to fashion the most intricate contrivances. Yet law could conceivably have this power if it were the extended arm of the Deity. As he put it in the essay:

> Doubtless it at first transcends our humble powers, to conceive laws capable of creating individual organisms, each characterised by the most exquisite workmanship and widely extended adaptations. It accords better with [our modesty] the lowness of our faculties to suppose each must require the fiat of a creator, but

> in the same proportion the existence of such laws should exalt our notion of the power of the omniscient Creator.[23]

The residual of these original considerations appears at the end of the *Origin*'s last chapter, which I have cited above (see text to note 16). For Darwin, natural selection was more than a blind force of nature. It functioned as the surrogate creator operating according to divine command.

As the above quotation from the 1842 essay suggests, this conception of natural selection as a stand-in creator had a facilitating role even in Darwin's convincing himself that a complex of natural causes could produce exquisite adaptations, such as the mammalian eye. From the time he read Paley's *Natural Theology,* Darwin never doubted that organs like the eye—Paley's favorite example—were adaptations of extreme perfection, hardly the sort of thing a machine could produce. Darwin later confessed that whenever he thought of the eye his blood ran cold. But if the agency producing an eye were a lesser God, then such production might well be intelligible.

Darwin's theologized conception of natural selection had a second important consequence. Because natural selection wore the mantle of the Creator, it also took on the wisdom of God: it daily and hourly scrutinized, throughout the world, every variation, even the slightest; not a sparrow fell that it was not aware. If it were like a Manchester spinning loom, the outcome would not have been fine damask. Only a skillful hand could spin that, or the fabric of the eye. Moreover, if like God, natural selection was always working to evaluate every minute variation, selecting always the best, no matter how small, then evolution had to proceed gradually and continuously, rather than hesitatingly and saltationally, the way it would have to proceed if cranked out by a nineteenth-century machine. Indeed, the machine image of natural selection—which comes so easily to us as we reflexively refer to Darwin's *mechanism*—played no role in his own formulation of the "law" of natural selection; the word "mechanism" does not even appear in the *Origin.* The law of natural selection radiated omniscience, power, and exquisite sensitivity. Machines clanked out products stupidly, fitfully, and crudely. They operated in the sink of factories, not on empyrean heights whence the law of natural selection was given.

Huxley thought his friend Darwin to have committed himself to too much by assuming gradual and continual evolution. Huxley was a mutationist—a punctuationist of an earlier vintage. Darwin had briefly entertained the possibility of evolution by sports, as macro-mutations were then called; but he rejected that option. Huxley would have been even more chagrined had he been aware of one of the important factors leading Darwin to reject saltationism for gradualism.[24]

A third contribution of theological presumption to natural selection theory follows hard on the first two. Since natural selection was comparable to a divine selector, Darwin understood it to work like artificial selection. That is, Darwin could use artificial selection as a model for natural selection—as he does in the *Origin*—only insofar as he conceived them both to be the work of intelligence at some level of expression. This link of artificial with natural selection, made through the assumption of a theologized process, explains one of the very curious claims of the *Origin*. This is a claim that no modern evolutionist would accept: namely, that large numbers of a given species in one location promote faster evolution. In the *Origin*, Darwin first remarks on an important condition for the success of artificial selection, namely, that breeding flocks be large, since "variations manifestly useful or pleasing to man appear only occasionally, the chance of their appearance will be much increased by a large number of individuals being kept; and hence this comes to be of the highest importance to success."[25] He then, some pages later, reintroduces this condition as one necessary for the success of natural selection: "A large number of individuals, by giving a better chance for the appearance within any given period of profitable variations, will compensate for a lesser amount of variability in each individual, and is, I believe, an extremely important element of success."[26] With large flocks, the absolute number of favorable variations will certainly increase, but the proportion of favorable to unfavorable (or neutral) variations should remain constant; and, indeed, large flocks will be even more subject to the phenomenon of swamping-out (when favorable varieties breed with unfavorable). Only if natural selection acts intelligently and with foresight, that is, like a divine selector, will large numbers avail.

The fourth real difference Darwin's theologized version of natural selection made to the general idea of species descent concerned his assumption about the utility of traits exhibited by organisms. The infinite wisdom accorded natural selection meant that it produced only traits that were useful, real adaptations. Darwin probably exaggerated, in that passage from the *Descent of Man*, the degree to which he found only useful adaptations in biological organisms. His conception of natural selection as having a wisdom that passeth understanding meant, nonetheless, that most all traits exhibited by an organism should work to that organism's benefit—all should be useful. In later editions of the *Origin*, Darwin had to retreat a bit from that presumption.

Fifth, not only would natural selection endow creatures with only useful traits, as befitting the surrogate creator, it would crown them with only perfect adaptations, adaptations that could not be better. This was the view, as Ospovat has cogently argued, that Darwin held up to the time he composed the *Origin of Species.*[27] For instance, in the large manuscript he was working on in the mid-1850s, when he received the letter from Alfred Russel Wallace announcing a theory virtually identical to his own, he still argued for perfect adaptation: "Can we wonder then, that nature's productions bear the stamp of a far higher perfection than man's product by artificial selection. With nature the most gradual, steady, unerring, deep-sighted selection,—perfect adaption to the conditions of existence,—the direct action of such conditions—the long-continued effects of habit & perfect training, all concur during thousands of generations."[28] In the *Origin,* Darwin moderated this conception. He recognized that "[n]atural selection tends only to make each organic being as perfect as, or slightly more perfect than, the other inhabitants of the same country with which it has to struggle for existence."[29] Yet, he still conceived of natural selection's work as really quite divine, as he maintained in a passage in the *Origin* comparable to the one just cited from his large manuscript: "Can we wonder, then, that nature's productions should be far 'truer' in character than man's productions; that they should be infinitely better adapted to the most complex conditions of life, and should plainly bear the stamp of far higher workmanship?"[30]

Finally, if natural selection really were tacitly identified by Darwin with nature's God, then we might expect it to work only for the good, for the moral good of creation. Man's selection of animals was vain and selfish; he chose whimsically and only to suit himself, whereas natural selection was unselfish, at least Darwin so judged in the *Origin,* where he observed: "Man can act only on external and visible characters: nature cares nothing for appearances, except in so far as they may be useful to any being. She can act on every internal organ, on every shade of constitutional difference, on the whole machinery of life. Man selects only for his own good; Nature only for that of the being which she tends."[31]

Theologians from very ancient times have had to conjure with the paradox of a morally good God creating a universe of pain and suffering. Darwin considered, as every thinking religious person must, the deep difficulty this presented. Initially he attempted to mitigate the moral responsibility of the Creator by shifting the blame to contingency: God created perfect and good laws, but an intractable matter introduced evil. Ultimately, of course, this explanation will not do. After all, it is an infinitely powerful God that triumphs in the Christian tradition. Darwin, like many theologians before him, attempted another kind of explanation of evil: namely, that a certain amount of evil was necessary to bring forth good; it was thus only apparent evil. At the end of his chapter on the "Struggle for Existence," chapter three of the *Origin,* Darwin splashed the oil of consolation on his reader: "When we reflect on this struggle, we may console ourselves with the full belief, that the war of nature is not incessant, that no fear is felt, that death is generally prompt, and that the vigorous, the healthy, and the happy survive and multiply."[32] It is a good universe after all.

Human Moral Nature

Darwin not only reinfused nature with value, he was especially keen not to leave man morally naked to the world, or so I have argued in a recent book.[33] But this is not the general view of Darwin's construction of morality, as the quotation from Susan Cannon (cited in the introduction, above) suggests. Michael Ghiselin also makes the easy and usual assumption about Darwinian morals. Darwin, he declares, recognized that "since it furthers the competitive ability of the individual and his family, an 'altruistic' act is really a form of ultimate self-interest."[34] This evaluation of Darwin's position certainly makes him a forerunner of Ghiselin's brand of sociobiology; but it completely misses what Darwin himself thought to be distinctive of his biology of morality, namely, that it overturned utilitarianism.

During his five-year voyage on H.M.S. *Beagle,* Darwin experienced the extremes of moral behavior, from the brutality he frequently observed among the South American gauchos to the nobility of the Indians whom they slaughtered. Darwin yet perceived within this moral diversity some common principles of conduct, principles that his Cambridge reflections on Paleyan moral theory helped him to recognize. While an undergraduate, Darwin had to get up for his exams William Paley's *Moral and Political Philosophy.* The position worked out in this book provided him with a framework through which to weave his emerging biological theory of behavior.

On 8 September 1838, about three weeks before he read Malthus, who inflamed the idea of natural selection in his imagination, Darwin reread Paley, who also sparked an idea. In his *Notebook M,* he considered "Paley's Rule." In *Moral and Political Philosophy,* Paley offers his rule of "expediency" as the central axiom of his ethics. He states it this way: "Whatever is expedient is right. But then it must be expedient on the whole, at the long run, in all its effects collateral and remote, as well as in those which are immediate and direct."[35] Darwin gave this rule a biological interpretation: "Sept 8th. I am tempted to say that those actions which have been found necessary for long generation, (as friendship to fellow animals in social animals) are those which are good & consequently give pleasure, & not as Paley's rule is then that on long run *will* do good.—alter *will* in all such cases to *have & origin* as well as rule will be given."[36]

In this construction, Darwin has done something quite typical. He has given a biological interpretation to a principle that seems, on the surface, adequately to capture a certain phenomenon. To understand this particular interpretation, though, it must be borne in mind that Darwin had not yet, in early September, formulated his principle of natural selection. Prior to reading Thomas Malthus, who supplied him the insight that coagulated his speculations into the form of his chief principle, Darwin had a different device by which to explain the evolution of species. He believed that the habits an animal might adopt to meet survival needs in its environment would eventually become instinctive, that is, innate and heritable. This conviction expressed an idea that Darwin harbored quite early on and never relinquished, namely, that acquired characteristics could be inherited. The usual medium for such inheritance he understood, like Lamarck, to be habit. In Darwin's early theory, he supposed that habits practiced with regularity over a life-time and by succeeding generations would gradually reform the inheritable substance and become expressed as instincts. In this version of Paley's rule, he suggested that those useful and expedient habits which have been necessary to preserve animal groups, allowing them over long periods to propagate and protect their young (such habits as friendship, parental nurture, etc.) were what we had come to call morally good. The continued practice of such useful behaviors would produce moral instincts that conformed to a temporally readjusted rule of expediency: what has been good will become interred in an animal's bones, and thus will continue to be what animals and their offspring, including man, regard as good. The Paley of his youth was preserved as the biologized Paley of his later evolutionary theory of morals.

The difference between the earlier theory and the later, however, was just natural selection, which Darwin believed eliminated any residual utilitarian selfishness from morals. In the *Descent of Man*, which he saw published in 1871, Darwin urged that the moral sense—the motive feeling which fueled intentions to perform altruistic acts and which caused pain when duty was ignored—would be considered a species of social instinct. He conceived social instincts as the bonds forming animal groups into social wholes. Social instincts comprised behaviors that nurtured offspring, secured their welfare, produced cooperation among kin, and organized the clan into a functional unit. The principal device of their evolution, in Darwin's view, was community selection—that kind of natural selection operating at levels of organization higher than that of the individual.

Darwin discovered the device of community selection—or what we would call kin-selection—while studying the social insects, the bees and ants that occupied his growing interest from the 1840s through the publication of the *Origin of Species.* Initially, before he quite understood how selection might produce apparently 'altruistic' behavior in bees—when, for instance, the soldier bees disembowel themselves in defending the hive—he was at a loss. How could it happen that a soldier bee might sacrifice its life for the group? Such behavior would not contribute to generating more offspring; and, of course, worker bees were neuters and could not produce offspring in any case. So the problem which Darwin faced in the mid-1840s was acute. After some conceptual struggle, however, he finally understood that natural selection could be applied to the whole community or hive, rather than only to the individual. Those hives that by chance had individuals whose traits varied in a beneficial direction would have the advantage over competing hives. And so it would be with men, as Darwin explained in the *Descent:*

> It must not be forgotten that although a high standard of morality gives but a slight or no advantage to each individual man and his children over the other men of the same tribe, yet that an advancement in the standard of morality and an increase in the number of well-endowed men will certainly give an immense advantage to one tribe over another. There can be no doubt that a tribe including many members who, from possessing in a high degree the spirit of patriotism, fidelity, obedience, courage, and sympathy, were always ready to give aid to each other and to sacrifice themselves for the common good, would be victorious over most other tribes; and this would be natural selection.[37]

Darwin's theory of moral sense was taken by some of his reviewers, as well as some modern biologists such as Ghiselin, to be but a species of utilitarianism, one that gave scientific approbation to the morality of selfishness. But this was expressly not Darwin's understanding. He believed his theory rather different from that of the utilitarians Bentham and Mill. Individuals, Darwin argued, acted instinctively to avoid vice and to seek virtue without any rational calculations of benefit. Bentham believed pleasure to be the sole motivating spring of human action, including moral action. But Darwin held that pleasure was neither the usual motive nor the purpose of moral acts. Rather, moral behavior, arising ultimately from community selection, was directed to the vigor and health of the group, not to the pleasure of its individual members. This meant, according to Darwin, that the criterion of morality—that highest principle by which we judge our and others' behavior in a cool hour—was not the general happiness, but the general good. The general good, though, he understood in distinctly biological terms; it was the welfare and survival of the group. This was no crude utilitarian theory of morality dressed in biological guise. Darwin's theory cast moral acts as bred in the bone, as intrinsically altruistic.

Conclusion

Should we regard morality as grounded in our biology rather than in soul? Today the idea that morality is ultimately based in biology strikes fear and loathing into the bowels of the most diverse groups. Contemporary creationists, for example, find this an horrendous notion; it appears to authorize all manner of monkey behavior in man. On the other hand, many biologists and philosophers of biology, such as Stephen Jay Gould, Richard Lewontin, and Philip Kitcher, are no less dismissive: deep in the bosom of every sociobiologist, they believe, beats the cold heart of a conservative Republican. Behind E. O. Wilson stands Newt Gingrich. Neither group perhaps suspects the historical roots of Darwin's conception, nor appreciates its strengths. But Darwin did understand, in that intuitive and poetically expressive way of his, the profoundly religious bearing of his theory. The last paragraph of the *Origin of Species* echoes those lines from Milton's *Paradise Lost,* which Darwin read constantly while on the *Beagle.* In one passage, Satan is approaching Eden, but is stopped by an entangled bank; he finally jumps the bank and alights in the Tree of Good and Evil, where he readies death for man. The passage from Milton reads:

> Now to the ascent of that steep savage hill
> Satan had journeyed on pensive and slow,
> But further way found none, so thick entwined,
> As one continued brake, the undergrowth
> Of shrubs and tangling bushes had perplexed
> All path of man or beast that passed that way. . . .

> Thence up he flew, and on the Tree of Life,
> The middle tree and highest there that grew,
> Sat like a cormorant, yet not true life
> Thereby regained, but sat devising death
> To them who lived, nor on the virtue thought
> Of that life-giving plant, but only used
> For prospect what, well used, had been the pledge
> Of Immortality.[38]

For Darwin, the pledge of immortality and morality was realized in the laws of evolution, which brought life out of death—a reversal of that Old Testament view that Milton depicts. Darwin's biology was Christian—out of death comes life more abundantly. In the last paragraph of the *Origin,* he reflects:

> It is interesting to contemplate an entangled bank, clothed with many plants of many kinds, with birds singing on the bushes, with various insects flitting about, and with worms crawling through the damp earth, and to reflect that these elaborately constructed forms, so different from each other, and dependent on each other in so complex a manner, have all been produced by laws acting around us [such as the struggle for life]. . . . Thus, from the war of nature, from famine and death, the more exalted object which we are capable of conceiving, namely, the production of the higher animals, directly follows. There is grandeur in this view of life, with its several powers, having been originally breathed into a few forms or into one; and that, whilst this planet has gone cycling on according to the fixed law of gravity, from so simple a beginning endless forms most beautiful and most wonderful have been, and are being, evolved.[39]

Notes

1. Andrew Dickson White, *History of the Warfare of Science with Theology in Christendom,* 2 vols. (New York: Macmillan, 1896), 1:70.

2. This version of the confrontation between Wilberforce and Huxley is given in "Grandmother's Tale," *Macmillan's Magazine* 78 (1898): 425–435. See the discussion of various accounts of the interchange in Robert Richards, *Darwin and the Emergence of Evolutionary Theories of Mind and Behavior* (Chicago: University of Chicago Press, 1987), pp. 549–551.

3. Edward Manning, quoted in White, *History of the Warfare,* 1:71.

4. Anonymous, "Dr. Draper and Evolution," *Catholic World* 26 (1878): 774–789, quotation from p. 782.

5. Richard C. Lewontin, Steven Rose, and Leon Kamin, *Not in Our Genes: Biology, Ideology, and Human Nature* (New York: Pantheon, 1984), p. 51. See also Richards, *Darwin and the Emergence of Evolutionary Theories of Mind and Behavior,* pp. 404–407.

6. Susan Faye Cannon, *Science in Culture: The Early Victorian Period* (New York: Science History Publications, 1978), p. 276.

7. James Moore has charted the accommodation of Protestant religious thinkers to Darwinian theory in his continuously interesting *The Post-Darwinian Controversies: A Study of the Protestant Struggle to Come to Terms with Darwin in Great Britain and America, 1870–1900* (Cambridge: Cambridge University Press, 1979). Jon Roberts focuses a like concern on Protestant America in his *Darwinism and the Divine in America* (Madison: University of Wisconsin Press, 1988). And Frederick Gregory charts comparable relationships in his *Nature Lost? Natural Science and the German Theological Traditions of the Nineteenth Century* (Cambridge: Harvard University Press, 1992).

8. I discuss these above mentioned evolutionary "spiritualists" in *Darwin and the Emergence of Evolutionary Theories of Mind and Behavior,* chaps. 8–10.

9. Charles Darwin, *The Autobiography of Charles Darwin,* ed. Nora Barlow (New York: Norton, 1958), p. 87.

10. Darwin, *Autobiography,* p. 85.

11. Charles Darwin, *Notebook B,* MS 84, in *Charles Darwin's Notebooks, 1836–1844,* ed. Paul Barrett et al. (New York: Cornell University Press, 1987), p. 192.

12. Charles Darwin, *Notebook D,* MS 36–37, in *Charles Darwin's Notebooks,* p. 343.

13. When he lived in London, in the years just after his return from his voyage, Darwin saw Babbage frequently, as he noted in his *Autobiography:* "I used to call pretty often on Babbage and regularly attended his famous evening parties. He was worth listening to, but he was a disappointed and discontented man; and his expression was often or generally morose" (p. 108).

14. Charles Babbage, *The Ninth Bridgewater Treatise, A Fragment,* 2d ed. (London: Murray, 1838), p. 46.

15. Charles Darwin, *Essay of 1842,* in *The Foundations of the Origin of Species,* ed. Francis Darwin (Cambridge: Cambridge University Press, 1909), p. 51. See also similar remarks made in *Notebook D,* MS 36–37, *Notebooks of Charles Darwin,* p. 343.

16. Charles Darwin, *On the Origin of Species* (London: Murray, 1859), pp. 488–489. See also the comprehensive and instructive discussion of Darwin's conception of natural law and its relation to God's design in Neal Gillespie, *Charles Darwin and the Problem of Creation* (Chicago: University of Chicago Press, 1979), pp. 54–55.

17. Charles Darwin to Asa Gray (22 May 1860), in Francis Darwin, ed., *Life and Letters of Charles Darwin,* 2 vols. (New York: D. Appleton, 1891), 2:105.

18. See Richards, *Darwin and the Emergence of Evolutionary Theories of Mind and Behavior,* pp. 199–200.

19. Charles Darwin, *The Descent of Man and Selection in Relation to Sex,* 2 vols. (London: Murray, 1871), 1:153.

20. Darwin, *Essay of 1842,* in *Foundations of the Origin of Species,* p. 6. Bracketed phrase was erased in original manuscript.

21. Darwin, *Essay of 1844,* in *Foundation of the Origin of Species,* pp. 86–87.

22. Darwin, *Origin of Species,* p. 84.

23. Darwin, *Essay of 1842,* in *Foundations of the Origin of Species,* p. 52. Bracketed phrse was erased in original manuscript.

24. As an historian who adopts an evolutionary-epistemological method, I would not wish to suggest that any aspect of Darwin's theory had only one cause. Darwin's ideas evolved within a complex conceptual environment, which had many intellectual pressures being applied simultaneously though with varying force. Another obvious source of Darwin's gradualism was his dependence on Lyellian uniformitarian theory. Lyell maintained that great geological changes could be explained by appealing to causes that operated insensibly and continually over vast stretches of time. Darwin was a Lyellian geologist and, I believe, brought the same principles to explain great zoological changes. But geology and zoology different in a significant way. Animal morphology exhibited, while rock morphology did not, minute and exquisite features of design. To explain the former, some agents of creative intelligence had to be implicitly assumed, which believe Darwin did in the case of natural selection.

25. Darwin, *Origin of Species,* p. 41.

26. Ibid., p. 102. See also similar claims on pp. 105, 107, and 125.

27. See Dov Ospovat, *The Development of Darwin's Theory* (Cambridge: Cambridge University Press, 1981).

28. Charles Darwin, *Charles Darwin's Natural Selection,* ed. R. C. Stauffer (Cambridge: Cambridge University Press, 1985), p. 225.

29. Darwin, *Origin of Species,* p. 201.

30. Ibid., p. 84.

31. Ibid., p. 83.
32. Ibid., p. 79.
33. The discussion in this section is based on my *Darwin and the Emergence of Evolutionary Theories of Mind and Behavior,* pp. 71–242.
34. Michael Ghiselin, "Darwin and Evolutionary Psychology," *Science* 179 (1973): 964–968; quotation on p. 967. See also Richards, *Darwin and the Emergence of Evolutionary Theories of Mind and Behavior,* p. 218.
35. William Paley, *Moral and Political Philosophy,* in *The Works of William Paley* (Philadelphia: Woodward, n.d.), p. 40.
36. Charles Darwin, *Notebook M,* MS 1 32e, in *Charles Darwin's Notebooks,* p. 552.
37. Charles Darwin, *The Descent of Man and Selection in Relation to Sex,* 2 vols. (London: Murray, 1871), 1:166.
38. John Milton, *Paradise Lost,* book 4, ll. 172–201.
39. Darwin, *Origin of Species,* pp. 489–490.

CHEMISTRY THROUGH INVARIANT THEORY?

JAMES JOSEPH SYLVESTER'S MATHEMATIZATION OF THE ATOMIC THEORY

*Karen Hunger Parshall**

In his *Cours de Philosophie positive,* the nineteenth-century French philosopher, Auguste Comte, detailed a philosophy crucially dependent upon science and scientific thought.[1] In his system, human comprehension and explanation of the natural world progressed through three stages: in the first, theological stage, divine will accounted for natural phenomena; in the next, metaphysical stage, interpretations of nature took an abstract, philosophical cast; and in the final, so-called "positive" stage, scientific truth characterized the world. According to Comte, the mind reached the positive stage by moving through a well-defined hierarchy of scientific thought. At its base lay mathematics, the most complex of the sciences in his view. Furthermore, since mathematics represented a body of scientific truths, it had already reached the stage of positive knowledge and, in fact, was historically the first science to attain this goal. Upon this base rested five sciences in ascending order of their historical achievement of the positive stage and thus, to Comte's way of thinking, in order of their decreasing complexity and increasing generality. They were astronomy, physics, chemistry, biology, and, at the top, a new science for which Comte coined the term "sociology." Of these, he held that all but the essentially brand-new science of sociology had reached the positive stage by the mid-nineteenth century. He thus took it as his task not only to establish this discipline in terms of scientific truths but also to achieve the synthesis of all of the positive sciences into one complete and unified positive philosophy.

While few thinkers, especially outside of France, accepted completely the positivist philosophy as articulated by Comte, that many in England embraced at least part of his philosophical vision indicated the persuasiveness of some of his underlying tenets.[2] In particular, the idea of the certainty and truth of mathematics, the concept of a hierarchy of the sciences, the desire for a synthesis of human knowledge, and the notion of an inherent progress in scientific development found expression in the work and writing of many nineteenth-century scientists. By extending these precepts along Platonistic lines, some further concluded that the sciences moved toward a surer and more perfect realization as they became more highly and successfully mathematized. (Of course, the latter position echoed that of the various brands of German idealists, most notably Immanuel Kant and Friedrich Schelling.) Consistent

P. H. Theerman and K. Hunger Parshall (eds.), Experiencing Nature, 81–111.

with this point of view, a synthesis of human—or at least scientific—knowledge hinged on its mathematization. Thus, the scientist animated by this modified positivistic philosophy of science (or modified idealistic philosophy of science, depending on the initial source of inspiration) strove to unify the sciences through mathematics. Most generally, and more consonant with Comte's initial formulation of positivism, the unification occurred through the application of mathematics *to* the science, which had the additional consequence of rendering the science more irrefutable. (In the nineteenth century, the mathematical linkage of the phenomena of electricity and magnetism represented a stunning example of just such an application.) In this light, though, how could a pure mathematician participate actively in a positivistic synthesis? Clearly, since the notion of the mathematization of mathematics was tautological, one possibility was to turn Comte upside down and apply the *sciences* to mathematics.

As early as 12 April, 1846, that colorful, flamboyant, and eminently talented British algebraist, James Joseph Sylvester, had betrayed his positivistic—as opposed to idealistic—leanings in response to a letter from his friend, the American physicist and soon-to-be Secretary of the Smithsonian Institution, Joseph Henry. Writing to Sylvester on 26 February, 1846, Henry had mentioned his recent attempts to read some of the German metaphysicists and had "concluded to keep pretty close to *positive philosophy* during the remainder of [his] life."[3] Sylvester concurred, writing that "I agree with you as to the preferability of Positive Philosophy—the other is well enough for an occasional excursion."[4]

Indeed, throughout his career, Sylvester dropped hints as to the positivistic—although by no means purely Comtian—nature of his philosophical leanings in his published works, before finally giving them their most complete expression at his hand in a speech before the British Association's Exeter meeting in 1869.[5] However, nowhere in his works did the ideals he expressed at Exeter and elsewhere receive a more explicit, technically oriented airing than in his attempt, in the late 1870s, to apply the atomic theory of chemistry to a major area of nineteenth-century research in pure mathematics, namely, invariant theory.[6] As he saw it, his work successfully realized the goal of unifying these two seemingly disparate areas of scientific thought. Yet of equal importance in his view, it provided an increasingly necessary practical justification—through the linkage with chemistry—for a main branch of pure mathematics, as well as an analogy useful in rendering pure mathematical research comprehensible to a popular audience. This perceived need reflected two key consequences of the push toward specialization in the latter part of the nineteenth century, namely, the increasing chasm between pure and applied mathematics and the growing distance between the researcher and the educated laity. In what follows, I motivate Sylvester's linkage of the atomic theory of chemistry with invariant theory through an examination of his positivistic philosophy of science and argue that his efforts to address the related issues of the legitimacy of abstract mathematics and its popularization represented a natural outgrowth of his philosophical views.

The Philosophy of Science of James Joseph Sylvester

In 1876, Sylvester accepted the first professorship of mathematics at the newly founded Johns Hopkins University in Baltimore, Maryland.[7] A rather dramatic step for an Englishman in his sixty-first year, the transatlantic move represented not only Sylvester's first opportunity for regular employment in the six years since his forced retirement from the Royal Military Academy at Woolwich but also his first chance to instruct students at the research level. As a Jew in Victorian society, Sylvester had been barred from the Cambridge or Oxford professorships which his international reputation and impressive research output should have earned for him. Even at these premier institutions, however, he would have found himself lecturing on relatively elementary subjects, far removed from his concurrent research interests, given the relatively slow response of the British universities to the strong research ethic which had developed in the German states. At Hopkins, a nonpartisan and nonsectarian university animated by these German educational ideals, Sylvester formed part of a close society of scholars dedicated to original research and to the training of future researchers. He had finally found an intellectual home.

Upon his arrival in Baltimore, Sylvester resumed the researches in invariant theory which his period of unemployment and accompanying mathematical despondency had interrupted. After having co-founded the theory around mid-century with his friend, Arthur Cayley, Sylvester had worked consistently to develop it over the next two decades. His efforts had resulted not only in numerous seminal contributions but also in national and international recognition. Indicative of the high regard in which his British colleagues held him, Sylvester was named to the presidency of the Mathematical and Physical Section of the British Association for the Advancement of Science at its Exeter meeting in August of 1869.[8] There, as was the custom, the President opened the proceedings with an address on some timely aspect of the discipline, in

Sylvester's case, a reaction to then-recent remarks of Thomas Huxley about the nature of mathematical inquiry as conducted by mathematicians. In the course of his Exeter address, Sylvester also gave a reasonably clear formulation of the personal, positivistic philosophy of science which had long been evolving in his thought[9] and which, I argue below, motivated the seemingly surprising foray he made into the atomic theory of chemistry not long after his arrival at the Hopkins.

Formulating the aim of his talk in characteristically longwinded yet entertaining terms, Sylvester explained that

> rather . . . than favour the notion that we Algebraists (who regard each other as the flower and salt of the earth) are a set of mere calculating-machines endowed with organs of locomotion, or, at best, a sort of poor visionary dumb creatures only capable of communicating by signs and symbols with the outer world, I have resolved to take heart of grace and to say a few words, which I hope to render, if not interesting, at least intelligible, on a subject to which the larger part of my life has been devoted.[10]

Having thus rightfully cast himself in the role of the expert mathematical witness, so to speak, Sylvester proceeded to attack Huxley's views on mathematics.

In an after-dinner speech on scientific education delivered and published in 1869, Huxley had characterized mathematical education as being "almost purely deductive" in nature.[11] He argued that "[t]he mathematician starts with a few simple propositions, the proof of which is so obvious that they are called self-evident, and the rest of the work consists of subtle deductions from them."[12] Extrapolating from this statement about mathematical pedagogy to the nature of research-level mathematical inquiry, Sylvester interpreted Huxley as trivializing mathematics to the point "that mathematical truth rests on the narrow basis of a limited number of elementary propositions from which all others are to be derived by a process of logical inference and verbal deduction."[13] He backed up this interpretation with a quote from yet another one of Huxley's published papers, where the biologist defined mathematics as "that study which knows nothing of observation, nothing of experiment, nothing of induction, nothing of causation!"[14] Sylvester roundly attacked this view in his address through a series of examples taken from the recent history of mathematics as well as from his own personal experience.

In addressing the question of the origins of his own subject, invariant theory, Sylvester explained that it had come about through

> the accidental observation by Eisenstein, some score or more years ago, of a single invariant . . . . This single result of observation (as well entitled to be so called as the discovery of Globerigerine in chalk or of the Confoco-ellipsoidal structure of the shells of the Foraminifera), which remained unproductive in the hands of its distinguished author, has served to set in motion a train of thought and to propagate an impulse which have led to a complete revolution in the whole aspect of modern analysis.[15]

Aside from its exuberant overstatement of the impact of his area on contemporary mathematics, this passage gave clear expression to Sylvester's conception of observation in mathematics through analogy with observation in the natural sciences. In short, Sylvester argued that many highly articulated and abstract mathematical theories owed their discovery and development to the observation of patterns emergent from specific examples. From the pattern came the conjectured theorem; then, through deduction, the conjecture became mathematical truth. For Sylvester, observation, and its associated activities of induction and experimentation, gave mathematics life, depth, and meaning. "When followed in this spirit," he wrote, "there is no study in the world which brings into more harmonious action all the faculties of the mind . . ., there is none other which . . ., like this, seems to raise [its followers], by successive steps of intuition, to higher and higher states of conscious intellectual being."[16] Through observation, mathematics transcended not only the dead and vacuous realm of mere deduction described by Huxley but also all other forms of human reasoning. For Sylvester, however, observation applied even more globally to his chosen discipline.

By observation and through analogy, he held, mathematical ideas which arose in one area might be seen to fit unexpectedly in other mathematical or physical contexts. Quoting one of his predecessors in the chair of the Mathematical and Physical Section, William Spottiswoode, Sylvester portrayed invariant theory as a subject that

> reaches out and indissolubly connects itself each year with fresh branches of mathematics, . . . the theory of equations has almost become new through it, algebraic geometry transfigured in its light, . . . the calculus of variations, molecular physics, and

> mechanics (he might, if speaking at the present moment, go on to add the theory of elasticity and the highest developments of the integral calculus) have all felt its influence.[17]

Thus, in Sylvester's view, the observation of ties and connections between a given body of mathematical thought and other areas served not only to underscore the seminal nature of the mathematics but also to unite seemingly disparate disciplines. This union or synthesis represented the highest aim and achievement of the mathematician, and it was by no means limited to the union of mathematics and the other sciences.

In 1864, five years earlier than the Exeter address, Sylvester had hinted at his positivistic ideas when he digressed in one of his purely mathematical papers to link mathematics and music on the basis of observation and analogy romanticized. He wrote that

> I think one clearly discerns the internal grounds of the coincidence or parallelism, which observation has long made familiar, between the mathematical and the musical [ethos]. May not Music be described as the Mathematic of sense, Mathematic as Music of the reason? the soul of each the same! Thus the musician *feels* Mathematic, the mathematician *thinks* music,—Music the dream, Mathematic the working life—each to receive its consummation from the other when the human intelligence, elevated to its perfect type, shall shine forth glorified in some future Mozart-Dirichlet or Beethoven-Gauss—a union already not indistinctly foreshadowed in the genius and labours of a Helmholtz![18]

In his somewhat Pythagorean view, music and mathematics were expressions on two different levels of the same abstractions. Moreover, Helmholtz's then recent mathematization of the theory of sound established their inextricable union in modern terms, thereby achieving the desired synthesis.[19] Likewise, Sylvester saw mathematics and poetry as essentially one. This motivated the "mathematization" of versification he presented in 1870 in his book, entitled *The Laws of Verse,* a work at once realizing his ultimate goal of synthesis and confirming his belief in the underlying unity—through mathematical laws—of all knowledge and human expression.[20]

Eight years later in 1877, Sylvester not only echoed many of these same ideas but also developed additional components of his philosophy of science. Again called upon to speak before a mixed audience, this time at the Commemoration Day celebration at his newly adopted Johns Hopkins University, he waxed lyrical in his efforts to convey the vital, unbounded nature of mathematics to his uninitiated listeners. As if reacting anew to Huxley's misperception, Sylvester passionately informed his listeners that

> Mathematics is not a book confined within a cover and bound between brazen clasps, whose contents it needs only patience to ransack; it is not a mine, whose treasures may take long to reduce into possession, but which fill only a limited number of veins and lodes; . . . it is limitless as that space which it finds too narrow for its aspirations; its possibilities are as infinite as the worlds which are forever crowding in and multiplying upon the astronomer's gaze.[21]

Moreover, he firmly stated the Platonistic view that "every science becomes more perfect, approaches more closely to its own ideal, in proportion as it imitates or imbibes the mathematical form and spirit."[22] The almost evangelical zeal inherent in this statement, coupled with the positivistic goal of the synthesis of knowledge, goes far in explaining Sylvester's own attempt to inject "mathematical form and spirit" into poetry in 1870. Finally, in his first attempt to bring his Hopkins address to a close, Sylvester professed his belief in the intellectual affinity between mathematics and the physical sciences, an affinity which should clearly render a positivistic synthesis more immediate. Indeed, he proclaimed "[h]ow rejoiced should I be, were I of less ripe years and under less peremptory obligations as to the disposal of my time, branching out from mathematics as my natural mental centre of gravity, to diverge into the physical and chemical studies which lie so near to it."[23] Some ten months later, he did indeed delight in the unexpected discovery of what he deemed a potent analogy between the atomic theory of chemistry and invariant theory.

Graph Theory and the Mathematization of Chemistry in the Nineteenth Century

According to the Comtian hierarchy of the sciences, chemistry lagged behind both astronomy and physics in achieving the positive stage. Nevertheless, for Comte, chemistry had already satisfied the criterion for a positive science by the early years of the nineteenth century, that is, it constituted a body of scientific truth. As should be evident from the quotations cited above, however, Sylvester (and others) held a positivistic view which hinged on a different criterion. For him, the sciences achieved greater and greater perfection through their increasingly successful mathematization. Thus, in a view which he shared with many of his contemporaries, chemistry was still far from perfect even at mid-century.

Historically speaking, while astronomy and physics were being remathematized during the period of the Scientific Revolution, many chemists remained ambivalent as to both the efficacy and the desirability of an abstract mathematization of their discipline. As Allen G. Debus has pointed out, "a significant proportion of the scientific community [namely, the Paracelsians] was moving away from the modern concept of mathematical abstraction in natural philosophy at precisely the time when that concept was beginning to show its power in the hands of the mechanical philosophers."[24] For the Paracelsians, mathematics was appropriate in chemistry only in the form of number mysticism or simple quantification. By the nineteenth century, although the number mysticism had been stripped from the discipline in the wake of the Lavoisieran revolution, chemistry's mathematization found its greatest fruit in gravimetric and volumetric analysis—basic weights and measures—even while the discipline's theoretical underpinnings remained in flux.[25] For example, while chemists such as August Kekulé, Alexander Butlerov, Archibald Couper, and Edward Frankland developed the notion of valence in the 1850s and 1860s to account for chemical combination, there was no general agreement as to the atomic weights and formulas characterizing different elements and compounds.[26] In 1860, however, the first international chemistry conference convened in Karlsruhe. As a result of the papers read and the discussions moderated there, the decade of the sixties witnessed not only the widespread adoption of fixed values for atomic weights but also the publication, in 1869, of Dmitry Mendeleev's periodic table.[27] Still, this agreement did not preclude protracted debates, in London and elsewhere in the late 1860s, as to the real nature of atoms: were they fundamental constituents of matter, or were they merely useful theoretical devices?[28] It also left open the matters of whether the theory inherent in the notion of fixed weights was mathematizable and, if so, how best to achieve its mathematization. By the 1870s, at least the first of these issues seemed to be settled.

Beginning in his Edinburgh thesis of 1861, Alexander Crum Brown, soon to become professor of chemistry at his *alma mater,* developed a useful schematization of chemical molecules.[29] By 1864, he had settled on a graphical notation in which he denoted a given atom by a circle containing the appropriate chemical abbreviation and depicted the atom's valence by line segments poking from the circle's circumference.[30] Thus, carbon, an atom of valence four, appeared in his notation as

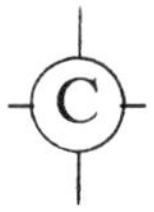 .

Crum Brown then denoted molecules as aggregates of these symbols linked at the spikes according to the laws of valence.

Both simple and suggestive, Crum Brown's system won approval almost immediately in Great Britain thanks to the publication in 1866 of Edward Frankland's widely used and influential textbook, entitled *Lecture Notes for Chemical Students.*[31] There, Frankland, one of the principal architects of the concept of valence, argued that Crum Brown's notation not only enjoyed "several important advantages over that first proposed by Kekulé" but also provided "most valuable aid to the teacher in rendering intelligible the constitution of chemical compounds."[32] For these reasons, he consistently used this graphical system throughout his text.

In presenting the notion of atomicity or valence of elements, for example, Frankland first described the concept in terms of chemical equations: "Thus an atom of zinc is equivalent to two atoms of hydrogen; for when zinc is brought into contact with steam at a high temperature, one atom of zinc expels from the steam two atoms of hydrogen and occupies their place—thus,

$$OH_2 + Zn = OZn + H_2."^{33}$$

He immediately followed this with Crum Brown's graphical interpretation of the notion of bond which underlay this equation. "To give a concrete expression to these facts," he wrote," the atom of hydrogen may be represented as having only one point of attachment or *bond* by which it can be united with any other element, zinc as having two such bonds, boron three, and so on. Thus the atoms of these elements may be graphically represented in the following manner:—

Hydrogen 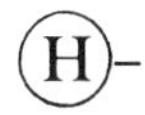

Zinc 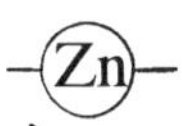

Boron  "[34]

Finally, these graphical "nodes" combined naturally to depict the chemical composition of molecules. Water, which Frankland represented symbolically as OH_2, thus appeared graphically as

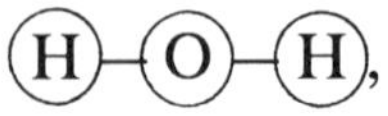

and sulphuric acid (or $SO_2(HO)_2$) was pictured as

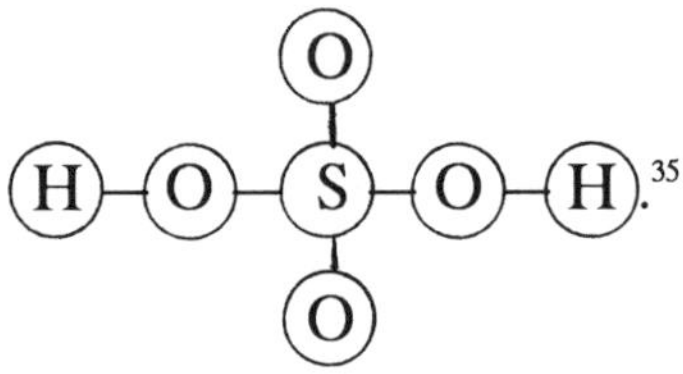

Despite his conviction of the pedagogical usefulness of the graphical notation, however, Frankland repeatedly cautioned his readers "that these graphic formulæ are intended to represent neither the shape of the molecules, nor the relative position of the constituent atoms."[36]

With this caveat, the new generation of British chemists learned its subject from the *Lecture Notes* in terms of Crum Brown's notation. Moreover, these "graphical formulæ," as Frankland called them, rather quickly gained acceptance internationally, winning out over the competing notational schemes developed by Kekulé and others.[37] Not only did they facilitate molecular representation, but they also accounted effortlessly for one puzzling chemical phenomenon, namely, isomerism or the existence of molecules with the same chemical composition yet different chemical properties.[38] The graphical notation immediately suggested that the phenomenon was due to different possibilities for the configurations of the constituent atoms.

Consider, for example, the two different molecules—then called propylic and Friedel's alcohol—which Crum Brown singled out in his paper of 1864. Sharing the chemical formula C_3H_7OH, these isomers may be represented graphically as

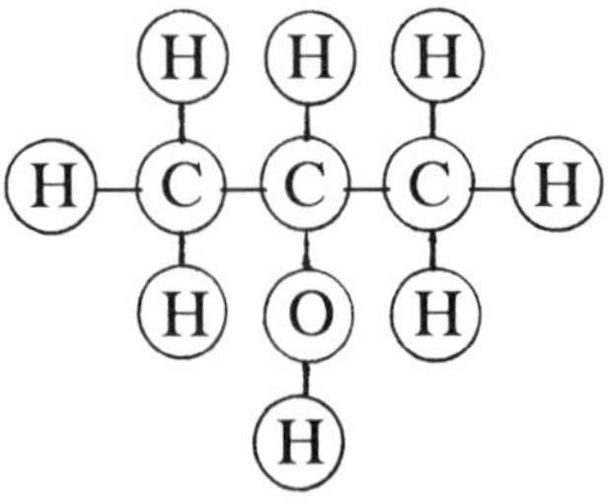

and

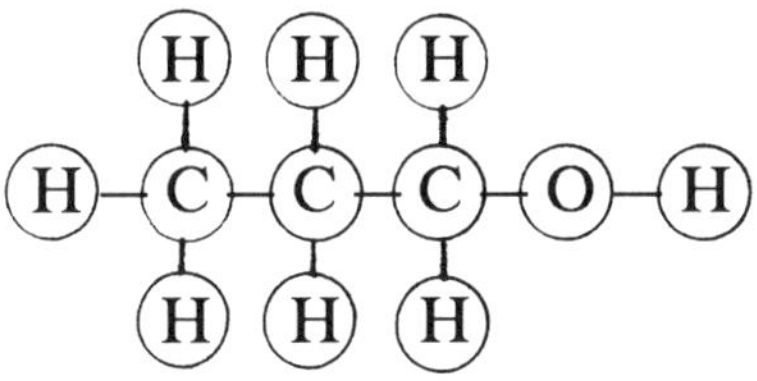

,

respectively.[39] The two different graphical representations thus reflect the distinctness of the two molecules, and they immediately and naturally suggest one question: how many isomers of a given chemical composition are possible? As Arthur Cayley realized in 1874, the mathematical work he had done in the late 1850s in what would now be called graph theory applied perfectly to answer this chemical query.

In 1857, Cayley's ongoing invariant-theoretic research had led him to investigate an abstract, combinatorial theory of what he called trees, that is, linear diagrams branching from one root with n knots and n–1 branches. As Cayley pointed out in his first paper on the subject, a work entitled "On the Theory of the Analytical Forms Called Trees," it is easy to see that if $n = 2$, 3, and 4, respectively, then the following are all of the possible (rooted) trees composed of n knots and n–1 branches:

.

The problem then clearly presents itself: for any n, determine the number of (rooted) trees with n knots and n–1 branches. In his paper of 1857, Cayley determined the generating function which produced these numbers and calculated them explicitly for $n = 2, \ldots, 11$.[40] By 1859, he had worked out a formula for determining the number of (shortest) trees with n terminal knots, which confirmed the following graphical results for $n = 1$, 2, and 3:

.[41]

Although Cayley did not pursue his study of trees any further in the 1860's, he returned to it in 1874 in a paper suggestively entitled "On the Mathematical Theory of Isomers."[42]

As already mentioned, the graphical representation of chemical molecules developed by Crum Brown had become institutionalized in the years after 1866 thanks to the influence of Frankland's textbook in Great Britain. As an active and respected participant in the British scientific community, Cayley could easily have learned of this new innovation as a result of his regular attendance at meetings of organizations like the British Association for the Advancement of Science and the Royal Society. Indeed, Crum Brown spoke on his graphical formulæ and their usefulness in chemistry in an address before the British Association in 1874, the very same year in which Cayley published his first chemico-mathematical paper.[43] Regardless of when he actually became aware of the contemporaneous developments in chemistry, however, by 1874 Cayley had clearly perceived the analogy between mathematical trees and chemical graphs, and he had recognized the utility of his mathematical results in attacking the problem of isomerism. He gave this analogy its full mathematical expression before the British Association at its meeting in Bristol in 1875.

In that work, Cayley focused on the case of the paraffins, that is, saturated hydrocarbon molecules of chemical composition C_nH_{2n+2}. As he explained, since carbon has valence four, these molecules "have n atoms of carbon connected by $n-1$ bonds, under the restriction that from each carbon-atom there proceed at most 4 bands (or, in the language of the papers first referred to [that is, his papers of 1857 and 1859], we have n knots connected by $n-1$ branches), in the form of a tree."[44] By way of example, he exhibited the three possible "carbon-trees" with chemical formula C_5H_{12}, namely,

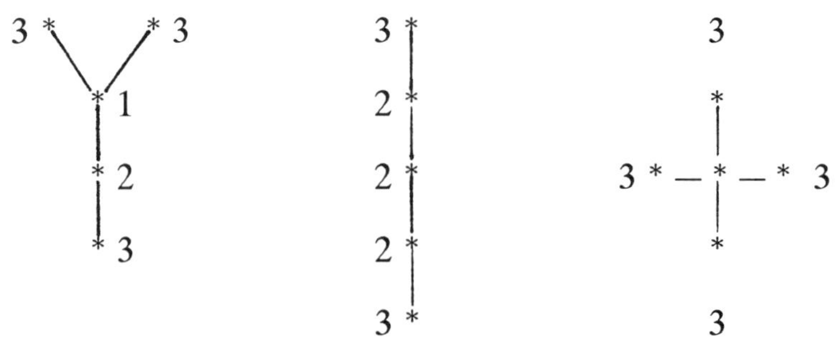

where each knot represents a carbon atom and where the numbers at each knot refer to the number of hydrogen atoms which bond to that carbon atom. He then stated the chemical problem in its equivalent mathematical terms: "the determination of the theoretic number of the paraffins C_nH_{2n+2} is consequently that of the determination of the number of the carbon-trees of n knots, viz. the number of trees with n knots, subject to the condition that the number of branches from each knot is at most 4."[45] In the remainder of the paper, Cayley mathematically analyzed this and the analogous problems of saturated graphs of elements of valence two and three, respectively. He succeeded in explicitly calculating the number of such trees possible for $n = 1, \ldots, 13$.

Late in 1877, Sylvester, who was familiar both with Cayley's research and with Frankland's text, also came to link chemistry and mathematics but in a very different way and for very different reasons. Whereas Cayley had applied his mathematical theory of trees to the chemical problem of isomerism, Sylvester applied the atomic theory of chemistry to his own special mathematical area of expertise, invariant theory. To appreciate his idea, we must pause briefly to examine the kinds of questions with which that mathematical theory deals.

A Brief Glimpse at the British Approach to Invariant Theory

In a paper published in the *Cambridge Mathematical Journal* in 1841, George Boole first laid out the phenomenon of invariance before the British audience by isolating an algorithmic technique for generating invariants.[46] To take the simplest example, consider the binary quadratic form $Q = ax^2 + 2bxy + cy^2$, that is, a second-degree (quadratic) form in two unknowns (binary). The discriminant of Q is defined as $b^2 - ac$, an expression exclusively in the coefficients of the form. Now define a linear transformation L of the variables x and y of Q as follows:

$$(1) \qquad \begin{aligned} x &\rightarrow mx' + ny' \\ y &\rightarrow m'x' + n'y', \end{aligned}$$

where $m, n, m', n' \in \mathbb{R}$ and $mn' - m'n \neq 0$. Applying L to Q, that is, forming $L(Q)$ through the substitution of x and y (as indicated in (1)) in the expression for Q, yields the new binary quadratic form

$$L(Q) = (am^2 + 2bmm' + cm'^2)x'^2 + 2(amn + bmn' + bm'n + cm'n')x'y' + (an^2 + 2bnn' + cn'^2)y'^2$$

in the variables x' and y'. Notice that by changing notation and replacing the coefficients in $L(Q)$ by A, B, and C, respectively, we have $L(Q) = Ax'^2 + 2Bx'y' + Cy'^2$, a binary quadratic form with discriminant $B^2 - AC$. Boole proved that

$$B^2 - AC = (mn' - m'n)^2(b^2 - ac),$$

that is, the discriminant of $L(Q)$ equals the discriminant of Q up to a power of the determinant $mn' - m'n$ of the linear transformation L. In terminology which would later develop in the pages of the correspondence of Cayley and Sylvester, any expression in the coefficients of a binary form (or binary *quantic* to use the nineteenth-century terminology) Q which satisfies this property is called an *invariant.* Any expression in the coefficients and the variables of Q which satisfies it is called a *covariant.*[47] (Notice that an invariant is just a special type of covariant, one in which the variables are absent.)

Although Boole did not vigorously pursue the line of research he had opened up in his 1841 paper, by 1844 it had come to the attention of the twenty-two-year-old Cambridge Fellow and Senior Wrangler of 1842, Arthur Cayley. In 1845, he published the first of his many papers on what would become a major area of nineteenth-century mathematical research, and by 1846 he had formulated one of its fundamental problems, namely, "[t]o find all the derivatives [i.e., invariants] of any number of functions, which have the property of preserving their form unaltered after any linear transformation of the variables."[48] Together with Sylvester, Cayley spent much of his mathematical career formulating a theory of invariants which would successfully resolve this and related problems. The calculationally oriented methods which emerged from their cooperative efforts came to characterize the nineteenth-century British, as opposed to the German, approach to invariant theory.[49]

In particular, in his 1856 paper, "A Second Memoir upon Quantics," Cayley formulated an algorithm for explicitly determining the independent covariants of a given binary quantic.[50] Without going into theoretical detail, Cayley's method generated the following complete (or minimal generating) set of independent covariants for the binary cubic $U = px^3 + 3qx^2y + 3rxy^2 + sy^3$ (or third-degree form in two unknowns):

$U = px^3 + 3qx^2y + 3rxy^2 + sy^3$, that is, the binary cubic itself,

$H = (pr - q^2)x^2 + (ps - qr)xy + (qs - r^2)y^2$,

$$\Phi = (p^2s - 3pqr + 2q^3)x^3 + (3pqs - 6pr^2 + 3q^2r)x^2y$$
$$+ (6q^2s - 3prs - 3qr^2)xy^2 + (3qrs - ps^2 - 2r^3)y^3,$$

and $\nabla = p^2s^2 - 6pqrs + 4pr^3 + 4q^3s - 3q^2r^2$.

Cayley also recognized that these four covariants satisfied the following algebraic relation or *syzygy:*

$$\Phi^2 - \nabla U^2 + 4H^3 = 0.$$

Furthermore, by virtue of the fact that these four covariants form a minimal generating set for the covariants of the binary cubic, every invariant of the binary cubic U is of the form ∇^n, for n a positive integer, and every covariant is either of the form $U^\alpha H^\beta \nabla^\gamma$ or $U^\alpha H^\beta \nabla^\gamma \Phi$, for positive integers α, β, and γ.[51]

While this example treats the case of finding the minimal set of covariants for one form, the binary cubic, Cayley and Sylvester also analyzed the much more complicated situation of calculating minimal generating sets for *systems* of binary forms of various degrees. Consider a system of forms comprised of the binary quadratic Q and the binary cubic U as defined above. The problem involved first determining covariants in the variables and coefficients a, b, and c of Q and p, q, r, and s of U, that is, calculating expressions which remained invariant in Boole's sense after the application of a linear transformation L as in (1) to Q and U. As in the simpler example above, it next required finding a minimal generating set of such covariants. In this particular case, the following expression in the coefficients of Q and U is one invariant of the system:

$$I = a(qs - r^2) - b(ps - qr) + c(pr - q^2)$$

$$= aqs - ar^2 - bps + bqr + cpr - cq^2, \tag{2}$$

and the minimal generating set contains I as one of its fifteen, distinct, independent covariants.[52] Here, although the final results about the system Q, U may be easy to write down *ex post facto,* the actual verification of these facts involved much calculation based on a highly articulated and non-trivial theory. Computationally nightmarish, this sort of more general situation indicates not only the technical difficulty of nineteenth-century invariant theory but also the seeming impossibility of conveying a sense of the subject in nontechnical terms. As on the occasion of his British Association lecture in 1869, Sylvester struggled with this latter problem after taking up his duties at the Johns Hopkins University in 1876.

Sylvester's "Chemistrization" of Invariant Theory

Invited to lecture before his new university's "Scientific Club" in the fall of 1877, Sylvester once again faced the problem of addressing a non-mathematical audience, this time one comprised mainly of active researchers in physics, chemistry, and biology, on his invariant-theoretic research. In his Commemoration Day speech earlier in the year, he had tried a suggestive but imperfect metaphor from spectroscopy, most probably with the work of his colleague in physics, Henry Rowland, in mind. On that occasion, Sylvester had explained that "[t]o every quantic," that is, to every homogeneous binary form

> is associated an infinite variety of other forms that may be regarded as engendered from and floating, like an atmosphere, around it— but infinite in number as are these derived existences, these emanations from the parent form, it is found that they admit of being obtained by composition, by mixture, so to say, of a certain limited number of fundamental forms, standard rays, as they might be termed in the Algebraic Spectrum of the Quantic to which they belong. And, as it is a leading pursuit of the Physicists of the present day to ascertain the fixed lines in the spectrum of every chemical substance, so it is the aim and object of a great school of mathematicians to make out the fundamental derived forms, the Covariants and Invariants, as they are called, of these Quantics.[53]

Thus, he hoped that from this analogy between the fixed lines, or the building blocks of spectra, and the minimal set, or the building blocks of covariants, his listeners would gain an appreciation of one of the invariant theorist's primary quests.

While the vagueness of the spectroscopic metaphor may have suited Sylvester's lay audience in February, it did not fit closely enough to satisfy his scientifically sophisticated auditors in the fall. The mathematician needed some effective means of communicating with them at their sophisticated level. Inspiration, instead of sleep, came late one night. As Sylvester dramatized it:

> Casting about, as I lay in bed one night, to discover some means of conveying . . . modern algebra to a mixed [scientific] society, . . . I was agreeably surprised to find, of a sudden, distinctly pictured on my mental retina a chemico-graphical image serving to embody and illustrate the relations of these derived algebraical forms to their primitives and to each other which would perfectly accomplish the object I had in view.[54]

Moreover, and perhaps more importantly, he saw that he might exploit his newly discovered analogy between the atomic theory and invariant theory both "as an instrument of investigation in purely algebraical investigations" and as means of providing mathematical justification for certain chemical phenomena.[55] Thus, while Sylvester envisioned a sort of "chemistrization" of his own special area of mathematical research, he also saw a symmetrical mathematization of certain aspects of chemistry. Just what was the analogy that achieved, at least to Sylvester's way of thinking, this synthesis of algebra and chemistry?

Writing to the British journal *Nature* on 1 January, 1878, Sylvester outlined the salient features of his new insight. First and foremost, he "compare[d] every binary quantic with a chemical atom. The number of factors . . . in a binary quantic is the analogue of the number of *bonds*, or the *valence*, as it is termed, of a chemical atom."[56] He then characterized "[a]n invariant of a system of binary quantics of various degrees" as analogous to "a chemical substance composed of atoms of corresponding *valences*. The order of such an invariant in each set of coefficients is the same as the number of atoms of the corresponding *valence* in the chemical compound."[57] After extending the analogy to the other basic concepts of invariant theory, Sylvester concluded that "[e]very invariant and covariant thus becomes expressible by a *graph* precisely identical with a . . . chemicograph."[58] In order to get a better sense of precisely what Sylvester had in mind here, reconsider the system of forms comprised of the binary quadratic form Q and the binary cubic form U defined above. (Recall that this system has I as expressed in (2) as an independent invariant.)

In referring to "factors" of the binary quantic, Sylvester meant the degree of the quantic or form. The binary *quadratic* Q has degree *two* and so is analogous, in Sylvester's system, to an atom of valence two, while the binary *cubic* U has degree *three* and so corresponds to an atom of valence three. Since I is an invariant of the system comprised of the binary *quadratic* Q and the binary *cubic* U, I represents a molecule composed of atoms of valence *two* and *three*. Consider any one of the summands in the expression of I given in (2), say, ar^2 ($= arr$). Since this and every summand of I is made up of precisely one coefficient of Q—a, in this case—and two coefficients of U—two r's, here—the invariant I is said to have *order* one in the coefficients of Q and *order* two in the coefficients of U. The chemical molecule which I represents thus consists of *one* atom of valence two and *two* atoms of valence three, and its corresponding chemicograph is

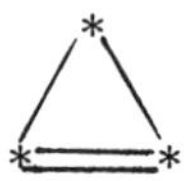
[59].

Exploiting this analogy, Sylvester provided a potentially powerful graphical schematization of invariant theory in terms given by the atomic theory and its schematization in the form of chemical graphs. In other words, Sylvester achieved a sort of chemistrization of invariant theory. The potential impact of the analogy, however, was not unidirectional. Sylvester further ventured the possibility that the invariant theory "whilst in nowise interfering with existing chemicographical constructions, accounts for the seeming anomaly of the isolated existence as 'monad molecules' of mercury, zinc, and arsenic—and gives a rational explanation of the 'mutual saturation of bonds.'"[60] While not neglecting these kinds of conclusions, Sylvester developed principally the implications of the atomic theory on invariant theory in his 1878 paper, entitled "On an Application of the New Atomic Theory to the Graphical Representation of the Invariants and Covariants of Binary Quantics—With Three Appendices."[61]

He began this work by establishing a notational equivalence between the chemical symbols and the invariant-theoretic entities. Since valence corresponds to the degree of a form and since hydrogen has valence one, let its symbol H denote a form of degree one (that is, a binary linear form $ax + by$). Similarly, let O (for oxygen) stand for a form of degree two (a binary quadratic form like Q above), and let N (nitrogen) represent a form of degree five (in other words, a binary quintic form $ax^5 + 5bx^4y + 10cx^3y^2 + 10dx^2y^3 + 5exy^4 + fy^5$). As Sylvester saw it, "[a]n invariant of a form or system of algebraical forms must thus represent a saturated system of atoms in which the rays of all the atoms are connected into bonds."[62] In this light, since nitric acid has as its graph

(3) 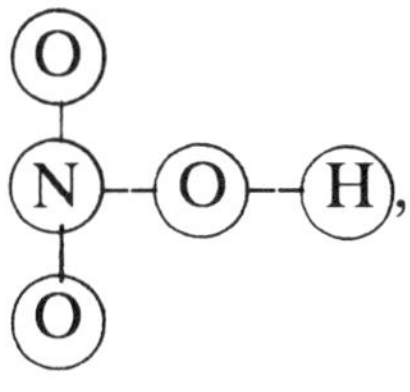

"[t]his will correspond to an invariant of a quintic, quadr[at]ic and linear form of the first order in the coefficients of each extreme and of the third order in those of the middle form."[63] Here, the invariant analogous to the molecule depicted in (3) corresponds to a system of binary quantics comprised of a *quintic, quadratic,* and *linear* form since nitrogen, oxygen, and hydrogen have valences *five, two,* and *one,* respectively. Since N and H appear only *once* in the graph, the invariant has order *one* in the coefficients of the quintic and linear forms, while it has order *three* in the coefficients of the quadratic form, since O occurs *three* times in the graph. By applying the established theory for this and numerous other systems of forms, Sylvester was able to provide chemicographical interpretations of their various theoretical components.

As a result of this success, Sylvester envisioned the potential translation of the computationally complicated theory of invariants into a presumably simpler, and more readily manipulable, graphical form. He thus concluded that "we see that the graphical method suggested by the theory of atomicity is a real instrument not merely for representation but also for the calculation and comparison of algebraical results."[64] Although he did not state it explicitly, Sylvester may have seen even more as at stake. Following Paul Gordan's proof in 1868 of the finiteness of the minimal generating set of covariants for any binary form, the British school's invariant-theoretic researches had markedly stalled. Through graph theory, it might be able to infuse new life into its approach to the area. Still, as mentioned above, Sylvester viewed his analogy as benefiting not only invariant theory but also chemistry. By seeking the physical realizations of the components of the mathematical theory, chemists might gain new insights into the composition and actual structure of chemical molecules.[65]

Although further details of this work, which extended to almost sixty quarto pages, need not concern us here, several non-technical aspects of Sylvester's paper do warrant mention. First, Sylvester chose to publish his novel ideas in the inaugural number of the *American Journal of Mathematics,* a publication underwritten by the Johns Hopkins University in its efforts to encourage and stimulate original research particularly on the part of its faculty and students.[66] As the journal's editor-in-chief, Sylvester felt pressed not only to help fill the pages of the early numbers but also to work to establish high regard for the venture as a whole. In short, he wanted to produce a first contribution worthy of the reputation he hoped his journal would soon earn. He ultimately submitted, with some trepidation, his new thoughts on invariant theory and chemistry.

Writing to Simon Newcomb, the Superintendent of the *American Ephemeris* and the *American Journal*'s cooperating editor in astronomy, Sylvester expressed his anxieties over the reception of his chemico-mathematical work. "[I]t will be thought by many strained and over-fanciful," he admitted. "It is more a 'reverie' than a regular mathematical paper. I have however added some supplementary mathematical matter which I hope will rescue the chemical portion from absolute contempt. It may at the worst serve to suggest to chemists and Algebraists that they may have something to learn from each other."[67] After briefly describing his analogy to Newcomb, Sylvester offered that "[t]his must appear to you very much of the nature of moonshine," yet he could not help wondering "what the chemists will think of my 'cheek.'"[68] He did not have to long to muse; almost immediately, reactions came in from both the mathematical and the chemical camps.

Sometime after conceiving of his new analogy, Sylvester had fired off a letter to his friend, the English mathematician William Kingdon Clifford. In it, he had described his novel approach to invariant theory and had asked for details about possibly related ideas which Clifford may have presented earlier before a meeting of the British Association. Clifford wrote back to tell Sylvester that, indeed, inspired by Sylvester's invariant-theoretic researches, he had hit upon the chemico-graphical representation of invariants and had spoken on it at that same meeting in Bristol where Cayley had applied the abstract notion of trees to the problem of chemical isomerism.[69] So excited was Sylvester to learn of this striking case of multiple discovery, a coincidence which seemed to validate his "reverie," that he published an extract of Clifford's letter immediately following his own paper in the *American Journal.* In a lengthy accompanying editorial note, Sylvester exuberantly asserted that

> [i]t seems to me in the highest degree improbable that our joint speculations should not eventually find their embodiment in chemical doctrine proper, and I think that young chemists desirous to raise their science to its proper rank would act wisely in making themselves masters betimes, of the theory of

> algebraical forms. What mechanics is to physics, that I think algebraical morphology, founded at option on the theory of partitions or ideal elements, or both, is destined to be to the chemistry of the future.[70]

Although the details of his vision of the future of mathematics in chemical research were wrong, Sylvester did seemingly portend the key role "algebraical morphology"—in the form not of nineteenth-century invariant theory but in that of twentieth-century group theory—would play in chemistry.[71]

If Sylvester felt absolved of the sin of undue "cheek" by Clifford's letter, he must have felt even more vindicated by the subsequent rumors which reached him from England. In a letter dated 7 September, 1878 to Daniel Coit Gilman, the President of the Johns Hopkins University, Sylvester reported that "[f]rom what I hear the chemico-algebraical theory has made a sensation among the English chemists."[72] These rumors soon proved well-founded when Sylvester received a lengthy and friendly letter from the author of the *Lecture Notes,* Edward Frankland. Dated 13 October, 1878 and published in part in the *American Journal,* Frankland's correspondence attested to the fact that he had carefully read and considered Sylvester's ideas, and, far from dismissing them out of hand, he found in them certain merit. Indeed, after admitting that he was "at first inclined to smile" at some of Sylvester's notions, "on further consideration, it seems to me that your conception of the constitution of atoms may prove of great value."[73] In light of this, he urged Sylvester to "go on with the consideration of chemical phenomena from a mathematical point of view, for I am convinced that the future progress of chemistry, as an exact science, depends very much indeed upon the alliance of mathematics."[74] Like Sylvester, Frankland clearly appreciated the positivistic goal of the progress of chemistry through mathematics. And, while he may not have committed his own research time to the pursuit of such an aim, he was more than willing to see Sylvester and others work toward it.

Empty Analogy or Positivistic Triumph?

Despite the initial flurry of interest in the chemico-mathematical theory as evidenced by this correspondence, no one, including Sylvester, carried forth the research initiative implicit in the linkage between the atomic theory and the theory of invariants.[75] This failure most glaringly reflected the impenetrability of invariant theory to all but the most adept. In the absence of a coherent, introductory treatment of the subject and given its emphasis on concrete calculation as opposed to true abstraction, few mathematicians and even fewer chemists would have been equipped to make meaningful connections between the two disciplines.[76] Thus, it would take Paul Gordan, the leader of the German school of invariant theorists and the principal mathematical rival of Sylvester and Cayley, to conceive—completely independently—of the same chemico-mathematical analogy in 1900, twenty-two years after the publication of Sylvester's paper and three years after his death. Like Sylvester, Gordan and his collaborator, the chemist W. Alexejeff, believed just as incorrectly that further advances in chemistry would follow through the application of invariant theory; unlike the Englishman, however, they did not see the benefit as mutual.[77]

As for the mathematician who had originally proposed the synthesis, Sylvester failed to pursue the implications of his analogy after its initial appearance in 1878 due to the quirks of his research habits but, more fundamentally, due to his broader philosophical objectives. Almost immediately after finishing his paper "On an Application of the New Atomic Theory . . . ," he became caught up once again in the actual calculation of complete sets of covariants for quantics of various general types. True to form, he threw himself single-mindedly into this task until some new mathematical passion flared up to consume him.[78] But redirected research interests accounted only superficially for Sylvester's effective abandonment of his then-new and seemingly fruitful theory. Quite simply, he had achieved his philosophical goals in his one, main pronouncement on the subject: he had unified two seemingly unrelated areas of human knowledge, and he had provided an increasingly necessary applied justification for the purely algebraic theory of invariants. He expressed his satisfaction and his feeling of closure quite succinctly at the end of the body of his *American Journal* paper. "Thus," he wrote, "we see that chemistry is the counterpart of a province of algebra as probably the whole universe of fact is, or must be, of the universe of thought."[79] As a mathematician, he might well have added "Q.E.D."

Sylvester's "chemistrization" of invariant theory ultimately proved devoid of substance as a real research tool, its promise going unfulfilled due to the superficiality of what was little more than a notational analogy.[80] Effectively, then, it did not mark the sort of positivistic synthesis which Sylvester so highly valued. Be this as it may, Sylvester's linkage of the atomic theory and the theory of invariants in 1878 and its initially enthusiastic reception underscored the strength and indeed the dominance of a positivistic approach to science in the last quarter of the nineteenth century. Moreover, Sylvester's quest for a practical application of his own special branch of pure, abstract mathematics marked not only a heightened sensitivity on the part of at least one mathematician to the growing distance between pure and applied mathematics but also a philosophical shift away from "mathematics for mathematics' sake" and toward applicability as a legitimate standard by which the subject should be measured. Like his younger contemporaries Felix Klein and Henri Poincaré, Sylvester believed that while it was legitimate to pursue pure mathematics, the discipline as a whole enjoyed a heightened vitality through the establishment of interconnections with other areas. Even more pragmatically, Sylvester viewed his efforts to link algebra and chemistry as a real attempt to communicate with non-mathematicians. In recognizing both the need and the desirability to answer those two most frequently asked questions "What do you do?" and "What is it good for?," he anticipated problems the mathematical community would increasingly come to face in the twentieth century relative to funding and public opinion.

Notes

*Departments of History and Mathematics, Mathematics/Astronomy Building, University of Virginia, Charlottesville, Virginia 22903–3199. The research for this paper was partially supported by NSF Grant # DIR 9011625. It is my pleasure to thank my colleagues, Peter Heath and Bruce Martin, as well as Allen Debus, Robert Kargon, Evan Melhado, and Paul Theerman for reading and providing insightful comments on a previous draft of this paper.

1. Auguste Comte, *Cours de Philosophie positive,* 6 vols. (Paris: Bachelier, 1830–1842).

2. Most noted among those who embraced as well as questioned many aspects of Comtian thought was John Stuart Mill. Even Thomas Huxley, in a blistering attack on positivism, allowed relative to Comte's ideas that "[n]othing could be more interesting to a student of biology than to see the study of the biological sciences laid down as an essential part of the prolegomena of a new view of social phenomena. Nothing could be more satisfactory to a worshipper of the severe truthfulness of science than the attempt to dispense with all beliefs save such as could brave the light, and seek, rather than fear, criticism." See Thomas Henry Huxley, "The Scientific Aspects of Positivism," *Fortnightly Review* 11 (1869): 653–670, on p. 653.

3. Joseph Henry to James Joseph Sylvester, 26 February, 1846, (Henry's emphasis) in *The Papers of Joseph Henry,* ed. Nathan Reingold, Marc Rothenberg, *et al.*, (Washington, D.C.: Smithsonian Institution Press, 1972–1992), 6:380–381. The manuscript retained copy of this letter is held in the Henry Papers, Smithsonian Institution Archives.

4. James Joseph Sylvester to Joseph Henry, 12 April, 1846, in ibid., 6:407–410. The original of this letter is also held in the Henry Papers, Smithsonian Institution Archives.

5. James Joseph Sylvester, "Presidential Address: Mathematics and Physics Section," *Report of the Thirty-ninth Meeting of the BAAS Held at Exeter in August, 1869* (London: John Murray, 1870); revised and reprinted as "A Plea for the Mathematician," *Nature* 1 (1869–1870): 237–239, 260–263. See, also, James Joseph Sylvester, *The Laws of Verse or Principles of Versification Exemplified in Metrical Translations: Together with an Annotated Reprint of the Inaugural Presidential Address to the Mathematical and Physical Section of the British Association at Exeter* (London: Longmans, Green, and Co., 1870); and "Presidential Address to Section 'A' of the British Association," *The Collected Mathematical Papers of James Joseph Sylvester,* ed. H. F. Baker, 4 vols. (Cambridge: University Press, 1904, 1908, 1909, 1912; reprint ed., New York: Chelsea Publishing Co., 1973), 2:650–661 (hereinafter abbreviated *Math. Papers JJS*). All specific page citations for Sylvester's papers will refer to the latter, most accessible, source.

6. James Joseph Sylvester, "On an Application of the New Atomic Theory to the Graphical Representation of the Invariants and Covariants of Binary Quantics,—with Three Appendices;" *American Journal of Mathematics* 1 (1878): 64–125, or *Math. Papers JJS,* 3:148–206, on p. 148.

7. Although there is as yet no full-scale biography of Sylvester, he has been the subject of many short biographical articles. See, for example, H. F. Baker, "Biographical Notice," in *Math. Papers JJS,* 4:xv–xxxvii. On Sylvester at the Johns Hopkins University, see Karen Hunger Parshall, "America's First School of Mathematical Research: James Joseph Sylvester at The Johns Hopkins University 1876–1883," *Archive for History of Exact Sciences* 38 (1988): 153–196; and Karen Hunger Parshall and David E. Rowe, *The Emergence of the American Mathematical Research Community 1876–1900: James Joseph Sylvester, Felix Klein, and Eliakim Hastings Moore* (Providence: American Mathematical Society and London: London Mathematical Society, 1994), Chapters 2 and 3.

8. David Knight discusses the prestige associated with this honor within the British scientific community in *The Age of Science: The*

Scientific World-view in the Nineteenth Century (Oxford: Basil Blackwell, 1986), pp. 128–132. See also Jack Morrell and Arnold Thackray, *Gentlemen of Science: Early Years of the British Association for the Advancement of Science* (Oxford: Clarendon Press; New York: Oxford University Press, 1981) for historical perspective on the organization, its members, and its influence.

9. Joan Richards approaches this speech from a different point of view in her book, *Mathematical Visions: The Pursuit of Geometry in Victorian England* (Boston: Academic Press, Inc., 1988), especially on pp. 134–136 and 163–166.

10. Sylvester, "Presidential Address to Section 'A' of the British Association," p. 652. (See note 5 above.)

11. Thomas Henry Huxley, "Scientific Education: Notes of an After-Dinner Speech," *Macmillan's Magazine* 20 (1869): 177–184, on p. 182.

12. Ibid.

13. Sylvester, "Presidential Address," p. 654.

14. Ibid., p. 654. See Thomas Henry Huxley, "The Scientific Aspects of Positivism," p. 667.

15. Sylvester, "Presidential Address," p. 656. Gotthold Eisenstein's researches were published in 1844 in "Allgemeine Auflösung der Gleichungen von den ersten vier Graden," *Journal für die reine und angewandte Mathematik* 27 (1844): 81-83. Here, Sylvester apparently forgot George Boole's isolation of an invariant in 1841. See note 46 below.

16. Ibid., pp. 657-658.

17. Ibid., p. 654.

18. James Joseph Sylvester, "Algebraical Researches, Containing a Disquisition on Newton's Rule for the Discovery of Imaginary Roots, and an Allied Rule Applicable to a Particular Class of Equations, Together with a Complete Invariantive Determination of the Character of the Roots of the General Equation of the Fifth Degree, &c.," *Philosophical Transactions of the Royal Society of London* 154 (1864): 579–666, and *Math. Papers JJS,* 2:376–479, on p. 419. Sylvester's emphasis.

19. In 1863, just one year prior to the publication of Sylvester's paper, Helmholtz systematized his findings in physiological acoustics in the book, *Die Lehre von die Tonempfindungen als physiologische Grundlage für die Theorie der Musik* (Braunschweig: F. Vieweg und Sohn, 1863).

20. Sylvester, *The Laws of Verse.* (See note 5 above for the full citation.) For example, relative to what Sylvester called the "principle of phonetic syzygy," he wrote "I conceive that the method of triadic analysis which I have employed in the genesis and distribution of the princi-

ples of lyrical poetry, is founded on the nature of things and not on an arbitrary subjective rule of classification. I can honestly aver that I was not on the look-out for any such an arithmetical law" (p. 12). He continued: "Of one thing I have no doubt, which is, that when the analysis of principles which I have here faintly indicated has been carried to its full term, we shall be in possession of a system of rules, by which the criticism of the technical part at least of lyrical poetry may be reduced to the form of propositions capable of being logically argued and debated, and entirely removed from that indefinite region of taste which, like the so-called discretion of a judge, does not admit of being made the subject of rational discussion" (p. 12).

21. James Joseph Sylvester, "Address on Commemoration Day at Johns Hopkins University 22 February, 1877," *Math. Papers JJS,* 3:72–87, on pp. 77–78.

22. Ibid., p. 78.

23. Ibid., p. 83.

24. Allen G. Debus, "Mathematics and Nature in the Chemical Texts of the Renaissance," *Ambix* 15 (1968): 1–28, on p. 3.

25. For a discussion of some of these developments, see Arnold Thackray, *Atoms and Powers: An Essay on Newtonian Matter-Theory and the Development of Chemistry* (Cambridge: Harvard University Press, 1970), Chapter 7, entitled "Quantified Chemistry: The Newtonian Vision," pp. 199–233.

26. On the complicated history of valency, see C. A. Russell, *The History of Valency* (Leicester: University Press, 1971); and J. R. Partington, *A History of Chemistry,* 4 vols. (London: The Macmillan Press, Ltd., 1970–1972), 4:500–565.

27. Knight, *Age of Science,* pp. 157–158; and Partington, *History of Chemistry,* 4:488 and 494.

28. Knight, *Age of Science,* p. 156.

29. Of course, others like Friedrich August Kekulé also developed notations around this time. See Partington, *History of Chemistry,* 4:533–565ff; and Russell, *History of Valency,* pp. 92–107. For a general treatment of the high points of the linkage between graph theory and chemistry, see Robin J. Wilson, "Graph Theory and Chemistry," *Colloquia Mathematica Societatis János Bolyai* 18 (1976): 1146–1164. I thank Robin Wilson for referring me to this work as well as to his book cited in note 37 below.

30. Alexander Crum Brown, "On the Theory of Isomeric Compounds," *Transactions of the Royal Society of Edinburgh* 23 (1864): 707-718. See, also, Partington, *History of Chemistry,* 4:552–553; and Russell, *History of Valency,* pp. 101–105.

31. Edward Frankland, *Lecture Notes for Chemical Students: Embracing Mineral and Organic Chemistry* (London: John van Voorst, 1866). Russell discusses the influence of this text in *History of Valence,* pp. 104–105.

32. Frankland, *Lecture Notes,* p. v. Kekulé had developed a graphical notation in which circles represented univalent atoms, but sausage-like shaded figures denoted 2–, 3–, and 4–valent atoms. Consult Partington, *History of Chemistry,* 4:540 for a discussion of the highly idiosyncratic system.

33. Frankland, *Lecture Notes,* p. 17. Here, Frankland employed boldface type to indicate that the element so symbolized "is directly united with all the active bonds of the other elements or compound radicals following upon the same line" (pp. 16–17).

34. Ibid., p. 18.

35. Ibid., p. 24.

36. Ibid., p. 25.

37. Partington, *History of Chemistry,* 4:552–553; Russell, *History of Valency,* pp. 92–100; and Norman Biggs, E. Keith Lloyd, and Robin J. Wilson, *Graph Theory 1736–1936* (Oxford: Clarendon Press, 1976), pp. 56–60.

38. Clearly, Crum Brown had realized this as early as 1864 given the title of the paper in which he introduced his new symbolism. Frankland also explicitly singled out this advantage of the graphical notation in his *Lecture Notes.* As he put it, "[i]t is also of especial value in rendering evident the causes of isomerism in organic bodies" (p. 23).

39. Crum Brown, "Theory of Isomeric Compounds," pp. 707–718; and Biggs, Lloyd, and Wilson, *Graph Theory,* p. 60.

40. Arthur Cayley, "On the Theory of Analytical Forms Called Trees," *Philosophical Magazine* 13 (1857): 172–176, or Arthur Cayley and A. R. Forsyth, ed., *The Collected Mathematical Papers of Arthur Cayley,* 14 vols. (Cambridge: University Press, 1889–1898), 3:242–246 (hereinafter cited as *Math. Papers AC*). All subsequent page citations refer to *Math. Papers AC.*

41. Arthur Cayley, "On the Analytical Forms Called Trees: Second Part," *Philosophical Magazine* 28 (1859): 374–378, or *Math. Papers AC,* 4:112–115. Cayley gave the following recursive version of the formula for the number of trees ϕn with n nodes: $\phi n = \phi 1 + (\frac{n-1}{1})\phi 2 + (\frac{n-1\, n-2}{1\cdot 2})\phi 3 + \ldots + (\frac{n-1}{1})\phi(n-1)$ (see p. 113). He also provided subsequent mathematical refinements of this formulation of ϕn. As his results showed, ϕn gets large quickly. For $n = 1, 2, 3, 4, 5, 6, 7, 8$, respectively, ϕn takes on the values 1, 1, 3, 13, 75, 541, 4683, and 47293.

42. Arthur Cayley, "On the Mathematical Theory of Isomers," *Philosophical Magazine* 47 (1874): 444–446, or *Math. Papers AC,* 9:202–204.

43. Charles S. Gillispie, ed., *The Dictionary of Scientific Biography,* 16 vols. 2 supps. (New York: Charles Scribner's Sons, 1970–1990), s.v. "Brown, Alexander Crum," by D. C. Goodman.

44. Arthur Cayley, "On the Analytical Forms Called Trees, with Application to the Theory of Chemical Combinations," *Report of the British Association for the Advancement of Science: Bristol 1875* (London: John Murray, 1875), pp. 257–305, or *Math. Papers AC,* 9:427–460, on p. 427.

45. Ibid., p. 428.

46. George Boole, "Exposition of a General Theory of Linear Transformations," *Cambridge Mathematical Journal* 3 (1841–1842): 1–20. See note 15 above.

47. In precise mathematical terminology, if L is a linear transformation of nonzero determinant Δ defined as in (1) which takes a binary form

$$f = a_0x^n + a_1\binom{n}{1}x^{n-1}y + a_2\binom{n}{2}x^{n-2}y^2 + \ldots + a_{n-1}\binom{n}{n-1}xy^{n-1} + a_ny^n$$

into

$$L(f) = A_0(x')^n + A_1\binom{n}{1}(x')^{n-1}(y') + A_2\binom{n}{2}(x')^{n-2}(y')^2 + \ldots + A_{n-1}\binom{n}{n-1}(x')(y')^{n-1} + A_n(y')^n,$$

then the expression $I(a_0, a_1, \ldots, a_n)$ in the coefficients of f is an *invariant* provided

$$I(A_0, A_2, \ldots, A_n) = \Delta^k I(a_0, a_1, \ldots, a_n),$$

for some positive integer k. The expression $K(a_0, a_1, \ldots, a_n; x, y)$ in the coefficients and variables of f is called a *covariant* provided

$$K(A_0, A_1, \ldots, A_n; x', y') = \Delta^k K(a_0, a_1, \ldots, a_n; x, y),$$

for some positive integer k.

48. Arthur Cayley, "On the Theory of Linear Transformations," *Cambridge Mathematical Journal* 4 (1845): 193–209, or *Math. Papers AC,* 1:80–94; and Arthur Cayley, "On Linear Transformations," *Cambridge and Dublin Mathematical Journal* 1 (1846): 104–122, or *Math. Papers AC,* 1:95–112. The quotation appears in the latter text on p. 95.

49. For discussions of the historical development of invariant theory both in Great Britain and on the Continent, see Tony Crilly, "The Rise of Cayley's Invariant Theory (1841-1862)," *Historia Mathematica* 13 (1986): 241–254; "The Decline of Cayley's Invariant Theory (1863–1895)," *Historia Mathematica* 15 (1988): 332–347; and Karen Hunger Parshall, "Toward a History of Nineteenth-Century Invariant Theory," in *The History of Modern Mathematics,* ed. David E. Rowe and John McCleary, 2 vols. (Boston: Academic Press, 1989), 1:157–206.

50. Arthur Cayley, "A Second Memoir upon Quantics," *Philosophical Transactions of the Royal Society of London* 146 (1856): 101–126, or *Math. Papers AC,* 2:250–281. Cayley enumerates the complete system for the binary cubic on pp. 271–272.

51. See George Salmon, *Lessons Introductory to the Modern Higher Algebra,* 5th ed. (Dublin: Hodges, Figgis, & Co., 1885; reprint ed., Bronx, N.Y.: Chelsea Publishing Co., 1964), pp. 183–187 for the enumeration of the complete system for the binary cubic form. Indicative of the level of complexity of the theory associated with these computations, Salmon required the first 175 pages of his text to reach the point where he could carry out this enumeration.

52. These are listed in ibid., pp. 187–188.

53. Sylvester, "Address on Commemoration Day," p. 76.

54. Sylvester, "On an Application of the New Atomic Theory," p. 148. (See note 6 above.)

55. James Joseph Sylvester, "Chemistry and Algebra," *Nature* 17 (1877–1878): 284, or *Math. Papers JJS,* 3:103–104, on p. 103.

56. Ibid. Sylvester's emphasis.

57. Ibid., pp. 103–104. Sylvester's emphasis.

58. Ibid., p. 104. Sylvester's emphasis. Sylvester was initially confused on the issue of who developed the so-called chemicographs and attributed them to Kekulé (see ibid.). His error was pointed out to him by an editorial remark in *Nature,* and he subsequently acknowledged and corrected his mistake in the *American Journal.* See *Nature* 17 (1877–1878): 309; James Joseph Sylvester, "Historical Data concerning the Discovery of the Law of Valence," *American Journal of Mathematics* 1 (1878): 282; and "On an Application of the New Atomic Theory," pp. 172–173.

59. Sylvester dealt with this example in greater generality in "On an Application of the New Atomic Theory," pp. 164–165. Notice that this figure is a chemicograph because it represents a saturated system of atoms. Sylvester made no claims about the necessary existence in nature of all possible saturated systems.

60. Sylvester, "Chemistry and Algebra," p. 104.

61. See note 6 above.

62. Sylvester, "On an Application of the New Atomic Theory," p. 149.

63. Ibid., pp. 149–150. The diagram appears as figure 5 on p. 163.

64. Ibid., p. 165.

65. See, for example, Sylvester's comments in ibid., pp. 165–166, relative to the possible physical implications of Hermite's law, an important technical result in the theory of invariants due to the French mathematician, Charles Hermite.

66. For details on the establishment of the *American Journal of Mathematics,* see Parshall and Rowe, *Emergence of the American Mathematical Research Community,* Chapter 2.

67. Raymond C. Archibald, "Unpublished Letters of James Joseph Sylvester and Other New Information Concerning His Life and Work," *Osiris* 1 (1936): 85–154, on p. 134.

68. Ibid., p. 135.

69. William Kingdon Clifford, "Extract of a Letter to Mr. Sylvester from Prof. Clifford of University College, London," *American Journal of Mathematics* 1 (1878): 126–128, or "Remarks on the Chemico-Algebraical Theory," *Mathematical Papers by William Kingdon Clifford,* ed. Robert Tucker (London: Macmillan & Co., 1882; reprint ed., Bronx, N.Y.: Chelsea Publishing Company, 1968), pp. 255–257.

70. Ibid., p. 126.

71. See, for example, David M. Bishop, *Group Theory and Chemistry* (Oxford: Clarendon Press, 1973); and David S. Sconland, *Molecular Symmetry: An Introduction to Group Theory and Its Uses in Chemistry* (London: Van Nostrand, 1965).

72. James Joseph Sylvester to Daniel Coit Gilman, 7 September, 1878, Daniel C. Gilman Papers Ms. 1, Special Collections Division, Milton S. Eisenhower Library, The Johns Hopkins University. As always, I thank the University for permission to quote from its archives.

73. Edward Frankland, "Extract from a Letter of Dr. Frankland to Mr. Sylvester," *American Journal of Mathematics* 1 (1878): 345–349, on p. 346.

74. Ibid., p. 349. Sylvester's paper on the atomic theory also attracted some critical attention in the United States. See, J. W. Mallet, "Some Remarks on a Passage in Professor Sylvester's Paper on the Atomic Theory," *American Journal of Mathematics* 1 (1878): 277–281. Mallet was Professor of Chemistry at the University of Virginia.

75. Cayley briefly pursued his research in graph theory after 1875, but from an abstract point of view and not with an eye toward its application to chemistry. See Arthur Cayley, "A Theorem on Trees,"

Quarterly Journal of Pure and Applied Mathematics 23 (1889): 376–378, or *Math. Papers AC,* 13:26–28. Likewise, Sylvester returned to graph-theoretic researches for several months in 1889–1890 as a result of his contact with the Danish mathematician, Julius Petersen. See Gert Sabidussi, "Correspondence between Sylvester, Petersen, Hilbert and Klein on Invariants and the Factorization of Graphs 1889–1891," *Discrete Mathematics,* 100 (1992): 99–155. I thank Gert Sabidussi for providing me with a preprint of his edition of this correspondence.

76. Both George Salmon and Edwin Bailey Elliott tried to fill this lacuna in the literature by presenting a systematic treatment of the theory of invariants. See Salmon, *Lessons Introductory to the Modern Higher Algebra;* and Edwin Bailey Elliott, *An Introduction to the Algebra of Quantics,* 2d ed. (Oxford: University Press, 1913; reprint ed., Bronx, N.Y.: Chelsea Publishing Company, 1964).

77. Paul Gordan and W. Alexejeff, "Übereinstimmung der Formeln der Chemie und der Invariantentheorie," *Zeitschrift für physicalischer Chemie* 35 (1900): 610–633. See, also, Max Noether's assessment of this work in "Paul Gordan," *Mathematische Annalen* 75 (1914): 1–41, on pp. 34–35.

78. On Sylvester's idiosyncratic working style, see Parshall, "America's First School of Mathematical Research," pp. 165–172; and Parshall and Rowe, *Emergence of the American Mathematical Research Community,* Chapters 2 and 3.

79. Sylvester, "On an Application of the New Atomic Theory," p. 166.

80. Biggs, Lloyd, and Wilson concur in this assessment in *Graph Theory,* p. 67.

Religion, Science, and the Public Imagination: The Restoration of Order in Early Modern France

*Terence D. Murphy**

Introduction

On the night of 17–18 October 1534, radical religious reformers posted broadsheets throughout Paris proclaiming war on public idolatry. This "affaire des placards" was a provocation that precipitated the prolonged and violent confrontation between Protestants and Catholics, the French Wars of Religion. The radical reformers attacked "the pompous and proud Papal Mass," denying that the words of the celebrating priest changed bread and wine into the body and blood of Christ. To them transubstantiation resembled a magical conjuration, a remnant of medieval ignorance; it was "a diabolical doctrine" that drove "these blind *sacrificateurs*" to idolatry, to the worship of Satan. As the civil authorities had failed to stop these offenses against religion, true believers had to alert the public to the danger; otherwise a divine vengeance "will soon totally destroy this world, plunging it into desolation, ruin, perdition, and the abyss."[1]

Previously, reformers had challenged the magical healing power of relics and the sale of indulgences. Now Protestant radicals extended this critique to the central Christian mystery of the Eucharist, rejecting the claim that the priest's words transformed bread and wine into the real body and blood of the Lord. From the radicals' point of view, the unique sacrifice foreshadowed in the Last Supper could only be commemorated; they believed that the words and actions of mere men could never recapture the redeeming power of that sacred moment. More generally, the radicals asserted that the Christian religion had no need of the repeated and miraculous intervention of the Creator in the affairs of men.

Jean Calvin (1509–1564) ably defended this interpretation of the relationship between God and his creation. While giving paramount importance to biblical testimony of Christ's miraculous works, Calvin denied such powers to ordinary men. In his thinking, God's providential hand guided the natural order, making miracles unnecessary.[2] To assert the contrary, he argued, challenged God's omniscience. Banishing the miraculous from history and containing nature within prescribed boundaries, Calvin sought to limit the effects of the popular tendency towards superstition and idolatry. He detected just such a movement in the popular enthusiasm for the adoration of the Host. This led him to

P. H. Theerman and K. Hunger Parshall (eds.), Experiencing Nature, 113–161.

decry the fertility of "the field of falsehood in the human mind," where "the smallest of seeds, when sown there, will grow to yield an immense harvest."[3] He explained this persistent failing of the human mind as a consequence of the fall from grace, through Adam's original sin. Emphasizing the Augustinian interpretation of the Fall, Calvin maintained that once driven from Paradise the heirs of Adam's fault could never repair by word or deed the consequences of the original sin. Adam's heirs, condemned to labor, suffering, and ignorance, were denied the mastery of nature that had been their original gift.

This stark and unforgiving understanding of the human condition struck hard against medieval beliefs, excluding from the realm of the possible all magical manipulations of natural forces. It was ultimately responsible, in Max Weber's analysis, for what he described as the disenchantment of the world. The Protestant critique, Weber maintained, drove miracles and magic from ordinary existence. In this way the reformers removed a serious obstacle blocking the rationalization of economic life.[4] Weber's appraisal has found its echo among historians of science who have documented the gradual triumph of the notion of immutable natural laws governing a mechanistic universe, a belief that established the grounds for the modern scientific attitude.[5] On the threshold of the modern age, Johannes Kepler (1571–1630) exemplified the new perspective emphasized by these historians: "My aim," he wrote, "is to show that the heavenly machine is not a kind of divine, live being, but a kind of clockwork . . . insofar as nearly all the manifold motions are caused by a most simple, magnetic, and material force."[6] And, at its center, the founder of modern thought, René Descartes (1596–1650), carried out the intellectual *tour-de-force* that would create an independent and sovereign domain of knowledge protected from its perennial enemies: mystical or occult forces, skeptical doubt, and the weight of traditional, university-bred learning.

Emphasizing the progressive and innovative in the scientific revolution, this rationalist historiography celebrates the achievements of the secular current of human history in overcoming the constraints of tradition. Not particularly interested in describing the local context of a particular belief, or in seeking explanations for its hold on the imaginations of contemporaries, this historical perspective invokes enduring standards of objectivity. It suggests a movement from the medieval to the modern that followed a process of purification and simplification, of liberation of human potential from the encrustations of tradition, of release of the individual from the tutelage of authority. Despite the compelling interest in establishing a lineage of human achievements that traces the steps leading to the release of the human spirit from the fears of the unknown, it takes great faith in the linear advance of history to seek some form of causal relation between the religious motivations of reformers obsessed with the impending blow of divine vengeance and the supreme confidence that inspired the Cartesian system and the dynamic for scientific discovery. Moreover, the religious inspiration that fostered Calvin's biblical literalism was ill-suited to the spirit of inquiry, the recourse to instrumental observation, and the ambition to know that characterize the modern world view.

Historians such as Walter Pagel, Frances Yates, Allen Debus, and other distinguished scholars working in the tradition of the Warburg-Courtauld Institute have found other sources for the revolution in scientific ideas. They are representatives of a school of historians who have discovered in Neoplatonism and Renaissance mysticism some of the origins of the modern world view. Through a series of studies on Robert Fludd (1574–1637), his work on the English Paracelsians, and the emergence of the chemical philosophy, Allen Debus clearly establishes the persistence of Neoplatonic themes and Renaissance mysticism in the scientific revolution. Most recently, he has demonstrated the importance of the Paracelsian corpus in early modern France and, like Frances Yates and William Shea, related René Descartes's early work to the inspiration of the Rosicrucian movement, a movement deeply indebted to the Paracelsian legacy.[7] Disseminated through the writings of Marsilio Ficino (1433–1499) and the hermetic tradition, he argues, Neoplatonism stood at the fountainhead of a general cultural revolt against the domination of the medieval synthesis.[8] Those who rejected the Scholastic system and ecclesiastical authority pursued experimental knowledge in the search for remedies against illness.[9] Heinrich Cornelius Agrippa of Nettesheim (1487–1535), foremost among these practitioners, suggested in his *Occult Philosophy* that the search for the universal medicine that would treat all sicknesses could be found through meditation on the system of cosmic sympathies that established correspondences between astral bodies and the virtues inherent in metals, plants, and animals.[10] In his view, the celestial hierarchies echoed their powers in the terrestrial hierarchies, as the whole of nature was organized under the ordering principle of the chain of being. Later, John Baptista della Porta (1540–1615) in his *Natural Magick* described the regime of resemblances that governed Nature:

> As with respect to its vegetation the plant stands convenient to the brute beast, so through feeling does the brutish animal to man, who is conformable to the rest of the stars by his intelligence; these links proceed so strictly that they appear as a rope

> stretched from the first cause as far as the lowest and smallest of things, by a reciprocal and continuous connection; in such wise that the superior virtue, spreading its beams, reaches so far that if we touch one extremity of that cord it will make tremble and move all the rest.[11]

This system of thought represented a fundamental tendency of Renaissance culture: the belief in the profound harmony between the science of the ancients and Christian revelation, in the ability to reconcile the domains of nature and grace and to see the history of the world as the preparation for the second coming of Christ. From this perspective, the intellectual ferment of the early sixteenth century testified to the desire for spiritual rejuvenation; it was evidence that a new era had opened. For some, like the religious reformers, the reign of the Anti-Christ had begun, and divine vengeance would soon destroy this world. They spoke and acted with the fervor of Savonarola, the Florentine preacher who in the previous generation had condemned the Pope as the Anti-Christ and had interpreted the invasion of Italy by Charles VIII as an apocalyptic moment. For others, like the advocates of natural magic—Agrippa, Porta, Paracelsus, and later, the followers of the "rosy-cross"—the apocalyptic moment was indeed at hand, but it preceded a golden age when the fire of faith would transmute death-ridden earthly matter into the quintessential form animating celestial bodies. Natural magic would transform the world, mitigating the effects of original sin.

Whether rationalist or mystical, these dominant interpretations of the origins of the scientific revolution pursue an internalist approach to the history of scientific ideas. They tend to neglect the fact that in the century between the affair of the placards and the *Discourse on Method* (1637), the peoples of France lived through one of the most murderous and anxiety-ridden periods of their history. The French heard their preachers rage against idolatry or heresy, denouncing superstition and pacts with the devil. They experienced a terrifying diabolic presence that roamed the world, taking possession of the faithful, tormenting them.[12] Witches and sorcerers reappeared, provoking the panic that brought torture and death to innocent victims of a collective fantasy. When the author of the placard spoke of the impending doom that awaited a sinful world, he reflected the dark vision of an elite convinced that the apocalyptic moment approached.[13] An obsession with the eschatological fixed the elite's attention on the confrontation with the forces of evil. Historians of social movements or collective mentalities, who have studied these manifestations of the diabolic as a form of social anxiety, have most frequently concluded that the panic ended in the middle of the seventeenth century as a consequence of the triumph of rationalism identified with the scientific revolution.[14] Social historians, in other words, have agreed with historians of science that a rationalist mode of thought and cultural expression eventually transformed or dominated the theocratic or mystical tendencies of an earlier, formative age.

While not disputing the persuasive effects on the educated public of the scientific achievements of a Galileo or a Descartes, I suggest that we reformulate the relationship between the revolution in scientific ideas and the decline of the satanic and the magical in ordinary life. In this essay, I will argue that the critique of magical influences was not so much a consequence of the development of the new science as a prerequisite for it. This critique imposed a strict interpretation to the division between the supernatural and the natural. Human affairs would be protected from the arbitrary intervention of the divine; they would also be set free to assume the awesome responsibility of a newfound freedom.

To present this interpretation of the shift in the early modern mentality, the first part of this essay considers the rupture in Christian culture provoked by the challenge to the Eucharist. There I argue that the critique of a central religious mystery provoked a challenge to the reigning political and religious myths that structured public life; it engendered a destructive element in the body politic, undermining secular authority and encouraging religious zealots to crusade against idolaters or heretics. The eucharistic topos focused their obsessions; it became the governing metaphor in the struggle against the satanic influences that appeared to threaten the community.

In this context, the restoration of order, which is the subject of the second part of the essay, required more than Henri of Navarre's victory over the forces of the Catholic League. It also required that the forces of the supernatural play for the king's cause. His conversion and scrupulous respect for the rituals of Catholic France demonstrated as much. In this effort to restore the monarchical foundations for civil society, he called upon distinguished men of the robe to make a significant contribution. These latter used their professional knowledge and capacities to defend the cause of order and, above all, to moderate the religious enthusiasm of zealots who threatened to subordinate the traditional corporate order to their religious obsessions. Yet, when the magistrates and doctors confronted the demonic they had to confront deeply-held convictions. Some of them were convinced that sinful creatures could enter into satanic pacts. Others preferred to argue in the tradition of Aristotelian naturalism that natural effects required natural causes.

But, was Satan's presence physical or only a fantasy? Was the witches' sabbath a real event or only an illusion? Could Satan, the witch, or the magician defy the general laws of nature? Accusations of witchcraft and malevolent magic encouraged reflection on the difference between fantasy and the real, and on the distinction between the supernatural and the natural. Above all, it demanded consideration of the functioning of the human mind, its ability for delusion and its capacity for knowing the truth. These were not abstract philosophical questions, but practical issues of everyday life faced by a professional elite in its efforts to uphold forms of social and cultural control. That these same issues prompted René Descartes's search for a new philosophy suggests that the effort of magistrates and doctors to understand and to master the effects of the supernatural in ordinary life contributed to the formation of the modern mentality.

Part I: The Rupture in Christian Culture

A. Eucharistic Imagery and Political Legitimation

Throughout the history of the Church, bloody feuds had accompanied debates over Christian doctrine. The losers, judged guilty of heresy, were either silenced or executed. During the medieval period, debates over the nature of the Eucharist had even ended with Church scholars submitting their arguments to the arbitration of either Church Council or the authority of the Pope. The celebrated painting of Raphael, *The Dispute over the Eucharist,* painted for the Vatican in 1505, glorified this effort at reconciliation. Or, more probably, the painting pointed to the papal desire for submission to its authority. Such was the rule of faith. A new generation of theologians, represented by Luther and Calvin, thought differently. It was their intention to purge the Christian religion of the accretions of the medieval legacy. In the affair of the placards more radical reformers moved beyond debate to violent confrontation.

After the blasphemous outrage of that night of October 1534, the municipal and corporate authorities of Paris organized a series of expiatory gestures to purge the community of the pollution of heresy. Over the next few weeks, twelve persons of markedly different social stations were condemned to death. The heretics were executed on different days in several sectors of the city, prolonging the expiatory gesture and distributing its effect over a wider area and population. The ritual burning of the victims followed. Reduced to ashes, their corporal remains were thrown to the wind, depriving the heretic of the right to burial in holy ground. Dispersed to the forces of chance, the bodily remains of the heretics were forever separated from the beneficent effects of the religious rituals of atonement regularly enacted in the nearby churches. This gruesome gesture, the immolation of the heretic's body, gave public form to the Christian community's understanding of how the body of each member mediated overriding religious values. Fearing the contamination of heresy, the municipal and ecclesiastic authorities ordered the destruction and dispersal of the contagion threatening the body of the visible Church with pollution. But this expiatory sacrifice was not enough.

Catholic preachers inflamed popular passions, saying that sedition and anarchy loomed over the kingdom. To hold public passions in check and to contain the fanaticism of these preachers, François I, king of Catholic France, took steps to reestablish the Christian and civil order. Upon his return to Paris he set up a commission for judging heretics, and on 21 January 1535 he led the nobility, clergy, the university faculties, the municipal government of Paris, and the urban corporations in an elaborate expiatory procession. The canons of the cathedral and the ordinary clergy carried the relics of the passion, those of Saint Geneviève, the patron of Paris, and those of numerous saints from each parish church. Carrying a lighted torch, the king, bareheaded and on foot, walked behind the monstrance, the receptacle designed in the form of a radiant sun and containing the Blessed Sacrament, the *Corpus Domini.*[15] The procession wound through highly-decorated city streets to Notre-Dame, where the canons of the chapter sung a High Mass.[16]

The king and his officers demonstrated to the public and to religious enthusiasts that they would defend the community from outrage to the doctrine of the real presence in the Eucharist. The public display of the Host and the celebration of the High Mass focused the community's emotions and energies on the sacred signs that pointed to their fundamental beliefs. The order and form of the procession represented a living image of those beliefs. The presence of the civil hierarchies and the urban corporations testified to their preeminent role in ordering social and culture life. The ecclesiastical presence spoke of their primary responsibility for presiding over the rites and beliefs of the true religion. The king, the embodiment of the community, walked at their head. The Host, the sign of the sacred presence in their midst, dominated the whole.

Already highly charged with political content, the eucharistic imagery of state ceremonies called attention to the claims for a sacred sanction that justified the monarch's authority. The monstrance, for example, was used in processions and state ceremonies. There the king's person represented the body politic. And in royal entry ceremonies the king rode beneath the dias, the covering held over the Host in religious processions. Since the fifteenth century, the Host had been increasingly displayed during liturgical ceremonies. This separation of the Host from its place in the ritual of the Mass suggested to the reformers a dangerous exploitation of public emotions. No longer locked within the ciborium, a heavy stone spire hidden off to the side of the altar, the Host was frequently shown in a monstrance high above the altar in the plain gaze of all. Over the course of the sixteenth century, another development reenforced this identification of kingship with eucharistic imagery. The monstrance designed as a radiant sun, the image of God in this world, came to symbolize the power and beneficence of the kings of France. All these iconographic statements shaped the public imagination. Going beyond the boundaries that feudal practice would have permitted, they eventually prepared the public imagination for claims to a kind of royal authority that was as absolute as God's own claims on His creatures.

What seems clear is that over the course of the early modern period, the Eucharist, the preeminent social sacrament, became the metaphorical representation of the body politic; it recapitulated the medieval understanding of national identity subsumed under the concept of the Mystical Body.[17] Just as Christ stood for the Christian community, the king stood for the nation. The Eucharist, a symbol for both, aided the real presence of the sacred in the midst of the profane. The transposition of sacred signs onto the social articulated the public belief that the king, the anointed minister of the nation, played the role of the pontiff, the bridge between heaven and earth. For the Christian imagination the king represented the community before the divinity; he stood as a Christ-figure, mediating between the domains of the sacred and profane, between the demands of eternal justice and the contingencies of history. Like Christ, he had a miraculous healing touch and dutifully punished those who harmed. Like the sinners who transgressed the laws of God, he made amends, carrying the lighted candle of penitence; he walked bareheaded behind the Host, a sign of submission.

When the ceremonies were ending, a witness reported François I's words: "If the arm of my body was infected with this disease, I would want it cut off, and if my children were tainted by It, I would want to immolate them myself."[18] The rhetoric bore a clear message: sworn to uphold the Catholic religion, the anointed king of France had to eliminate heresy even at the price of what was most precious to him. Earlier, Louis IX, Saint Louis, had said: "Je suis la France"; and he persecuted the heretic. Later, Louis XIV claimed: "L'etat c'est moi"; and he subjected the French to the Catholic religion. A sacred mission explained this imposition, for the king identified his legitimacy with the obligation to impose religious conformity. Guided by the ordering principle: "one king, one law, and one religion," the sole guardian of national unity could not do otherwise.

Proof of this commitment to subject the bodies of the Christian community to the established order again claimed its victims. To close out the ceremonies of expiation several accused heretics were burned at the stake. This murderous manifestation of collective anxiety, the annihilation of the heretic's body, parodied the celebration of the Body of Christ. The destruction of the alien's body enhanced the honors rendered to those authorities who represented the body politic. This collective celebration of the body politic and the immolation of its infected members demonstrated the importance of corporal imagery for the mediation of the desires and anxieties of this community of believers.

More than an image of a hidden presence, in this belief system the eucharistic signs became what they signified, the body and blood of the expiatory victim who died for the sin of Adam. Participation in the sacrificial act identified the community of believers as those who won their salvation through the ritual sacrifice of the sacred victim. This act gave meaning to the phrase: "Outside the Church there is no salvation," for communion with the sacrificial victim earned believers access to heavenly grace, to salvation. This theological principle argues in favor of the validity of Northrop Frye's illuminating analysis of Christian mythology. In his persuasive presentation, the eucharistic symbol articulates the human desire for salvation from pain and suffering, from the miseries of the human condition and from the servitude imposed by the immutable hierarchy of the natural and social orders. Within the mythological structure of the Christian belief system, the Eucharist, the bread and the wine, was the archetype for communion with the fruits of human labor, the harvest and the vine. Eating the sacrificial victim, the Lamb of God, signified the redemption of each person from the fatal consequences of the Fall. Partaking of that sacrificial victim bound the human community not only to the divine, but also to nature, for the signs of the Eucharist stood for the categories of the chain of being—mineral, vegetable, and animal—giving the communicant possession of all of the natural world.[19] The supreme metaphorical expression of the human desire to possess the world, the Eucharist signified at once the promise of the Paradise to come and a longing for the Garden of Eden lost through Adam's original sin.

The radical reformers' interpretation of the Christian myth had an original and fatal quality, for they believed that popular piety and the abusive reasoning of theologians had violated the spiritual meaning of the Eucharist. Unable to accept the gross materialism that they read into Rome's interpretation of the Eucharist, they denounced the notion of the real presence. In their view the bread and wine retained their physical natures during the ceremony that commemorated the Last Supper. They could not accept then the Scholastic doctrine of transubstantiation, wherein the words of the minister transformed the substance of bread and wine into the body and blood of the Lord. The appeal to Aristotelian categories, like the appeal to popular emotions, only served to prove that the purity of the original act of atonement foreshadowed in the Last Supper had been tarnished by the obfuscation of theologians and the superstitions of the people.

Rather than this appeal to human emotions or philosophical ploys, the radicals emphasized the message of faith spoken in the Scriptures. For them the word of God would transform the hearts of believers. God's free gift of faith alone could save without the mediation of priest or bishop. This exaltation of the spiritual was reminiscent of the gnostics who, in the early history of the Church, had decried public forms of worship presided over by the hierarchy of the visible Church.[20] Convinced that the Fall of Adam had radically altered human nature and nature itself, the radicals could never accept that the manipulation of bread and wine could transformation them into the body and blood of the Lord. To affirm as much was blasphemous. To adore a piece of bread, claiming it was the body of the Lord, was mere idolatry. The worshipers of Baal now threatened the Christian community. The call of the *sacrificateurs* for the multiple repetitions of the redemptive act distorted the meaning of Christ's words at the Last Supper: "Take and eat; this is my Body." This momentous phrase had a metaphorical sense; otherwise, the Christian ritual would be a horrible form of cannibalism. In view of this understanding of what was happening to true religion, it is no wonder that the radical reformers feared that divine vengeance would soon plunge the world into an abyss.

Catholic zealots shared this apocalyptic vision, for they believed that the denial of the real presence in the Eucharist signified the active presence of Satan among men. The reign of the Anti-Christ, a theme sounded in the pamphlets of Protestant reformers, found its echo in the sermons of Catholic preachers. As the affair of the placards showed, the popular fears and anxieties released by the controversy over the signification of the Body of Christ suggested that the unity of Christian community could not withstand the challenge to this central mystery; its fragmentation prepared for horrifying patterns of violence: the desecration of the dead, the torture and defilement of those accused of breaking with the true religion, even cannibalism, both metaphorical and real, moved to the forefront of the public imagination. And, as if to demonstrate the parodic nature of the religious mentality, the rupture in Christian culture provoked the rediscovery of the satanic presence among men. The devil took possession of the human body just as the communicant took possession of Christ's body. In the witches' sabbath Satan mimicked the rituals of the Mass. That the Devil, the Ape of God, demanded sacrifice and adoration only validated the Sacrifice of the Mass as the primordial social sacrament. Within this frame of reference, dominated by the eucharistic topos, no other form of expression could give form to the desires and fears of a people facing destruction.

B. The Demonic Parody and the End of Politics

Threatening the reigning political and religious mythologies, the affair of the placards thus disrupted the cultural consensus supporting civil society. The critique of the radical reformers enforced a kind of self-awareness and demanded a personal commitment to doctrinal purity. Where religious practice once enjoyed a kind of unreflective and unconditional support of the community, it now became a matter of conscience. Now, religious enthusiasts measured the legitimacy of king or magistrate against the uncompromising standard of ideological orthodoxy. Taking God as the source of all authority, they refused obedience to a king who did not punish idolaters or execute heretics. In this way the attack on a central mystery of the Christian religion became inextricably linked in the public imagination to the legitimacy of the king.

In 1559, Anne Du Bourg, counselor to the Parlement of Paris and nephew of a former Chancellor to François I, drew a radical conclusion from Calvin's mandate to the magistrates. On 10 June, the Parlement of Paris was nearing the end of a protracted *Mercuriale* begun in late April. Instituted in 1499, the *Mercuriale* was an occasional gathering of the magistrates "to review their moeurs and manner of living both public and private."[21] On this occasion, Anne Du Bourg, provoked by the Edict of Ecouen, Henri II's most recent effort to contain the expansion of the reformed religion, spoke out against the king's protection of false religion. The presence of King Henri II, François I's successor, heightened the significance of Du Bourg's impassioned plea for a respect of scriptural purity. The magistrate remarked that the court accused some of "sedition, of disobedience to the king as long as we do not offer gifts to Baal." In response he refused obedience to a king who tolerated idolatry. "Who has given the king our prince authority over so many people? Isn't it the Seigneur of all kings? Would God have invested the king with such authority simply to countermand His will, to exempt the king from the commandments that he has given to all nations in heaven and on earth?" And daring even to raise the threat of regicide, he said: "it is our king and those who follow him who commit the crime of lese majesty if they work against the will of our common Seigneur. And by this they are guilty of death if they persist in an error that should be condemned." Turning to the magistrates present, he warned that "your dissimulation is equivalent to penury by which you miserably enchant your king and abuse the people."[22] Unable to pardon such a direct offense, Henri II insisted on the punishment of the young magistrate, a decision that had momentous consequences, for the condemnation and execution of Anne Du Bourg prepared the ground for rebellion.[23]

Catholic preachers echoed Du Bourg's appeal to the law of God as the standard for political fidelity. These zealots of the Catholic faith cried out against the failure of the civil authorities to punish the heretic. Unable to contain the advance of the Protestant movement, Catholic preachers had come to perceive the disavowal of the Mass as a sign of a more general moral disorder. The violence of Catholic polemicists cultivated a sense of righteousness that set aside the procedures of ordinary justice and recommended popular vengeance against the reformers in the name of God's law. Leniency or hesitation in matters of religion rang of complicity. On Easter of 1559, just before Anne Du Bourg's plea, the Franciscan Jean de Han, "using the parable of the adulterous woman brought before Christ, declared it no wonder that judges refused to cast the first stone against the Lutherans, since they were Lutherans themselves."[24] Such collaboration suggested, in the zealot's eyes, the widespread corruption that had seized the country. The charisma of the youthful Jean de Hans projected into his sermons a prophetic quality. Hearing his impassioned plea for the purification of the community, the populace could easily believe that they heard God speaking directly to them. The very embodiment of God's will, a kind of incarnation of his justice, preachers like Jean de Han came to challenge directly the authority of both the king and his officers, the magistrates.[25]

The liberal corporate elites, traditional mediators of social conflict, lost ground; they were displaced from the center of social life as the exigencies of ideologues dominated the public imagination and focused national passions. Sensing the dangers of popular passions that fed on ideological differences, the chief magistrates tried to contain the disruptive effects of religious fanaticism. At the moment of Du Bourg's execution, the Parlement of Paris condemned those "preachers of this town and faubourg, who abusing the authority of the pulpit, instead of restraining the people in the fear of God and obedience to the king and his magistrates, pronounce scandalous commentaries supporting sedition and encouraging popular emotion rather than edification."[26]

At the end of 1559, the premature death of Henri II set the stage for a national tragedy, for subsequent attempts at compromise and reconciliation proved futile. Early in 1560, François II, the young successor to Henri II, promulgated an edict of toleration, pardoning all those accused of heresy. For the first time a king of France tolerated religious differences, a policy strongly supported by the Queen Mother, Catherine de Medici. Intending to defend this compromise, Michel de l'Hospital, François II's new Chancellor, recognized in the rule of conscience a challenge to the rule of the prince. An Erasmian figure, he opened the Estates General at Orleans with an appeal for tolerance: "The knife means little against the spirit and its use only risks losing both spirit and body. . . . Let us banish these diabolic words, names of party, factions and seditions, Lutherans, Huguenots, Papists . . . and bear only the name of Christian."[27] When the Edict of April 1561 mandated that ideal into law, Catherine and the magistrates manifested their intention to confine the extremists and to limit their power over the populace. The preachers of Paris, Jean de Han in particular, reminded the public authorities that the people of God would carry out the justice that the magistrates refused to exercise.

After the untimely death of François II brought Charles IX, a minor, to the throne, national reconcilation became more urgent. At the Colloquy of Passy (September–October 1561), the Chancellor and the Queen Mother brought together Catholics and Huguenots. Again, the question of the real presence in the Eucharist frustrated compromise. Théodore de Bèze (1519–1605) could not concede the principle for which Anne Du Bourg died. "My conscience is taken captive by God's word." In these terms Luther had justified his revolt against Rome. Anne Du Bourg's similar appeal justified his revolt against king and Parlement. Bèze, the Huguenot theologian, persistently defended the same cause. And Catholic theologians and preachers were as adamant as their Protestant critics. In the Chancellor's appeal to the preeminence of law and respect for civic order, religious enthusiasts read only the submission to evil. The framework of the law failed to restrain the demands for religious purity. In fact, it won only the resentment of Protestants and Catholics alike. Such appeals to civic loyalties divested of religious doctrine only reenforced the sentiment that the authorities protected the agents of Satan.

Provoking widespread anxiety, the breakdown of religious unity not only threatened the civil order but also deranged the public imagination. As if to give adequate measure to the distortions troubling the community, the phenomenon of demonic possession reentered the annals of Christian history. Religious zealots exploited this phenomenon for partisan purposes. As exorcising Satan fell to the Church, the rituals now made public could only dramatize its power over the forces of the supernatural; it drew public attention to the indispensable place of the Church in protecting the community. The pressing need to mobilize the Catholic majority against the Protestants gave preachers an opportunity to mold the popular imagination. Itinerant preachers brought the message to the villages. They took over the pulpits of parish churches. Everywhere they reiterated the same message: compromise with the heretics confirmed the action of diabolic forces. Unfortunately for the cause of civil order, the terrifying and captivating effects of demonic possession and the process of exorcism united *les curieux* of aristocractic or bourgeois origin with the urban and rural populous. It was the perfect baroque theater prepared and talked about through contemporary channels of communication. Heralds brought the date of the execution of a magician or a witch to neighboring villages and towns, inciting the country people to swell the crowd of urban spectators. And those unable to attend could hear visiting preachers describe the crimes and their punishment. In these moments, normal civil relations, ordinary criteria for judging, were suspended. The community was made over in the image of its fantasies.

There were notable examples of this manipulation. In 1566, the exorcism of the young maiden, Nicole Aubray, in Laon, a provincial capital torn by religious divisions, provided an excellent opportunity for ecclesiastics to demonstrate the power of the rites of the Catholic religion over evil.[28] The moment could not have been more propitious. Excluded from the councils of Catherine de Medici, opposed to her tendency towards compromise and toleration, and fearful of the advance of the Huguenots' cause, Catholic zealots found their cause waning. Occurring during a fragile peace in which Protestants and

Catholics heightened their propaganda, the exorcism of Nicole Aubray focused public attention on the central issues of the religious struggle. While the Protestants denied the validity of the case, each day the exorcists brought Nicole in procession to the cathedral. Displayed on a large scaffold, similar to those used for the theatrical performances of the mystery plays, she was exorcised. Constrained to obey the priests, the Prince of the devils, Beelzebuth, lectured the crowd on the violence and infidelity of the Huguenots. Following the script elaborated in the sermons of Catholic preachers, s/he spelt out the Counter Reformation program. Over a two month period, the symbolism of the Host dominated the scene. Beelzebuth, speaking with Nicole's voice, told how Huguenots had desecrated the Host; it was the Host that finally drove the Beelzebuth from Nicole's body. According to witnesses, a converted Huguenot cried out: "I believe. Because I have seen it. I will no longer be a Huguenot. Damned are those who fooled me. Now I know the Mass is good and that the priests are virtuous."[29]

In this way Catholic preachers fed the persuasive fear of a satanic presence. These clerics, reenforcing the deeply held convictions of the people, spoke with a defining voice on contemporary issues. To insure their dominance over public debate, they sanctioned and encouraged the purported intervention of the supernatural in daily life. In so doing they only expressed in comprehensible terms the deeply felt anxieties of the people, anxieties, it must be said, that they shared and magnified. In so doing they also set themselves up as rivals of the normal civil authorities. This development had momentous and tragic consequences. At decisive moments of popular sedition, riot, and revolt, the civil authorities and liberal professions became the victims of religious fanaticism and simple greed. During the St. Bartholomew massacres, the men of the robe who ruled over the intellectual life of the town, who claimed responsibility for forming public opinion, paid the price of resisting the exigencies of the zealots. Pierre de La Place, First President of the *Cour des Aides,* and numerous men of the law were killed. Another victim was Petrus Ramus (1515–1572), the leading humanist of his day. Ambroise Paré (1509–1590), celebrated surgeon trained at the Faculty of Medicine, was spared only by the intervention of Charles IX, the hapless king who tried to excuse the violence by explaining the massacre as the consequence of a Protestant plot. To save his life, Prince Henri of Navarre feigned conversion. Rather than restoring order, these distortions only fed disorder. In the 1570s, a literature of resistance flourished. Théodore de Bèze's *On the Right of Magistrates* (1574) took up a theme sounded in the personal sacrifice of Anne Du Bourg. In 1579, the most radical expression of the need to resist tyranny in the name of an individual's covenant with God came from Phillipe du Plessis-Morney (1549–1623) in his *Vindiciae contra Tyrannos*. This development within the literature of political and religious theory exemplified the singular breakdown of the traditional construct, the sacrality of the king's person, as a compelling national myth.

Another measure of the displacement of legitimate and traditional imagery and mythology and their corporate interpreters was the intensification and transformation of beliefs in witchcraft. Viewed as an opportunistic development, it can be argued that this tragedy exploited common fears to punish, torture, and often execute social deviants, marginal personalities, most frequently aging women, or religious dissidents. Outside of France Protestant leaders participated as actively as their Catholic counterparts in this murderous pursuit of the witch. The command of Exodus 22:18 motivated both communities: "Thou shalt not suffer a witch to live." Together theologians and jurists fabricated the literature of demonology. The prototype for this literature was the *Malleus maleficarum* (1486), a treatise written by Jakob Sprenger and Heinrich Kraemer (Institor), two Dominican fathers.[30] Translated as the *Hammer of Witches* and edited several times over the course of the sixteenth century, the text served as a referential code for inquisitorial proceedings and for the witchcraft trials. The witches' sabbath, discussed in the early demonological literature as a folk superstition and judged by the Church in the *Canon episcopi* to be the fantastic imaginings of the credulous, under the pressures of religious conflict, now became a reality. Previously, Saint Augustine, who had judged the supposed transportations of witches as illusions, informed Church teaching. Unfortunately, Saint Thomas (c. 1225–1274) and Scholastic teaching came to accept the absolute reality of these transformations: the witches are not merely deluded; their transportation to the sabbath and their intercourse with the devil was a reality.[31]

Just as the medieval Church used this ideological innovation to justify the persecution of its enemies, in the sixteenth century, the enemies of the faith and their accusations focused the theological and judicial imagination.[32] In this way, the crime of idolatry and the denial of the real presence in the Eucharist, were assimilated into the literature of demonology. Just as the exorcist demonstrated that the Host empowered the priest to command the devils, so too the witchcraft trial was an opportunity to demonstrate that Satan parodied the true religion and its rituals. The eucharistic ritual being central to the true religion, it was not surprising that Satan demanded a similar form of adoration, a similar sacrifice of innocent victims in the witches' sabbath.

The diffusion and exploitation of such ideas accelerated the persecution of accused witches and magicians. The nature of the crimes committed–the transportation to the sabbath, intercourse with the devil, the transmutation of the witch into a wolf–defied normal juridical procedures and accepted rules of evidence. A physician and student of Agrippa, Johan Wier, defended the victims, offering melancholy as the cause of the illusions: "The purpose of this work," he wrote, "is to prove that the crimes attributed to sorcerers are imaginary; that these women are not criminals, but are afflicted with an illness of their mental faculties; that they should not be brought before priests, monks, or judges; they should not be imprisoned, tortured, nor delivered up to the flames at the stake. On the contrary, they should be confided to the care of doctors."[32a] Wier represented the skepticism of the trained physician and the credulity of a true believer. His text, widely diffused and commented on, certainly encouraged medical practitioners to look carefully at gesticulations of the possessed or the evidence of witchcraft.

In 1580, the year following the French translation of an enlarged version of Wier's original Latin text, Jean Bodin (1530–96), the author of *The Six Books of the Republic,* ventured into the domain of the demoniac with the *De la Démonomanie des Sorciers,* in part an effort to refute Wier's work.[33] Significantly, Bodin practiced law at the center of a region plagued with witchcraft accusations: Laon. He was perfectly aware of the exorcism of Nicole Aubray that had occurred in that town. In fact, just before Bodin wrote his response to Wier, Jehan Boulaese, a local Catholic polemicist, refreshed the public's memory by publishing a book on the celebrated exorcism.[34] Bodin's rambling text can be read as a response to priests who would appropriate the demonic for their use. While he believed in the reality of demonic possession, Bodin lamented the spectacular exploitation of the possessed by the Catholic clerics of his town.[35] Nor could he accept the authority of medical practicioners to annul the trial of accused witches. And while he clearly sets down the rules of evidence for trying the accused witch, his book, like his *Republic,* argues for the establishment of a well-regulated society under the absolute authority of the sovereign. Critical of magistrates who would restrain the sovereign, Bodin urged the relaxation of any restraining influence on his will to order society. Fearing the continued civil war, he urged respect for the Salic Law and the absolute authority of the prince. Fearing the spread of satanic influences, he defined witchcraft as *crimum exceptum.*[36]

The magistrates of the high courts did not share Bodin's views. They, however, were an exception. Lesser magistrates and clerics saw in Bodin's text the complement of the learned *Republic.* In ordinary times the king's will ruled over civil society, uniting the various interests and estates in his person. Satanic threats to the public order could not, however, be repulsed and punished in the same manner as threats to the body politic. Satan could penetrate the body and take hold of it. He also possessed the mind of his willing victim. He actually carried them off to the witches' sabbath where they performed their abominations. Like the *Malleus maleficarum,* Bodin's collection of tales and his repeated appeals to the authority of the ancients had a formative effect on demonologues and the educated public alike.

Just as Bodin had found in the tales of human sacrifice practiced in the Americas the prime example of the corruption of the human imagination, Jean de Léry, the author of *History of Voyage to the Land of Brazil,* found in Bodin's tale of the witches' sabbath an explanation for his own discovery twenty years earlier. In the 1585 edition of his book Léry noted that Satan operated in the New World in a manner similar to his behavior in the Old. Speaking of the demonic behavior of the Tupinamba performed at a witches' sabbath, he wrote: "I have concluded that they have the same master: that is, the Brazilian women and the witches over here were guided by the same spirit of Satan; neither the distance between the places nor the long passage over the sea keeps the father of lies from working both here and there on those who are handed over to him by the just judgment of God."[37] Léry could also understand the literal cannibalism of the Brazilians, because French Catholics practiced a metaphorical cannibalism in his own day. The obsessive attention of Catholics and Protestants to the rituals of blood sacrifice and cannibalistic practices among the Amerindian peoples tells us little of the significance of those customs for the peoples of the New World, but it contributes to our understanding of the intensity of meaning that the European gave to the Eucharist, their central sacrament; it also explains the ease with which the Spanish *conquistadors* could justify their pillage and murder of the native peoples in the name of the true religion.

In this context, it is not surprising that the eucharistic topos articulated communal anxiety. It provided a locus which fixed the central antagonisms dividing the nation. From the very outbreak of the religious wars, the defense of the true nature of the Eucharist encouraged and justified popular violence. Religious riots occurred especially during processions such as those of Corpus Christi Day when the monstrance provoked attack or justified violence.[38] In Lyon, the painter Denis de Vallois threw himself on the "God of paste" in order "to destroy him in every parish in the world." And Catholics shouted: "For the flesh of God, we must kill all the Huguenots."[39] The defense or attack on the

Eucharist and the mystery of transubstantiation had already colored the representations of Europeans' observations in the new world. Sensitive to the charges of idolatry from the Moslems and Jews, the Spanish conquistadors violently criticized the bloody human sacrifices made to the Aztec gods.[40] And, the Calvinist Jean de Léry readily identified the cannibalism he discovered in Brazil with the Catholics who "wanted to eat the flesh of Jesus Christ raw."[41] Accused of being cannibals and *sacrificateurs,* the Catholic masses defended their faith while their preachers articulated the explanation for collective violence, even cannibalism, in cogent if chilling terms:

> The people's vengeance against this traitor [Ravaillac] and the holy rage [were] . . . taken against a body torn into a thousand pieces. But to eat its raw flesh imposes silence on us. This image of hatred is worth more than all the discourses of the world . . . . The iron, the sulfur, the fire, the hooks—a thousand deaths rolled into one, the body of the patricide quartered, these quarters torn into a thousand other pieces cannot satisfy the just suffering of this people. The people's vengeance imitates its love . . . . They carry out with an infinite affection what a barbarous and denatured cruelty imposes upon the cannibals.[42]

Montaigne lamented the contemporary barbarity carried out in the name of the syllable "hoc." His *Essays,* an effort to distance the magistrate from his former responsibilities, contained a memorable reflection on the troubles of his time. "On Cannibals" is particularly pertinent as it is at once a commentary on Léry's discovery and on Montaigne's own experience:

> I am not sorry that we notice the barbarous horror of such acts, but I am heartily sorry that, judging their faults rightly, we should be so blind to our own. I think there is more barbarity in eating a man alive than in eating him dead; and in tearing by tortures and the rack a body still full of feeling, in roasting a man bit by bit, in having him bitten and mangled by dogs and swine (as we have not only read but seen within fresh memory, not among ancient enemies, but among neighbors and fellow citizens, and what is worse, on the pretext of piety and religion), than in roasting and eating him after he is dead .[43]

Montaigne's protest recalled the admonition of earlier magistrates and foreshadowed that of a later generation. These officials of the monarchy argued that the value of citizenship should carry more value than religious enthusiasm, that loyal to the civic order should influence action more directly than the appeal of personal conscience. Henri de Navarre would heed their advice, for to secure his throne military victories alone were not enough. To reign he had to bring the struggle over culture to an end; he had to make the supernatural play for the king's cause.[44] The liberal elites would assist him in this task, for they understood that appeals to a conscience respecting only a personal interpretation of the law of God could threaten civil authority. Above all, they witnessed the popular violence generated by zealots who cultivated the popular imagination and its fascination with the supernatural.

Part II: The Restoration of Order
A. Ritual and the Reconstruction of Political Mythology

Henri of Navarre's accession to the throne of St. Louis promised a new beginning. Compromising the demands of conscience for the purposes of civil order, Henri IV promised in July 1593 to take instruction in the Catholic faith. Success against the armies of the League was an essential first step towards the throne, but to recreate the nation out of the chaos of civil war and ideological conflict Henri tried to demonstrate that the demands of conscience could be reconciled to the need for order. Yet, concessions in matters of faith could equally compromise political legitimacy in the mind of the religious enthusiast. In September 1593, a Catholic zealot and a well-educated member of the legal classes wrote of Henri's attendance at Mass as the symbolic gesture that had secured his crown. Had the new king failed to do so, nobles would have had the duty to rebel.[45] The same author identified the true conditions for inheriting the crown of St. Louis:

> The true inheriters of the Crown are those who are worthy of carrying the imprint of God. If it pleased God to give us a king of the French nation, blessed be His name; if he be of Lorraine, blessed be His name; if Spanish, blessed be His name; if German, blessed be His name. From whatever nation he might be, being catholic and filled with piety and justice, as coming from the hand of God; it does not make a difference to us. We are not moved by the nation, but only by religion.[46]

The orison denied the authority of the Salic Law, pointing beyond traditional usage to religion as the test of legitimacy and to God as the source of all authority. Part of the legacy of ideologically inspired politics, the text refused the ambiguities of dynastic power and personal ambition; it denied the separation of religious belief from political ends; it subordinated temporal purposes and interests to the requirements of eternal salvation. Religion, not national ambitions, moved this author.

Henri, the victorious hero, could be portrayed as Hercules, the mythical representation of an ordering principle,[47] as the king who dominated his rivals on the battlefield. But to reign, Henri IV had to end the ideological confrontation that religious propagandists had encouraged for two generations. That was the purpose of his conversion, the first gesture of identification with the beliefs of the majority religion. Others soon followed, for Henri, the Christian king, had need of the symbols of Catholic France to sanctity and legitimize his authority.[48] A well-established political-religious liturgy alone could confirm legitimacy upon Henri. The Basilica of St. Denis, symbol of the unity of France, offered the perfect place for the ritualization of the conversion. It was there that Louis VI had identified his defense of the realm against the threat of foreign invaders by appealing to the Apostle of France for protection and by identifying in the popular imagination his cause with that of Charlemagne's. On 25 September 1593, Henri IV appealed to the national memory and abjured his former beliefs with solemnity before twelve prelates. An immense crowd celebrated the recognition of the king. Such close identity with the national past help legitimize his claim to kingship readily proclaimed by the people present. Kept out of Rheims, which remained hostile to him, Henri chose Chartres for his consecration, and it was there that he was crowned king of France on 27 February 1594. Elaborate public pronouncements informed the nation how scrupulously the ceremony respected the liturgy of past coronations. On 22 March, after brushing aside the last resistance of the League forces, the king entered Paris, his presence in the capital symbolizing the union of the king with his people. At the close of the day, a *Te Deum* was sung, the people's prayer asking God's protection for their king: "Domine, Fac sauve le roi."[49] The following week, on 29 March, a magnificent procession under the authority of the king took place. Modeled on the expiatory procession of 1535, the Blessed Sacrament again was honored, numerous relics from the royal chapel and from the parish churches demonstrated the union of the king with the people. At Notre-Dame, the bishop preached on the theme of clemency. Afterwards, each of the representations of the city's constituted corporations begged the king's grace and benevolent assistance in the restoration of order.

If the conversion and anointment of Henri of Navarre was the symbolic healing of the wounds opened in the terrible civil and religious wars, this triumphant procession finally exorcised the blasphemous acts committed in the affair of the placards. On Easter Sunday, 10 April, Henri brought to a climax the rituals celebrating his kingship. After the Mass, "empowered by a eucharistic fire," he touched the victims of scrofula, the ulcerous tumors provoked by tuberculosis. Believers knelt before the king and heard him repeat the traditional prayer: "Le roi te touche, Dieu te guerit." When the news spread through Paris that the anointed king had touched the victims of scrofula, the virulent cries of Catholic preachers calling for continued resistance to the heretic ceased to move the people of Paris. The royal touch, with its magical healing power, belonged to them once again. Having conquered the throne by force of arms, Henri de Bourbon legitimated his power by his conversion; his acknowledgement of the religious grounds for political authority confirmed the people's belief that their king and their church partook of a single communion.[50] Interrupted for five years, the ceremony of the royal touch, fittingly celebrated on the feast of the Resurrection, regenerated the nation under the signs of Catholic France. The restoration of the king's real presence mediated once again the divine benevolence for the people of France.

Unfortunately for the royal cause, the appeal to the rituals of the Catholic religion might win Paris for the king, but it failed to convince Henri's erstwhile correligionists. Théodore de Bèze spoke for the disenchantment of the committed Protestant when he said that the religion of Machiavelli now reigned in France. Like the phrase, "Paris is well worth a Mass," that summing up challenged the religious foundations for the rituals of the state. It would not be excessive to conclude that the explicit appeal to the supernatural and its exploitation for ideological purposes had exhausted the credibility of the supernatural in the political domain. After the killings in the name of religious purity, Henri IV's attempt to master the rituals of public ceremony struck a false note. At best the appeal to tradition spoke to a sense of civic order animating the action of higher magistrates and other men of the robe. At worst the mask of state ceremony covered up the design to manipulate the public imagination for the purposes of personal and dynastic ambition. In the end, the religious wars exposed all appeals to the supernatural to the scrutiny of the skeptic or cynic who might accept such usage as necessary for the maintenance of public order. Rarely would the force of faith ever again move men to see the supernatural working for the cause of the nation. It is here, at the level of faith, that we must see how the rituals of state became the monopoly of the person of the king. Begun under Louis XIII and fully developed under Louis XIV, this process eventually stripped the monarchy of its representational function in the domain of the sacred. In this context the restoration of the civic order and the maintenance of the structures of ordinary life required more than ceremony. To assure the functions of state and society, the nation, like the sick, required more than religious enthusiasm. In this context it was imperative that the supernatural be mastered for the cause of civil society. To assure this development leading members of the liberal elite actively propagated a message destined to reconcile religious beliefs with the new monarchy. They did so in the name of a particular competence, won through university study, developed through daily practice, and sanctioned through public recognition. Their overriding objective remained social and cultural control exercised in the name of the restored monarchy and in the name of the public welfare. André Dulaurens, Henri IV's First Physician, and Guillaume Du Vair, President of Parlement, exemplified this effort.

B. Professional Knowledge and the Corporate Order: The Royal Touch

Henri IV's First Physician wrote a treatise on the magical healing power of the king's touch, "on this obscure but beautiful subject that nobody has yet undertaken to explain."[51] Prepared by a night of prayer, the king confessed, heard mass, and armed himself with "the celestial and eucharistic bread." Burning with "the fire of charity," he entered into a vast, spacious place where the sick turned toward the king, "the royal sun, followed by his doctors as the dawn is followed by the day. Then, looking at the sick, with a peaceful and benign eye, his sacred and medical touch reached out to cross the forehead of each, saying: 'The king touches you; and God heals you.'"[52] Separately, the engraver, Pierre Firens, issued a print depicting the ceremony. The physician and engraver celebrated an image of kingship embedded in tradition.[53] Behind the king stands his chaplain giving his benediction, representing the ecclesiastical sanction given to the ritual. The captain of the Scottish guards, relaxed but vigilant, looks on while in the outer semicircle his men form a barrier between *les curieux* and the participants in the ceremony; they create a frame for the engraving, suggesting the need for secular vigilance and force in human affairs even if at this particular moment they are peripheral to the ceremony. And while the king is a central figure in this engraving, the print highlights the preeminent role of the First Physician whose gaze engages the spectator's and leads it into the ritual of healing. And first among the observers was the medical profession itself. The First Physician reminded them that peace could be restored to the imagination of their war-torn culture only if trained and recognized professionals mediated the effects of the supernatural in the lives of men.

Standing opposite the king, the titular head of the medical profession places his hands on the neck, on the sores, of the sick, as the king makes the sign of the cross on the pilgrim's forehead. The meaning was clear enough: between royal benevolence and popular suffering, between health and sickness, the medical professional projected an active and necessary mediating power. Beside him stands another physician who points to the next person selected for royal beneficence. Other physicians stand behind the sick, pointing to the extensive participation of medical professionals in the rituals of royal order. Labeled a *représentation au naturel,* the engraving resembled, like the text itself, a prayer seeking to transform a desire into a reality. Dulaurens ended his text with: "Amen."

Coherent with his professional responsibilities, Dulaurens did not leave all to the religious imagination. In the second part of his treatise, the First Physician set down other remedies for healing the victims of scrofula. The order and manner adhered strictly to the conventions of medical tradition. Most importantly, he arranged the treatments in a precise order that recalled the social and professional hierarchy familiar to medical practitioners. The king's cool eucharistic fire healed without harming. The physician's insight into the cycle of the disease was more beneficial than the surgeon's industrious hand; the doctor's cool touch was better than the surgeon's burning iron. An implicit assumption governed this hierarchy of treatments: the king's reception of the Eucharist empowered his medical act; an act of faith disposed the pilgrim to a healing virtue; the doctor's liberal education nurtured his understanding; the surgeon's manual skills prepared his hand for cutting out the malignant tumors. All other remedies, especially those hinting of any occult property, he dismissed as superstitious.

The appeal was credible, for Dulaurens structured his argument in terms that were recognizable and acceptable to the literate elites of his day. He wrote in the context of the Hippocratic and Galenic medical traditions, demonstrating the synthetic quality of medical knowledge and extending its interpretive force beyond the domain of human illnesses to comment on the significance of the human condition. Typical of the rationalist tendencies among his colleagues, Dulaurens typified the attachment to the Aristotelian model and its search for natural causes. In his essay on the recognition of the critical moment in the cycle of an illness Dulaurens wrote:

> She [Nature] moves and operates according to a certain, constant order which never varies such that she seems to govern by counsel and reason. She contains, says Hippocrates, the fatal necessity of life and death. And, according to Galen, her movements are regulated by certain terms and fixed, determined cycles. Nature imposes on herself certain laws and ordonnances which she can never go beyond. Thus . . . she respects them without inconstancy, and without changing anything concerning their order. It is she who is the unique doctor of the sick; it is she who brings about the crisis when Nature separates the harmful humors from those which are useful and prepares them for excretion. She finds hidden paths unknown to us to purge sicknesses from the body.[54]

Their limited ability to intervene effectively in the cycle of diseases imposed a healthy skepticism upon medical practitioners.[55] Trained in the Galenic tradition and tested by bitter experience, they hesitated to base their professional and corporate authority on a chance success and preferred instead to follow, to the best of their ability, the Hippocratic admonition: "Do no harm." This understanding of the natural order confirmed what the Fathers of the Church had taught, encouraging medical professionals to accept sickness and death as fatalities. They were consubstantial with all things temporal and created. Hence Galenic doctrine reenforced Christian teaching and the transposition of the Christian understanding of human fatalities to those of the Galenic mode was easily achieved: "Death and the ruin of things does not depend on heaven, but on the condition of elementary matter, and on its vices and defects."[56] If Saint Augustine explained the frailty of the human condition through the doctrine of original sin, before him Galen had taught that "the malignancy of time" wore down the particular virtue or force of every living thing. The synthesis of these two traditions underwrote medical doctrine and explained in part its inherent skepticism; it also served to soften doctors' frustration before sicknesses that they could not effectively treat. Like the priest presiding over the ailing soul, they presided over the sick and the cycle of disease.

If Dulaurens celebrated the king's special healing powers, he did not celebrate royal absolutism. Rather, he defended a broad notion of political power that included a vital role for the liberal professions, the corporate elites, in the definition and exercise of political responsibilities. In this he was much closer to his contemporary Charles Loyseau (1564–1627), eminent magistrate and defender of corporate power, than to Jacques-Bénigne Bossuet (1627–1704), the theoretician of the divine right of kings.[57] His essay can rightly be read alongside a text published in the following year, Loyseau's *Traité des Ordres et des simples Dignitez* (1610), where the corporate responsibilities of the liberal professions are justified in terms of a special education which prepared them for their task of guardianship. An Aristotelian model of social and political organization served both Dulaurens and Loyseau. Professional competence, earned through a special training of the mind, formed the elites of a nation and gave them the duty and privilege of a specialized function within the body politic. If Dulaurens sought to rehabilitate the mythical foundations of the monarchy, he consciously endeavored to reenforce the notion that professional knowledge gave the men of the robe a privileged place in corporate society; it granted them certain privileges but required of them a special service. Furthermore, these liberal professionals provided essential services to their contemporaries. Responsible for the continuities of everyday experience, they tended to act as a conservative element in a society sorely tested by the demands of religious enthusiasm. As corporate leaders obliged by office to respond to the needs of daily life, they accommodated change with difficulty. Responsible for the transmission and the development of traditional learning, they could not be indifferent to a system of thought that believed in the regular intervention of the supernatural in ordinary life.[58] In this context the professional elites participated in the struggle to circumscribe the purported influence of the supernatural on public life; they participated in an effort to restructure the public imagination in terms of beliefs and practices which would encourage wide support and participation in the restored monarchy. The severe cultural and political crisis of the previous generations made this task an essential feature of the reconstitution of a unified nation. The professional elites, armed with classical education and a revitalized Christian learning, were well-suited to the undertaking.

Formed in the university, these professionals of liberal learning feared the subordination of a social system dependent on the highly differentiated qualities for public office to a homogeneous order founded on the criteria of religious practice. Like their monarch, they had an interest in making the people believe in the society of orders based upon respect for the rule of quality in public office. If a common purpose bent them to the service of the nascent absolute monarchy, their training fostered a dedication to the rule of law. If they assumed the role of the guardians of orthodoxy, they did so in the name of professional knowledge and experience. For these reasons, it would be wrong to reduce them to mere henchmen of the king or to confuse them with the propagandists of royal absolutism. And if they were faithful to the patterns of traditional usage and encouraged the ceremonials that formed their public image and ratified their claims to social place, they tried to set boundaries around acceptable ritual and to encourage its propagation. Fearing the threat of sedition, they tried to purge Christian culture of its excessive and uncontrollable elements. Fearing the threat of eternal damnation, they warred against idolatry, the cause of superstition. As interpreters of orthodoxy, they had to make the people believe that the forces of the supernatural worked for the purposes of public order. As believers themselves, they had to reconcile their professional knowledge and standards of conduct with their own understanding of the true religion.

Taken together, the liberal professions constituted what can be referred to as the elite of the dominant literate culture. The gatekeepers of tradition, they exercised their formative authority over the culture as teachers, censors, doctors, judges, priests. Through the Faculties of the University of Paris, through the law courts of Parlement, they carried out social functions in the name of a professional training that gave them privileged access to a body of knowledge and custom beyond the reach of the layman. In their respective domains, they acted as the licensed practitioners responsible for regulating the affairs of daily life. A kind of fourth estate, they served the community neither by arms nor by production, but rather through their interpretive abilities, their power to recognize the just or right order of things, to discern threats to the equilibrium of the body politic.

In this sense Dulaurens's project concurred with the humanist motivation of the men of the robe, particularly ennobled magistrates, the successors of Michel de l'Hospital and Christian De Thou. The most preeminent among them was Guillaume Du Vair (1556–1621). Before his fellow magistrates he persuasively advocated the respect of the Salic Law. The defense of national traditions, personified in the king, reasserted the crucial function of the magistrates themselves to enforce these traditions. "What we want to make them (the people) believe is that this is done according to the laws of the nation."[59] Erase this belief from the spirit of the people," he continued, and you will violate your charge, for "you have the guardianship of the laws and the tutorship of the kingdom in your hands." "If this evil is to be remedied, only you, the guardians of the law, can do so."[60] Without the commitment of the corporate elites to the customs and symbols of monarchical culture, the forces of sedition and disorder would threaten the civil order.

Professing these beliefs and persuading others to assent to them, Dulaurens asked that civic virtue and Christian values which permeated the doctors' professional training shape their judgment and incline their intellect to the defense of tradition. Where religious ideologues and political realists saw the hand of Machiavelli at work, Dulaurens and Du Vair sought to move the wills of the liberal professions through an appeal to intellectual constructs coherent with their education and social function. The same metaphors structured the discourse on sickness, sedition, and sin. Judge and priest assimilated their functions to medical imagery. Magistrates and lawyers "entered the public theater of justice" as "doctors of the spirit." "In that the priest confessor bears the double charge of judge and doctor, he must exercise the responsibilities of both." He judges and heals.[61] Understood in this manner, the texts of the men of the robe synthesized a coherent religious mythology and a political ideology which identified corporate functions as vital to the restoration of the monarchy of Catholic France. By appealing to traditional learning and accepted practice, doctors, like Dulaurens, and magistrates, like Du Vair, tried to shelter the persuasive force of custom from the reductionist formulations of religious purists who would strip away the cumbersome accumulations of history and expose the works of mere men to the harsh light of a purified and militant faith.

The rationalist mode of medical explanation and the assumptions of Dulaurens's Christian faith structured his discourse. But when he reviewed the causes of superstition, the most serious affliction of the human mind, his fundamental beliefs compelled him to conclude: "these . . . are nothing other than the external signs of a pact and alliance contracted with the devil."[62] In the contemporary context of the witch-craze, this judgment demonstrated the inclusive reach but also the disabling effects of medical discourse. For the action of Satan among men could be explained in terms totally coherent with Galenic learning: melancholic humors, it was argued, provided the devil with an opportunity to conjure up the illusions of a false power to heal or to harm—an explanation that concurred with the widespread sentiment nurtured by the religious enthusiasts. Like them, Dulaurens too believed that human history tended towards the accumulation of sin and the risk of idolatry.

Regardless of their religious or political caste, men of liberal learning viewed the popular mind as a fatally flawed instrument incapable of independent and trustworthy reflection. Du Vair articulated this same understanding of the human condition when he spoke to his fellow magistrates. He warned them of the corruptions of mind and body: As doctors found themselves among "the ulcers, abscesses, rot, and the corruption of the body, they [the magistrates] must live continually among the most pestilent illnesses of the spirit."[63] Notwithstanding the skepticism and doubt about the human condition, the magistrate, like the doctor, had to preside over the decisions affecting the lives of their fellows. How can the magistrate dare break the silence and pronounce his verdict when the anguishing diversity of customs, of manners, and of laws alone serve to guide his fallible judgment? Only an interior vision of the divine justice and truth permits the magistrate to speak and to judge. Doubt only ends when a source of truth prior to human events and words makes itself felt in the depths of the soul, dictating its righteousness amidst the inconstancy of human words and things.

When he spoke at a *remonstrance d'ouverture* (a ceremony like the *Mercuriale*), Du Vair recalled the elevated function of the magistrature, defining their efforts against the measure of the highest values of their culture. Offering an implicit response to Anne Du Bourg's idealism, Du Vair recalled that the service of law and the exercise of royal justice mirrored the service of divine law and a universal justice. Strongly affected by the contemporary effort to wed civic humanism to Christianity, the president of the Parlement represented the office of the magistrature as a kind of priesthood.[64] Imitating the Jewish Neoplatonist Philo of Alexandria, Du Vair's model of civil virtue and rectitude, the judge had to purge himself of self-interest. He had to assume the role of a purified victim capable of dying to personal choice before he could preside over the crimes and punishment of others. In Du Vair's terms, the judge became a *sacrificateur* who carried out his ministry only after having sacrificed his person to the needs of the community.[65] Unlike Anne Du Bourg, however, Du Vair never asserted the prerogatives of conscience. On the contrary, as a representative figure of the Catholic renaissance he assimilated the reformers' critique, and in a parodic turn he used its energy to generate a heightened awareness of the need to check the impervious demands of self. By dramatizing the sacerdotal quality of the magistrate, he drew on a theme of an emerging baroque theology, a theology of the holocaust wherein the Christian, like Christ, became the victim or the *hostie,* annihilating the self in a sacerdotal act of adoration. His contemporaries, Father Coton, a Jesuit, the friend and confessor of Henri IV, and Pierre Bérulle, the founder of the Oratorians and the confessor to Maria de Medici, formulated in mystical terms the same intense devotion to the cause of sacrifice. The charge of *sacrifacteur,* an accusation that had sustained the hatred and violence of the civil and religious wars, was transformed into a central trope or metaphor sustaining the Catholic renaissance.

C. Baroque Theology and the Demonic in the Restored Monarchy

Coherent with traditional learning and constructed upon the Augustinian understanding of a flawed human nature, the conventions of magistrate and doctor drew on those of a revitalized Catholicism. This latter identification strengthened a desired tendency to reunite political authority with a reinvigorated culture. Yet, Catholic zealots were not willing to accept a return to the dispassionate exercise of political power. Religious enthusiasm, represented by Coton or Bérulle, still tried to play a vital role in the affairs of the state. These religious zealots, particularly Bérulle, had no intention of subordinating the revitalized religion to the cause of political order or the affairs of the state. He preferred instead to give new life to the forces of religion.

In January 1599, just as the Parlement of Paris was to consider the registration of the Edict of Nantes, Martha Brossier, a victim of satanic possession, appeared in Paris with her father and sister. The young woman performed convincingly in Parisian churches. Ecclesiastical authorities exorcised Martha Brossier before spectators of rank, *les curieux,* who rewarded the family generously. The Parisian crowd also assisted. The demonstration of religious enthusiasm was a troubling reminder of the force of the supernatural working in the city. Ecclesiastical authorities took advantage of this development to remind the civil authorities of their responsibilities to defend the true religion, a gesture the king failed to appreciate. Fearing a mobilization of public opinion to oppose the registration of the Edict of Nantes, the king ordered the municipal authorities to challenge the veracity of the possession. Michel Maresicot, doctor regent of the Faculty of Medicine, deemed the possession a masquerade and blamed the churchmen "who claimed this cause of the supposedly possessed woman as their own."[66] Pierre Bérulle took up their defense, accusing Maresicot of being isolated and unrepresentative of the medical profession. As his witness, the cleric cited an eminent physician, none other than André Dulaurens as his ally. Bérulle defended the prerogative of the Church and "the judgment of an ecclesiastical body, the first of France, on a preeminently ecclesiastical subject which is beyond his [Maresicot's] merit and power."[67] For him, demonic possession demonstrated the power of true religion over Behemot, the satanic beast that roamed the world. As such, it was a persuasive force in the conversion of the heretic.

Maresicot argued differently. He appealed to the authority of the Faculty of Medicine. Interestingly enough, he too invoked the support of Dulaurens and other eminent physicians who had examined the woman: "they saw nothing beyond the common laws of nature." That Dulaurens's authority could be invoked by both sides suggests not only his preeminence but also the ambiguity of public perceptions of medical authority. Was the doctor's judgment valid only if it concurred with the priest's? Or, could the doctor exercise an independent authority and challenge the conclusions of the cleric? If Bérulle represented the former opinion, Maresicot's conclusion accepted the latter. The doctor believed that the Brossier family was exploiting public credulity. The problem, he wrote, was "the exaggerated credulity" of the spectators which, he explained, "derives from the imbecility of the spirit of men and the suggestions of the devil." Like Wier before him, Maresicot never denied the action of the devil, but he noted that without the guidance of medical professionals, churchman, nobles, and the profane could fall victim to the deceptions of their imaginations too easily seduced by satanic wiles.

Maresicot was grateful for the intervention of the Parlement: "Fearing that her influence on the people who wanted to see the devil might engender a new monster of sedition," the Parlement ordered Martha Brossier out of the city. For the medical professional, the Parlement was "the true protector and defender of piety and justice."[68] The message was clear: unless the elite of the liberal professions checked religious enthusiasm in the name of their specialized knowledge, the monster of sedition would raise its head again.

This judgment reflected the common opinion of civic and religious leaders who understood that the public fascination with the supernatural could effectively terrorize public or church authorities into taking actions that lacked justification in more traditional terms. Yet, fascinated themselves by the supernatural and firm believers in the fatal weaknesses of the human mind before temptation to enter into pacts with the devil, magistrates, doctors, and civil authorities had no firm ground to resist the intrusions of zealots in civil affairs. Soon after the assassination of Henry IV, Du Vair, like Dulaurens during the affair of Martha Brossier, fell victim to his fears.

After serving in the Paris Parlement, Du Vair held the office of President of the Parlement of Provence. In this capacity he presided over the trial and execution of a priest from Marseille, Louis Gaufridy, accused of seducing and bewitching a young girl, Madelaine Demandois de la Palud. In the fall of 1610, in Aix-en-Provence, Sebastian Michaelis, reputed exorcist and Father General of the Reformed Order of

Preachers, began his attempts to drive the devils from the girl. At first reticent to perform publicly, Madelaine remained silent. Under the pressure of Michaelis, of a fellow nun who was the real mistress of ceremonies, and of the eager crowd, Madelaine eventually complied. She acted out "the venereal act," her intercourse with Beelzebub, and finally accused the priest, Gaufridy, of being "The Great Magician." Similar to the case of Nicole Aubray, Michaelis's history of the tragedy reads like a Counter Reformation tract with the possessed nun, under the sign of the Host, preaching to the assembly of the dangers of laxity among the clergy and laymen alike.

Unfortunately, a doctor of Maresicot's integrity and intellect was lacking. Instead, Jacques Fontaine, a doctor from the faculty of Aix-en-Provence found the crucial evidence of Gaufridy's pact with the devil. He and a surgeon examined the priest for the marks of the devil and discovered several.[69] "Then, the word spread throughout the people that he was a sorcerer and that he was marked." The news only confirmed what the people had already learned while assisting at Michaelis's exorcisms of Madelaine. Fontaine, who wrote a short piece on the investigation to describe his role, lingered on the people's reaction: "The voice of the people is the voice of God, that is, of the truth, as if it is accompanied by a kind of divine inspiration, by an assured capacity for recognizing the truth of these crimes. So too, it was filled with contentment and rejoicing when they saw the punishment come."[70] That the people's enthusiasm reassured the doctor and confirmed his specialized judgment underlined the treacherous quality of the widespread belief that the devil did indeed mark his own. But the terrible consequences of credulity infected even the victim, for when Gaufridy was informed of the evidence he accepted the doctor's findings, only pleading that "this was done without his consent,"[71] a phrase that judge, priest, and doctor repeated in their reports.

For all their skepticism with regard to the fantasies of the religious mentality, some doctors too easily submitted their critical judgment to the more imaginative claims of biblical conventions. As Behemot marks his own, he had marked Louis. Perhaps, Fontaine, like many of his medical colleagues, had come to accept the Paracelsian doctrine of signatures which assumed that even though God has hidden certain things, He "has allowed nothing to remain without exterior and visible signs in the form of special marks."[72] In any event, the human fascination with leaving traces or tags to mark a fictive possession or to affirm an identity carried terrible consequences for the true believers of the early modern period.[73]

In the face of the evidence, Gaufridy's bishop lifted his ecclesiastical immunity, allowing the court at Aix to bring him to trial on the charges of rape and sorcery. He vigorously denied both. The court, following common practice, ordered that he be tortured. When put to the question and asked to tell the truth, he responded that "he will speak the truth, but he will not be believed . . . and fears that he will despair under torture." He did. When he failed to confirm the expectations of his judges, he was "admonished." When he persisted in his denials, he was "severely admonished." Breaking into tears, he begged for a pause in the interrogation. The record notes that Gaufridy said he would try "to satisfy his interrogators." To escape torture, the accused priest decided to tell the story that he knew would be believed. "Let them continue the interrogation; he will respond and tell the truth." Playing to their fantasies, Gaufridy told of the sabbath, the adoration of Satan, the blasphemies, and profanation of the consecrated body of Our Lord. Du Vair, the presiding judge, recognized "a tale filled with truth and illusion." He seemed unable to distinguish where one left off and the other began. Nevertheless, he ordered the death sentence.[74] On 30 April 1611, the people rejoiced when Louis went to the stake.

Several years later, Madelaine was brought before the court on charges of sorcery. She was declared innocent, but condemned to remain close to her relatives and not to trouble the public order with her stories. When asked about her earlier relationship with Father Michaelis and about his account of the exorcism and trial, she responded: "Those good Fathers who wrote those memoirs or that book believed all too easily. They were wrong, for that book contains only illusions."[75] That Madelaine finally admitted what Du Vair had suspected might suggest that Michaelis, the priest, had been duped. Or, did judge, priest, doctor, and the victims, Louis Gaufridy and Madelaine, all suffer under a kind of collective illusion that infected the people? If so, Michaelis never admitted it. He wrote of the proceedings with the passion of the advocate, defending his behavior by citing the work of Pierre Delancre, whose pursuit of the Basque witches of Labourd just a few months early had created a sensation throughout the nation.

Pierre Delancre (1553–1631), a magistrate from Bordeaux, a relative of Montaigne, delighted in exploiting the theme of inconstancy. In *Tableau de l'Inconstance et Instabilité de toutes Choses, où il est montré qu'en Dieu seul gît la vraie Constance* . . . (1607), Delancre informed his text with Neoplatonic fantasies. He highlighted the conviction of an age that saw the world as theater, a stage upon which life struts like "a walking shadow." He affirmed that the imagination could control matter, that conjuring up images had the effect of recreating a kind of reality as valid as the material world known to human touch and sight. Disappointed with the reception of his text, Delancre sought fame with *Tableau de l'Inconstance des mauvais Anges et Démons où il est amplement traité des Sorciers et de la Sorcellerie* (1613), a text written after he had presided over the witchcraft trials of the women of St. Pé in the region of Labourd, Pays Basque. Updating his theme with confessions, Delancre told the story of satanic invasion. The engraving, *The Witches' Sabbath,* accompanied his text. It is a mix of images, a recapitulation of demonic literature spun out in the programed interrogations. As the devil presides, a sorcerer presents a child she has seduced. After feasting, others dance with their demonic partners. In the center a brew of ointments throws up a cloud of smoke, the symbol of inconstancy. Through the night sky witches ride their proverbial broomsticks as masked women of all rank ready themselves for the festivities. Written as a contrapuntal movement to the Counter-Reformation obsession with human sexuality, the new genre revelled in the obscene and manifested a baroque fascination with the perverse. Inconstancy is the theme. In Delancre's credulous eye the tales of the witches' sabbath painted a *tableau,* a series of scenes acting out what judge, public, and the victim understood to be the passing shadows of a life already condemned to death. The clouds, the smoke, the flights through the night sky—all these images portrayed for the eager spectators not so much the illusions of a distorted imagination. On the contrary, they played to a belief that words and gestures could transform matter. "Hoc est corpus meam," the sacred formula of the priest could be mimicked by Satan, the Ape of God. Like the priest, the magician had the power to transform men into wolves, women into witches.

Taken as a whole, demonology projected a grotesque distortion into the natural, generating a sense of awesome power in some and overwhelming fear in many. Yet, on the other side of the Pyrenees, Alonso de Salazar Frias, who was the compassionate and skeptical priest in charge of the Spanish inquisition, acquitted his Basque witches and decried the delusion that incited the judges and victims alike to believe in the fantasies of the sabbath.[76] What was truth for one was illusion for the other. The fantasies of an age made it difficult to sustain the distinction between truth and illusion, between signs and reality, between the true religion and superstition.

The overwhelming evidence of the demonic in ordinary life dramatized public fears and encouraged a kind of collective hysteria; it testified eloquently to the hold of the magical and mystical on the public imagination. Full participants in this system of belief, religious reformers again testified to the parodic quality of the religious mentality. To rid the world of diabolical forces and to prepare for the Second Coming, they encouraged devotion to the mystery of the incarnation. Bringing Christ into the world, having him take possession of it, just as he had taken possession of the mystics, became their obsession. Having seen the stigmata on those marked by a divine possession, the parodic quality of the religious mentality anticipated finding Satan's mark on "The Great Magician." A full participant in this system of belief, the leaders of French spirituality, Pierre Bérulle among them, welcomed the Carmelites of Theresa of Avila to Paris. Madame Acarie, their distinguished patron, soon bore the same signs of a divine possession.

These religious zealots, however, were not content with prayer and meditation. A new militancy, more disciplined than the prophetic preachers of the League, worked for the elimination of heresy "a body, which disturbs, agitates, and infects all the bodies of France." Appealing to Louis XIII as the successor to Saint Louis, Bérulle urged that the young king wage war against the heretics. For him, as for the enthusiasts of reform, the clearly defined social and civil hierarchies founded on the qualities of mind and person held little attraction. Rather than accepting the fundamentals of Aristotelian politics as interpreted through the French tradition of the civic order, Bérulle appealed to a Neoplatonic inspiration. "An excellent mind of this century," wrote Bérulle of Copernicus, "maintained that the sun was the center of the world and not the earth . . . . This new opinion, little followed in the science of the stars, is useful and must be followed in the science of salvation. For Jesus is an immobile Sun in his grandeur and moves all things."[77] This desire for illumination and participation in the states of Jesus took precedence over civic obligations. Called to contemplate the moments, the passing scenes of the life of Jesus, his states of being in the world, the Christian could judge as ultimately futile all efforts to give permanent form to the illusions of this world.

D. Scientific Optimism and the Disenchantment of the World

In the decade immediately following the most spectacular and highly publicized witchcraft trials, other students of Plato made a concerted effort to master illusion and to translate the world of appearances into the constant laws of nature. To do so they narrowed the Neoplatonic tradition to its mathematical mode of expression. Galileo did as much in *The Assayer* (1623). There he wrote:

> Philosophy is written in this great book of the universe, which stands continually open to our gaze. But the book cannot be understood unless one first learns to comprehend the language and to read the alphabet in which it is composed. it is written in the language of mathematics, and its characters are triangles, circles, and other geometric figures, without which it is humanly impossible to understand a single word of it; without these, one wanders about in a dark labyrinth.[78]

This effort to read a mathematical order beneath the movements and shapes of the visible world assumed that uniform laws, a commanding order, governed the whole of nature. These assumptions allowed Galileo to distinguish his methodical and verifiable study of nature from those phrased in terms of "sympathy," "antipathy," "occult properties," and "influences."[79] Similarly, Kepler set himself off from Robert Fludd: "His method is the business of alchemists, hermeticists and Paracelsians, mine is the task of mathematicians."[80]

That Kepler's own mother was subjected to the terrifying interrogations of a witchcraft trial convinced him of the dangers of excessive credulity. For over two years, he participated in the defense of his mother, her courage and fortitude as well as her son's persistence saved her from condemnation.[81] Against the background of anxiety and personal anguish experienced during the trial, Kepler wrote *Harmonices mundi* (1619), an essential text underlying the future Newtonian revolution. Prominently displayed in the appendix to Book 3 was a long and sustained attack on the cosmological metaphor developed in Bodin's *Republic.* It is tempting to read Kepler's extensive aside as a thinly disguised effort to discredit the author of the *Démonomanie,* an authority who had nourished the fantasies of his mother's accusers. The discrediting of the *Démonomanie* became, in fact, the hallmark of a generation reaching towards a definitive separation of illusion from the real.

The mathematicians of the new sciences were not alone in this effort to counter the extraordinary credulity of previous generations. In fact several philosophical traditions joined in the critique of the occult, of the invisible and supernatural powers acting in the world. Doctors imitated the earlier example of Agrippa, Wier, and Maresicot. Medical critics, obeying the constraints imposed by Aristotelian naturalism also attacked Jean Bodin and those who affirmed that the transformation of men into animals was possible. Typical of this reaction was Jean de Nynauld, a doctor and author of *De la Lycanthropie, Transformation et Extase des Sorciers* (1615). That text demands that "the order of nature" and the "nature of things" be respected.[82] Critical of the fantasies that encouraged Bodin or Delancre to see the world in a constant process of change, Nynauld, like other Aristotelians, sought to purge nature of demonic spirits, astral influences, and occult forces. Gabriel Naudé, (1600–1653) also educated to join the medical profession, exposed Bodin's credulity to the skeptic's critique. In 1625, he wrote of Bodin:

> This great man of France, Jean Bodin, who with a marvelous vivacity of spirit and solid judgment, treated of everything divine, natural, and civil which were unknown to man; and, we would have been regarded him as an infallible intelligence, if he had not left a few signs of his humanity in the *Démonomanie.*[83]

Two years earlier, Naudé, certainly writing under the tutelage of his teachers at the Faculty of Medicine of the University of Paris, had condemned the Rosicrucians, a sect that had suddenly surfaced in Paris promising "to show and to teach without books or marks how to speak all languages of the countries where we wish to be, and to draw our fellow men from the error of death."[84] Schooled in the skepticism of the medical practitioner, Naudé rejected the Rosicrucians' excessive claims. As an alternative to the speculations on satanic action or human prowess to master the forces of death, Naudé situated these pretensions within a historical frame of reference and offered a contextualized understanding of cultural constructs such as "magician" or "witch." Refusing to grant supernatural powers to human agents or agencies, he reflected on the social and political processes that might have effected the evolution of religion in various cultures.

Naudé's reading of history was also a warning to those, like the Rosicrucians, who would claim to add new gods to the temple of learning. He saw in this adventure a repetition of Luther's critique, which "like a cloud of those locust from the Apocalypse ruined the harvest of our belief and dimmed the brilliant light of the sun of religion." Ready to ridicule the demonologues, like Delancre and Bodin, Naudé saw in the reformers' refusal to submit to the unity of faith an unremitting and devastating challenge to the civic order. The multiplication of the reforming churches, "this mark of division, so essential to the lie," mirrored in the multiplication of the new sciences, recalled for Naudé the processes that led to the multiplication of the Roman gods. For him Paracelsus, "the Luther of Medicine," was an innovator who introduced the forces of division and error.

The skepticism and reserve that characterized medical writings such as Nynauld's and Naudé's encouraged their critique of Bodin's diabolic fantasies as well as their suspicion of any hint of a universal medicine. Attached to a corporatist understanding of knowledge, they looked to the traditional interpreters of science, the university faculties, for sound judgment. The innovators have "everywhere caused such confusion, particularly in the sciences, that it is very difficult, if not impossible, to discern those who profess to enlighten the spirit from the ignorant and foolhardy who . . . obscure [the sciences] by an infinity of frauds and superstitions."[85] Writing under the influence of his corporate mentors at the Faculty of Medicine, the youthful Naudé identified knowledge with those whose social function was its profession. In this context the Rosicrucian claim "to draw our fellow men from the error of death," their attempt to discover "a universal Medicine, the philosopher's stone," smacked of a dishonest manipulation of the public imagination.[86] For Naudé the defense of the civil order and the public good required the exposure of these charlatans. The same conviction animated Gui Patin, a scholar of liberal learning and aggressive defender of the corporate control of medical practice. In the 1630s, he condemned Théophraste Renaudot (1583–1653) with the same arguments used by his friend, Naudé, a view years earlier. For these defenders of orthodoxy, of the corporate professional, the outsider was an unwelcome element in the corporate community of learning.

The most prominent and most revolutionary of such intruders was René Descartes. Having returned to Paris just as the Rosicrucians made their appearance, Descartes attracted the support of Bérulle who became, it appears, his spiritual mentor. That the founder of French spirituality and the father of mechanism could have found a common ground for intellectual exchange is indicative of their common opposition to the corporate and university monopoly of learning. Their common dedication to the renovation of Catholicism also motivated their spiritual union. That Aristotelianism suffered from such an alliance disturbed neither the mystic nor the founder of modern philosophy. On the contrary, it was another bonding force that united the elderly Bérulle to the youthful Descartes.

Despite their common front against the philosophical grounds of university training, Descartes distinguished himself in a significant manner from religious reformers. He did not accept the fatalism of the radical Protestant reformers. Nor did he identify fully with the religious enthusiasm animating Bérulle. In fact, he avoided religious controversy, fearing that it would impede reception of his scientific discovery. This is a significant aspect of the Cartesian effort. Above all, Descartes had supreme confidence in human abilities to know the world and to attain certain knowledge of its workings. So inspired, he rejected any hint of the apocalyptic vision and championed instead the sovereignty of life over death; he fostered the conviction that reasoned effort would regain for humankind their rightful place as the masters of nature. Unlike religious reformers who feared that God's vengeance would soon plunge the world into desolation and the abyss, Descartes sought to remake the world and to defend a privileged place in it for humankind.

Recently convinced that he had discovered the principles of a universal science, Descartes wrote to Marin Mersenne (1588–1648) of his search for just such a remedy that Agrippa, Paracelsus, and the Rosicrucians had pursued: "I truly regret your erysipelas and the illness of M. M. [M. Montais]; try to bear up, at least until I know if there is a way of finding a Medicine which is based on infallible demonstrations. That is what I am working on at the moment."[87] The same scientific optimism that inspired the Rosicrucian movement was reiterated in Descartes's *Discourse on Method:* "All we know is almost nothing compared to what remains to be known; we could be freed from innumerable sicknesses of the body and the mind alike and perhaps even from the infirmities of old age, if we had sufficient knowledge of their causes and of all the remedies which nature has provided us."[88] To achieve this end Descartes dispensed with the categories of his Scholastic training and, like Galileo, chose to describe the world in terms of the structure and movement of its parts. Separating body and mind and giving them an independent existence, Descartes rejected the Thomistic doctrine of their substantial union.[89] In so doing he struck down the ontological grounding of Aristotelian science.

But to reach his goal Descartes also refused Neoplatonic fantasies or religious aspirations. Instead, he achieved a revolution in the sciences because he wanted to bring an end to illusion as well as to the pessimism nourished by the skeptic's doubt. His way out of the contemporary dilemma was to order the world of appearances according to the principles of a mechanistic philosophy. Having desubstantialized organized beings, Descartes made them so many automata or moving machines. Having taken a life principle or form from the essence of things, Descartes could take up anatomical research as a necessary first step in his search for a universal remedy.[90] The dissection of their parts and the study of their movements convinced Descartes that he could prolong life by understanding the causes of death. Perhaps he even attended the anatomy lessons of a Leiden physician made famous in Rembrandt's "The Anatomy Lesson of Dr. Trupp." Like the Rosicrucians, he wished "to pull men from the error of death," a desire signed in his metaphorical reconquest of the Garden of Eden. By discovering a medicine based on infallible demonstrations, Descartes intended to cure the sicknesses of the body and so push back the moment of death. He also sought to heal those illnesses of the mind, the source of error and illusion. It was as if the young philosopher wanted to recapture the tree of knowledge for human purposes and in this way carry out his promise to make humankind the master of nature. That message was the essence of scientific optimism, and would eventually become the hallmark of modern science. It points to the strange confluence of philosophical, religious, and humanistic elements that disciplined the contemporary fascination with magic and supernatural forces, transforming imaginative fantasy into habits of mind normally associated with the sciences.

To begin this essay, I suggested that the debate over the Eucharist provoked a rupture in Christian community. The ambition of the Cartesian enterprise, the attempt to create a universal science capable of ending the reign of error and illusion, generated enthusiastic support among contemporaries. Mersenne and others understood the importance Cartesian geometry and the liberation of algebra from geometrical representation. Yet, the corpuscular physics that underpinned Cartesian mechanism created a significant obstacle to wider acceptance of his new system. Persuasive opponents exposed the Cartesian assumptions to critical scrutiny. For different reasons, skeptics, political philosophers, medical doctors, and theologians hesitated to embrace the new philosophy. Among the theologians, the mystery of the Eucharist once again proved divisive. Having overturned the ontological basis for Aristotelian physics, Descartes had identified substance with extension. He thus banished occult qualifies, sympathies, and the magical forces, whether beneficent or diabolic, from operating in the world. His system, however, threatened the doctrine of transubstantiation. He worried about the problem and in 1630 wrote to Mersenne of his obligation to explain "the way in which the whiteness of bread persists in the Blessed Sacrament," for his system could not accept the persistence of "real accidents" after the priest spoke the words of consecration.[91] Later, turning theologian, he suggested that Christ's soul informs the particles of bread and wine without mingling with them. Friends and enemies alike warned Descartes of the heretical implications of his physics, a warning confirmed when in 1663 the Congregation of the Index condemned his system.[92] Descartes's attempt to safeguard the mysteries of faith and open nature's secrets to the methods of a human way of knowing was a singular effort at synthesis. That he failed to convince the religious hierarchy or even friends of the possibility to reconcile his corpuscular philosophy with church teachings prompted a reflection that came to define the attitude of the coming epoch. In a letter to

Huygens Descartes wrote: "whatever the teaching of religion, I must admit that I share an infirmity common to most men, namely, that even though we want to believe and even firmly believe all that religion teaches us, we are more accustomed to being touched and persuaded by the evidence of natural reason."[93]

If Descartes heralded an era of scientific optimism that opened the modern age, it is tempting to suggest that he did so in the name of the age-old desire to reclaim humanity's rightful entrance to the Garden of Eden. Framed in terms of the Christian mythology, his ambition was to correct for the legacy of original sin, a task that would once again make humankind the masters of nature. Acted out in the domains of science and medicine, the primary areas of Cartesian concern, his ambition reflected a similar spirit of conquest that encouraged his European contemporaries to project their political and economic force throughout the world. In sum, if the sixteenth-century critique of tradition and the efforts of religious reformers to disenchant the world played a formative role in the creation of the modern world, the currents that fed the Cartesian spirit and nourished scientific optimism found their source beyond the limited domain of biblical primitivism and a fatalistic reading of the allegory of Paradise lost. Unlike Augustine or Calvin who believed that with the Fall "man has become like nothingness," or "who wills to do what he cannot," Descartes read Christian mythology differently. He, along with the inventors of the new science, worked to erase the consequences of the Fall. Uncovering the secrets of nature, humankind would have full knowledge of the world and mastery over its creatures. They would win back the Garden of Eden, leaving them without need or desire. This perhaps was the real threat of scientific optimism to the religious mentality. Promising to satisfy every desire, the Cartesian model unwittingly challenged the essence of all religions—communion with the ritual victim, sacrificed and eaten in the name of the human desire to possess the world.

Notes

*This is a revised and developed version of papers read at the Society for French Historical Studies, April 1991, the University of British Columbia, and at a conference in honor of Allen G. Debus, "Experiencing Nature," October 1991, The University of Chicago. I am grateful for Keith Michael Baker's critical suggestions.

1. *Chronique du Roy François premier de ce Nom,* ed. Georges Guiffrey (Paris: Jules Renouard, 1860), p. 464.

2. Jean Calvin, *Institutes of the Christian Religion,* trans. Ford Lewis Battles, 2 vols. (Philadelphia: The Westminster Press, 1960), 1:96.

3. Carlos M. N. Eire, *War against the Idols: The Reformation of Worship from Erasmus to Calvin* (Cambridge: Cambridge University Press, 1986), p. 225.

4. Max Weber, *General Economic History,* trans. Frank H. Knight (New York: Collier Books, 1961); and Marcel Gauchet, *Le Désenchantement du Monde: Une Histoire politique de la Religion* (Paris: Gallimard, 1985).

5. Herbert Butterfield, *The Origins of Modern Science* (New York: G. Bell & Sons, 1957); Rupert and Marie Hall, *The Scientific Revolution 1500–1800: The Formation of the Modern Scientific Attitude* (London: Longmans, Green and Co., 1954); Charles C. Gillispie, *The Edge of Objectivity: An Essay in the History of Scientific Ideas* (Princeton: Princeton University Press, 1960); Robert Lenobel, *Mersenne ou la Naissance du Mécanisme* (Paris: Vrin, 1971); and I. Bernard Cohen, *Revolution in Science* (Cambridge, Massachusetts: The Belknap Press of Harvard University, 1986).

6. Kepler to Herwart, in Arthur Koestler, *The Sleepwalkers* (New York: Grosset and Dunlap, 1963), p. 340.

7. Allen G. Debus, *The French Paracelsians: The Chemical Challenge to Medical and Scientific Tradition in Early Modern France* (Cambridge: Cambridge University Press, 1991), pp. 65–74. See also William R. Shea, "Descartes and the Rosicrucians," *Annali dell'Istituto di Storia della Scienza di Firenze, [Nuncius]* 4 (1979): 29–47; and Frances A. Yates, *The Rosicrucian Enlightenment* (London: Routledge & Kegan Paul, 1972), pp. 103–117.

8. Ronald A. Knox, *Enthusiasm: A Chapter in the History of Religion, with Special Reference to the XVIIth and XVIIIth Centuries* (Oxford: Clarendon Press, 1950).

9. Walter Pagel, *The Smiling Spleen: Paracelsianism in Strain and Storm* (Basel: S. Kargen, 1984), pp. 1–5.

10. Heinrich Cornelius Agrippa, *Three Books of Occult Philosophy or Magic: Book 1, Natural Magic,* ed. Willis F. Whitehead (Chicago: Hahn and Whitehead, 1898). There is also a 1971 London reprint of this work.

11. Quoted in Michel Foucault, *The Order of Things: An Archaeology of the Human Science* (New York: Vintage Press, 1973), p. 19. On the idea of the chain of being, see Arthur O. Lovejoy, *The Great Chain of Being* (Cambridge, Massachusetts: Harvard University Press, 1936). On the use of analogy in Renaissance science, see Brian Vickers, "Analogy versus Identity: The Rejection of Occult Symbolism, 1580–1680," in

Occult and Scientific Mentalities in the Renaissance, ed. Vickers (Cambridge: Cambridge University Press, 1984), pp. 95–163.

12. Jean Delumeau, *La Peur en Occident (XIVe–XVIIIe Siècles): Une Cité assiégée* (Paris: Fayard, 1978), chap. 7.

13. Johan Huizinga, *The Waning of the Middle Ages: A Study of the Forms of Life, Thought, and Art in France and the Netherlands in the Fourteenth and Fifteenth Centuries,* trans. F. Hopman (1924; reprint ed., London: Penguin Books, 1985). More recently, *The Darker Vision of the Renaissance: Beyond the Fields of Reason,* ed. Robert S. Kinsman (Berkeley: University of California Press, 1974).

14. H. R. Trevor-Roper, "The European Witch-Craze of the Sixteenth and Seventeenth Centuries," in *Religion, the Reformation, and Social Change* (London: Macmillan and Co., 1967), pp. 90–192; Julio Caro Baroja, *The World of the Witches* (1961, reprint ed., Chicago: University of Chicago Press, 1964); Keith Thomas, *Religion and the Decline of Magic* (London: Weidenfeld & Nicolson, 1971); Robert Mandrou, *Magistrats et Sorciers en France au XVIIe Siècle: Une Analyse de Psychologie historique* (Paris: Seuil, 1980); E. William Monter, *Witchcraft in France and Switzerland: The Borderlands during the Reformation* (Ithaca: Cornell University Press, 1976); Lynn Thorndike, *History of Magic and Experimental Science,* 8 vols., (New York: Columbia University Press, 1923–1958); and Daniel Pickering Walker, *Spiritual and Demonic Magic from Ficino to Campanella* (London: The Warburg Institute, 1958). For an ethnographic perspective, see Mircea Eliade, "Some Observations on European Witchcraft," *History of Religions* 3 (February 1975): 149–172; and Carlo Ginzburg, *Ecstasies: Deciphering the Witches' Sabbath* (New York: Pantheon, 1991).

15. *Journal d'un Bourgeois de Paris sous le Règne du François I (1515–1536),* ed. V. L. Bourrelly (Paris: Picard, 1910), p. 359; Eire, *War Against the Idols* (Cambridge: Cambridge University Press, 1986), pp. 189–193; Gabrielle Berthoud, *Antoine Marcourt: Réformateur et Pamphlétaire, du "Livre des Marchands" aux Placards de 1534* (Geneva: Urn, 1973), pp. 157–244; Jean Jacquart, *François Ier* (Paris: Fayard, 1981), pp. 268–273, 365–367; and R. J. Knecht, *Francis I* (Cambridge: Cambridge University Press, 1982), pp. 248–252, 390–407.

16. Ralph E. Giesey, *The Royal Funeral Ceremony in Renaissance France* (1960; reprint ed., Geneva: Droz, 1983); Sarah Hanley, *The* Lit de Justice *of the Kings of France* (Princeton: Princeton University Press, 1983); and Lawrence M. Bryant, *The King and the City in the Parisian Royal Entry Ceremony* (Geneva: Droz, 1986).

17. Ernst H. Kantorowicz, *The King's Two Bodies: A Study in Medieval Political Theology* (Princeton: Princeton University Press,

1957), and Laudes regiae: *A Study in Liturgical Acclamations and Medieval Ruler Worship* (Berkeley: University of California Press, 1946).

18. Jean-Pierre Babelon, *Nouvelle Histoire de Paris: Paris au XVIe Siècle* (Paris: Hachette, 1986), p. 411.

19. Northrop Frye, "Archetypal Criticism: Theory of Myths," *Anatomy of Criticism* (Princeton: Princeton University Press, 1957), pp. 143–150.

20. Hugo Rahner, *Servir dans l'Eglise: Ignace de Loyola et la Genèse des Exercises,* trans. Guy de Vaux (Paris: Epi, 1959), p. 74. Pertinent for this discussion is Elaine Pagels, *Adam, Eve, and the Serpent* (New York: Vintage Books, 1989). On the Renaissance as a second encounter with gnosticism, see Hans Blumenberg, *The Legitimacy of the Modern Age,* trans. Robert M. Wallace (Cambridge, Massachusetts: M.I.T. Press, 1983).

21. Louis de Bourbon, Prince de Condé, *Memoires de Condé ou Recueil pour servir à l'Histoire de la France, contenant ce qui s'est passé de plus mémorale dans ce Royaume sous les Regnes de François II et de Charles IX,* 5 vols. (Londres: Claude du Bosse, J. Nillor, 1739–1743), 1:218.

22. Anne Du Bourg, *Oraison au Senate de Paris pour la Cause des Crestiens, à la Consolation d'iceux: d'Anne du Bourg Prisonnier pour la Parole* (n.p., 1560), p. 11. See Jean Crespin, *Histoire des Martyrs persecutez et mis à Mort pour la Verité de l'Evangile, depuis le Temps des Apostres jusques à Present,* 3 vols. (1619, reprint ed., Toulouse: Société des Livres religieux, 1885–1889), 2:675–705; and Donald R. Kelley, *The Beginning of Ideology: Consciousness and Society in the French Reformation* (Cambridge and New York: Cambridge University Press, 1981), pp. 171–177.

23. Condé, *Mémoires de Condé,* 1:217, note 1.

24. Quoted in G. Wylie Sypher, "'Faisant ce qu'il leur vient à plaisir': The Image of Protestantism in French Catholic Polemic on the Eve of the Religious Wars," *Sixteenth-Century Journal* 9 (1980): 77. See Edouard Maugis, *Histoire du Parlement de Paris de l'Avènement des Rois valois à la Mort d'Henri IV,* 3 vols. (Paris: Picard, 1914), 2:14 ff.

25. See Denis Crouzet, *Les Guerriers de Dieu: La Violence au Temps des Troubles de Religions vers 1525–vers 1610,* 2 vols. (Paris: Champ Vallon, 1990), 1:411–489.

26. *Arret du Parlement au Sujet des Prédicateurs de la Ville et Diocèse de Paris* (16 décembre 1559). Quoted in Condé, *Mémoires de Condé,* 1:314–315.

27. Babelon, *Nouvelle Histoire de Paris,* p. 427.

28. Jonathan Pearl, "'A School for the Rebel Soul': Politics and Demonic Possession in France," *Historical Reflections/Réflections historiques* 16 (Winter 1989): 294.

29. Ibid. See Henri Weber, "L'Exorcisme à la fin du XVie Siècle, Instrument de la Contre-Réforme et Spectacle baroque," *Nouvelle Revue de seizième Siècle* (1983): 79–101; and François Azouvi, "Possession, Revelation et Rationalité medicale au Debut du XVIIe Siècle," *Revue des Sciences philosophiques et théologiques* 64 (1980): 355–362.

30. Sprenger, the prior of the Convent of Saint André of Cologne, had been a disciple of Alain de la Roche (d. 1475) who had mobilized popular piety in the fifteenth century. A member of the Inquisition since February 1479, Sprenger's authority was expanded by Pope Innocent VIII in the bull *Summis desiderantes effectibus* (1484).

31. Charles Edward Hopkin, *The Share of Thomas Aquinas in the Growth of the Witchcraft Delusion* (1940; reprint ed., New York: AMS Press, 1980).

32. Julio Caro Baroja, "Witchcraft and Catholic Theology," in *Early Modern Witchcraft: Centres and Peripheries,* ed. Dengt Ankarloo and Gustav Henningsen (Oxford: Clarendon Press, 1990), pp. 19–43.

33. Jean Bodin, *De la Démonomanie des Sorciers* (Paris: Jacques Du Puys, 1580). See Walker, *Spiritual and Demonic Magic,* pp. 152–156; E. William Monter, "Inflation and Witchcraft: The Case of Jean Bodin," in *Action and Conviction in Early Modern Europe,* ed. T. K. Rabb and J. E. Seigel (Princeton: Princeton University Press, 1969), pp. 379–384; and Sydney Anglo, "Melancholia and Witchcraft: The Debate between Wier, Bodin, and Scot," *Folie et Déraison à la Renaissance: Colloque internationale* (novembre 1973) (Brussels: Editions universitaires de Bruxelles, 1976).

34. Jehan Boulaese, *Le Tresor et entière Histoire de la triomphante Victoire du Corps de Dieu sur l'Esprit malign Beelzebuth, obtenue à Laon l'An mil cinq cens soixante six* (Paris, 1578).

35. Jonathan Pearl, "Le Role énigmatique de la Démonomanie dans la Chasse aux Sorciers," *Jean Bodin: Actes du Colloque interdisciplinaire d'Angers, 1984* (Angers: Presses de l'Université de Angers, 1985), pp. 407–408.

36. Anglo, "Melancholia and Witchcraft," pp. 209–228.

37. Stephen Greenblatt, *Marvelous Possessions: The Wonder of the New World* (Chicago: University of Chicago Press, 1991), pp. 15–16.

38. Natalie Zemon Davis, "The Rites of Violence," *Society and Culture in Early Modern France* (Stanford: Stanford University Press, 1975), pp. 163–169.

39. Ibid., p. 171.

40. Carmen Bernard and Serge Gruzinski, *De l'Idolotrie: Une Archéologie des Sciences religieuses* (Paris: Seuil, 1988).

41. Greenblatt, *Marvelous Possessions,* p. 8; see Jean de Léry, *History of a Voyage to the Land of Brazil, Otherwise Called America,* trans. Janet Whatley (Berkeley: University of California Press, 1990).

42. Sara Mamone, *Paris et Florence: Deux Capitales du Spectacle pour une Reine* (Paris: Seuil, 1987), pp. 196–199.

43. Michel Eyquem de Montaigne, "On Cannibals," *The Complete Essays of Montaigne,* trans. Donald Frame (Stanford: Stanford University Press, 1965), pp. 292–293.

44. See Emanuelle le Roy Ladurie, *L'État royal de Louis XI à Henri IV: 1460–1610* (Paris: Hachette, 1987), pp. 95–200.

45. John H. M. Salmon, "A Note on the *Dialogue d'entre le Maheustre et le Menant* and Its Variants," in *Renaissance and Revolt: Essays in the Intellectual and Social History of Early Modern France* (Cambridge: Cambridge University Press, 1987), pp. 264–266.

46. George Livet, *Les Guerres de Religion* (Paris: Presses universitaires de France, 1962), p. 62.

47. Françoise Bardon, *Le Portrait mythologiques à la Cour de France sous Henry IV et Louis XIII: Mythologie et Politique* (Paris: Picard, 1974).

48. Yves-Marie Bercé, *Le Roi caché* (Paris: Fayard, 1990), pp. 176–185.

49. Michèle Fogel, *Les Cérémonies de l'Information dans la France du XVIe au XVIIIe Siècle* (Paris: Fayard, 1989), pp. 181–182.

50. Marc Bloch, *Les Rois thaumaturges: Étude sur le Caractère surnaturel attribué à la Puissance royale particulièrement en France et en Angleterre* (1924; reprint ed., Paris: Gallimard, 1983; Jean-Claude Schmitt, *La Raison des Gestes dans l'Occident médiéval* (Paris: Gallimard, 1990).

51. André Dulaurens, *Les Oeuvres de M. André du Laurens,* trans. Théophile Gelée (Paris: Jean Petit-Pas, 1639), p. 93.

52. A provincial lawyer embellished Dulaurens's account. See Jean Barbier, *Les miraculeux Effets de la sacrée Main des Rois de France très-chrestiens: Pour la Guarison des Malades, et Conversions des Herétiques* (Paris: Jean Orry, 1618), p. 26.

53. Bloch, *Les Rois thaumaturges,* pp. 342–344; Robert Muchembled, *Popular Culture and Elite Culture in France, 1400–1700,* trans. Lydia Cochrane (Baton Rouge: Louisiana State University Press, 1985), pp. 183–186.

54. André Dulaurens, *Des Crises, auquel sont expliquées toutes les Differences des Jours critiques, & les Vertus d'un chacus d'iceux,* in *Les Oeuvres,* p. 71.

55. See also Jean-Paul Pitton, "Scepticism and Medicine in the Renaissance," in *Scepticism from the Renaissance to the Enlightenment,* ed. Richard H. Popkin and Charles B. Schmitt (Wiesbaden: Otto Harrassouitz, 1987), pp. 103–132.

56. Ibid., p. 71.

57. See Brigitte Basdecant-Gaudement, *Aux Origines de l'État moderne, Charles Loyseau 1564–1627: Théorecien de la Puissance publique* (Paris: Economica, 1977).

58. On the development of Aristotelianism in the Renaissance, see Charles B. Schmitt, *Aristotle and the Renaissance* (Cambridge: Harvard University Press, 1983).

59. Guillaume Du Vair, *Suasion de l'Arrest donne au Parlement, pour la Manutention de la Loy salique* (1593), p. 71. See Marc Fumaroli, *L'Age de l'Éloquence: Rhétorique et "Res literaria" de la Renaissance au Seuil de l'Époque classique* (Geneva: Droz, 1980).

60. Du Vair, *Suasion de l'Arrest,* p. 69.

61. Del Rio, *Les Controverses et Recherches magiques . . .* (Paris: R. Chaudière, 1611), p. 956.

62. Dulaurens, *Des Crises,* p. 94.

63. Guillaume Du Vair, *Ouverture du Parlement de Aix-en-Provence (1604)* in *Les Oeuvres du Sr. Du Vair vivant Garde des Seaux de France* (Rouen: Louys Du Menil, 1636), p. 325.

64. See the perceptive discussion of the Neoplatonic influences on Du Vair in Fumaroli, *L'Age de l'Éloquence,* p. 517.

65. Ibid., p. 480.

66. Michel Marescot, *Discours veritable sur le Faict de Marthe Brossier de Romorantin, prétendu Démoniaque* (Paris: M. Patisson, 1599), p. 12.

67. Pierre Bérulle, *Traicté de Energumenes, suivy d'un Discours sur la Possession de Marthe Brossier* (Troyes, 1599), p. 3.

68. Marescot, *Discours veritable,* pp. 2–3.

69. Jacques Fontaine, *Discours des Marques des Sorcières et de la réelle Possession que le Diable prend sur le Corps des Homes sur le Sujet de l'abominable et détestable Sorcier Louis Gaufridy . . .* (Paris, 1611).

70. Ibid., p. 5.

71. Ibid. Sébastien Michaelis, *The Admirable History,* trans. W. B. (London: W. Aspley, n.d.), p. 378. The same admission appears in his confession, *Procès verbal de l'Ordre tenu depuis le 23 fevrier jusqu'à 24 avril 1611.* Bibliotéque Nationale MS. 673(148) FF165V°.

72. Foucault, *The Order of Things,* p. 27.

73. Janet Whatley, "Savage Hierarchies: French Catholic Observers of the New World," *The Sixteenth-Century Journal* 17 (1986): 325.

74. Mandrou, *Magistrats et Sorciers en France,* p. 202.
75. Blouch, *L'Histoire chronologique de Provence,* (Aix: C. David, 1664), 2:850–852.
76. Gustav Henningsen, *The Witches' Advocate: Basque Witchcraft and the Spanish Inquisition, 1609–1614* (Reno, Nevada: The University of Nevada Press, 1980).
77. Pierre Bérulle, *Discours de l'Estat & les Grandeurs de Jésus* (Paris: Antoine Estiene, 1623), p. 41.
78. Galileo Galilei, *The Assayer,* quoted in Stillman Drake, *Galileo* (Oxford: Oxford University Press, 1980), p. 70.
79. Ibid.
80. Koestler, *The Sleepwalkers,* p. 397.
81. Max Caspar, *Kepler,* ed. and trans. C. Doris Hellman (New York: Collier Books, 1959); J. V. Field, *Kepler's Geometrical Cosmology* (London: Athlone Press, 1988).
82. Jean de Nynauld, *De la Lycanthropie, Transformation et Extase de Sorciers* (Paris: Frénésie Editions, 1990), p. 77.
83. Gabriel Naudé, *Apologie pour les grandes Hommes soupçonnez de Magie* (Amsterdam: Pierre Humert, 1712), p. 82.
84. Gabriel Naudé, *Instruction à la France sur la Verité de l'Histoire des Frères de la Roze-Croix* (Paris: François Ivlliot, 1623), p. 23. The phrase reads: "pour tirer les hommes nos semblables d'erreur de mort."
85. Ibid., p. 50.
86. Ibid., p. 64. See Yates, *The Rosicrucian Enlightenment,* p. 103.
87. René Descartes to Marin Mersenne, January 1630, *Oeuvres philosophiques (1618–1637),* ed. Ferdinand Alquié, 3 vols. (Paris: Gallimard, 1963–1973), 1:235.
88. René Descartes, *Discourse on Method,* trans. F. E. Sutcliffe (Harmondsworth: Penguin, 1968). See T. D. Murphy, "The Transformation of Traditional Medical Culture under the Old Regime," *Historical Reflections/Réflections historiques* 16 (1989): 318–319.
89. Étienne Gilson, *Études sur le Rôle de la Pensée médiévale dans la Formation du Système cartésien* (Paris: J. Vrin, 1984), p. 9.
90. On the relation of the anatomical theater and the literature of the Baroque to the Cartesian imagination, see Jean-Pierre Cavaillé, *Descartes: La Fable du Monde* (Paris: J. Vrin, 1991).
91. René Descartes to Marin Mersenne, 25 November 1630, in *Oeuvres,* 1:285–286. On the theological controversy, see Espinias, "L'Histoire de Cartesianism," *Revue métaphysique et moral* (1906): 265–293; Étienne Gilson, *La Liberté chez Descartes et la Théologie* (Paris: Félix Alcan, 1913); Henri Gouhier, *La Pensée religieuses de Descartes* (1924; reprint ed., Paris: J. Vrin, 1972); Gouhier, "La Crise de

la Théologie du Temps de Descartes," *Revue de Théologie et de Philosophie* 4 (1954): 19–54; Gouhier, *Cartesianisme et Augustinisme au XVIIe Siècle* (Paris: J. Vrin, 1978); J. R. Armogathe, Theologia Cartesiana: *L'Explication physique de l'Euchariste chez Descartes et Dom Desgahats* (La Haye: Nijhoff, 1977). On the question of transubstantiation in the work of Descartes and Galileo, see Pietro Redondi, *Galileo Heretic (Galileo eretico),* trans. Raymond Rosenthal (Princeton: Princeton University Press, 1987), pp. 280–286.

92. René Descartes to Marin Mersenne, March 1642, in *Oeuvres,* 2:923–926.

93. René Descartes to Christiaan Huygens, 10 October 1642, in *Oeuvres,* 2:936–938.

94. Augustine, *The City of God,* trans. Henry Bettenson (London: Penguin Books, 1984), book 14, chapter 15, 575.

DANCING WITH SPIDERS:

TARANTISM IN EARLY MODERN EUROPE

Martha Baldwin

In the early modern period learned physicians not only engaged in diagnosing and treating the ailments of the suffering but also evinced strong commitment to explaining how the forces of nature operated within the sick body to produce both diseases and cures. While instruction in natural philosophy and in medicine remained distinctly divided between university faculties, in practice such a division was blurred. Early modern learned physicians commonly felt themselves fully qualified to make theoretical pronouncements about a multitude of natural phenomena which were directly or indirectly observable. Moreover, if these phenomena manifested themselves in diseased human subjects, little deterred the early modern physician from regarding them as within the legitimate bounds of his profession. In this paper I will examine discussions of one peculiar disease, tarantism, to probe the character of early modern medicine.

Such an examination reveals the physician as engaged in matters far more exotic than his routine cases of gout, indigestion, and catarrh, and as keenly interested in the exotic world of nature well beyond his geographical confines. An investigation of inquiries into the disease inflicted by the tarantula's bite reveals the physician as closely linked to his professional *confrères* by various forms of learned communication. While extremely few physicians had the opportunity to treat or to observe firsthand the disease which so piqued their curiosity, far greater numbers addressed the subject in their medical discourses.

Discussion of tarantism started in 1621 when a physician committed to print a case history he had followed in his long-established medical practice in southern Italy. One Pietro Simone, a youth living on a country estate, suffered a tarantula bite. Suddenly the young harvester felt a strong pain on his leg where he had been stung. In the time in which it took to say three Ave-Marias, he fell on the ground with chills, and endured strong pains in the chest, priapism, and convulsions. He breathed heavily, and cried out to bystanders that he was suffocating. At dawn the physician called in a musician; slowly young Pietro took pleasure in one particular melody and began to dance until he was covered in sweat. During the following week, the patient spent eight to twelve hours a day dancing strenuously. When not dancing, he was wakeful, ate little, drank wine modestly, and slept lightly; he yearned for lullabies, wished to plunge into water, and fondled objects which

P. H. Theerman and K. Hunger Parshall (eds.), Experiencing Nature, 163–191.

were red. One week after the spider had injected its poison into his body, the patient returned to full health, and all traces of his mad behavior vanished.[1]

The author of this case history was a university-trained physician, Epiphanio Ferdinando, who had a well-established practice in Lecce, in the province of Otranto in southern Italy. While he was not the first to speak of tarantism,[2] Ferdinando gave the fullest extant description of a disease which would come to absorb the attention of learned physicians and natural philosophers in other parts of Europe. Although incidences of the malady were quite geographically specific, case histories interested physicians and natural philosophers living in such faraway places as London, Paris, Ulm, Jena, Marseilles, and Frankfurt. Books issuing from the printing houses of Würzburg, Ulm, Amsterdam, London, and Leyden cited Ferdinando's case history.

While the interest in tarantism widely permeated medical discourse, a review of the extant seventeenth- and eighteenth-century literature similarly reveals that interest in the affliction among learned physicians eroded with time. Historians of medicine have long noted that some diseases have higher visibility among patients and physicians than others at certain periods of time. Incidences of the disease by no means vanished by 1800 (indeed, twentieth-century anthropologists and folklorists have documented a wide range of dancing, singing, and social rituals accompanying its outbreaks in certain remote regions of southern Italy), but interest in tarantism on the part of learned physicians had notably evaporated by the end of the eighteenth century.[3] Thus, in tarantism the historian of medicine encounters a disease which is essentially culturally, geographically, and temporally located. In the early modern period, it became transformed from an unknown folk disease with an operating folk remedy into an openly addressed subject in the learned discourse of natural philosophy and medicine. By the end of the eighteenth century, it moved from being a hotly discussed medical topic to an almost invisible malady.

Shared Beliefs about Tarantism in the Early Modern Period

Physicians who investigated this disease shared certain assumptions about it. A review of all the available treatises reveals that doctors viewed it as real, not imaginary. Although some learned men would come to concede that in certain cases the disease was feigned, no physician ever strongly challenged, let alone overturned, the assumption that a poison injected into the bloodstream by the tarantula caused all of its baffling symptoms.

Following from their conviction that direct observation of natural species would reveal the secret and mysterious workings of nature, early modern naturalists and physicians increased their efforts to identify the spider whose toxin inflicted the disease and to follow its life cycle and breeding habits. Although their conclusions differ from those of modern taxonomists, the attention paid to the spider itself suggests how committed early physicians were to viewing the etiology of the disease as toxicological. Thus, in many early modern medical treatises discussions of tarantism are found alongside those of rabies and snakebite. Despite the interest shown by some Italian physicians in the actions of toxins on live animals and despite the heightened interest in chemical explanations of natural phenomena,[4] no chemical analysis of the tarantula's poison was undertaken in a systematic way until the twentieth century. Although the modern pharmacologist and medical historian believe that the toxin of the *Lycosa tarantula* is relatively harmless to man, the early modern audience clearly did not.

In addition to perceiving the disease as essentially toxicological, all physicians and natural philosophers interested in tarantism agreed that the *disease* did not exist outside the confines of Apulia, a region of southeastern Italy. While many conceded that the species did inhabit nearby regions of Calabria, Sicily, and the Roman countryside, all accepted the ocular evidence and testimony of observers that tarantulas living in these milder climates had no toxicity. Early modern physicians and natural philosophers concluded that topographical and climatic particularities of Apulia alone provoked the tarantulas' toxicity.

Because the spiders believed to inflict the disease evinced no toxicity in experimental trials on humans and were, in fact, rarely viable outside the confines of Apulia, physicians relied heavily upon textual accounts provided by a handful of local observers and foreign travelers in their written descriptions. That they were so willing to accept second-hand reports in place of their own first-hand observations and that they then so blithely constructed their own theoretical explanations may strike the modern reader as odd. But the early modern literature nevertheless reveals an acute concern for the accuracy of observations and for scrutiny of the testimony of witnesses providing information about the patients, the spiders, the climate of Apulia, and the customs of its natives.

Finally, there was universal agreement that music was the only effective remedy for the disease. While some physicians advocated the scarification of the wound, the application of topical ointments to the bite, or the ingestion of herbal sudorifics, virtually all of them regarded these standard remedies at best as *supplemental* to the playing of zithers, lyres, drums, or flutes. Musical therapy was certainly not unprecedented in Western medicine, but it was most unorthodox among learned physicians in the early modern period. Undoubtedly, the bizarre disease and its equally odd therapeutic regimen contributed greatly to arousing the attention of the learned community of scholars. Yet, although learned physicians clearly felt it their prerogative to explain the efficacy of the musical remedy, they agreed that the music-making itself, like the preparation of pharmaceuticals, fell outside the boundaries of their professional duty and training. They had the challenge and responsibility of explaining the disease and how it operated inside the human body; musicians who were experienced in curing the afflicted with particular songs and melodies had the task of performing the appropriate music. The narrative accounts of Apulian witnesses suggest that itinerant musicians earned their livelihoods wandering through the province playing for individual patients or for small groups of the afflicted. Reportedly, the musician came upon a suitable melody and musical instrument after empirically testing the effects of his repertoire on his patient.

The Accretion of Case Histories

Since incidences of this disease occurred only in Apulia, physicians living outside the region necessarily relied completely upon written reports of case histories. Ferdinando's account provided such necessities. In addition to the case of young Pietro Simone, he cited other vivid and detailed cases of several more victims of the tarantula's bite, including a ninety-four-year-old man confined to crutches. At the sound of music, the elderly cripple had danced and jumped about with the agility of a mountain goat. Ferdinando reported additional case histories which presented a variety of physical symptoms including stupor, paralysis of the limbs, spasms, nausea, and stomach-ache. He added that the disease was a serious one, and, if not treated quickly with music, patients suffered a relapse of symptoms for several years. He had seen women have annual relapses for ten, fifteen, seventeen, twenty, and thirty years when the musicians had been called in too late. Moreover, he had known deaf patients bitten by tarantulas whom music had induced to dance and jump about.

In addition to the array of somatic symptoms, Ferdinando documented types of mad behavior he had observed in bite victims—virgins flaunting their private parts; matrons ripping out their hair and ululating; and men jumping into the sea, brandishing swords like gladiators, or wallowing in the mud like swine. He noted that such behavior was concomitant with the physical symptoms and that patients cured by the music had no memory of their actions during their affliction.[5]

While virtually all later writers on tarantism cited Ferdinando, they also slowly built up an accretion of case histories which became standard fare in their medical treatises explaining the efficacy of the musical cure. When the Jesuit natural philosopher, Athanasius Kircher (1602–1680), first discussed tarantism in 1641, he was well versed in the few cases available in the medical literature. But Kircher enlarged significantly the somewhat skimpy collection of case histories by drawing on the written reports sent to him from Apulia by the Jesuits living there, Paolo Nicolelli, rector of the college at Nardò, Giovanni Baptista Galliberto, rector of the college of Taranto. As Jesuits, after all, Kircher's local informants were *sine disputatione* educated and credible witnesses. Kircher's case histories included three important additions which came to be widely cited in the later medical literature. One related the case of a member of the local nobility, Roberto Santorio, who had been bitten unawares and whose physician had been quite slow to diagnose his symptoms. Santorio almost perished as a result of the lapse of time before musicians were called in, and he lay in agony in his bed for several days until the fortuitously chosen melody of a musician excited him out of his lethiferous torpor. Slowly rousing himself, he indicated to the musician that the music pleased him. As the musician began to play the zither more quickly, Santorio got to his feet and began to dance vehemently and to sweat profusely. Soon restored to perfect health by the exercise of dance, he thereafter lived free from all infirmity. From this one carefully documented case, Kircher added important new evidence that the spiders did not make social distinctions about their victims.[6] While Kircher agreed with Ferdinando that peasants laboring in the fields were the most common *tarantati,* he argued that this was because they worked half-naked in the fields and napped outdoors where they exposed their bodies to spiders burrowing in the soil.

Kircher also recorded further cases of *tarantati* who suffered delusions and who took particular delight in certain colors. Here, he again enlarged on Ferdinando's evidence that delusions and mad behavior were symptomatic of the disease, and concluded that forbidding a mad victim's indulgence of his fantasies only begged for tragic consequences. As illustration, he related the case of a Capuchin monk whose superior, in accordance with the rules of the order, forbade him to listen to music or to bathe. Driven with an insane desire for water, the monk escaped his cell and dashed into the sea where he drowned. As further proof, Kircher related a case where a victim was seized with an insane desire to fondle anything purple. When an ecclesiastical official wearing a purple cloak appeared for a visit at a local monastery and realized what the poor mad man yearned for, he offered him his fine vestment. After stroking the cloth for some time, the patient returned to sanity. Drawing on the accounts of the Apulian Jesuits, who had interrogated both local patients and physicians,[7] Kircher detailed yet other cases where patients had thought themselves ducks, covering themselves with reeds and cattails and submerging themselves in water. Still others impersonated governors and military leaders, speaking pompously and gravely, while some, imagining themselves beasts, climbed and hung from trees. Moreover, in far greater detail than anyone in his century, Kircher recorded the songs and melodies which natives used in the therapeutic process.

While Kircher focused on the mad behavior accompanying the poisoning, most accounts paid equal attention to the somatic symptoms. At the end of the seventeenth century, the distinguished physician, Giorgio Baglivi (1668–1707), enlarged upon the physical symptoms associated with the disease. He verified that the tarantula, venomous only in summer and only in the Apulian region, attacked without provocation, often at night when the victim was asleep. Leaving a welt like a bee-sting on the skin, the tarantula's bite brought on difficulty of breathing, heavy anguish of heart, loss of appetite, headache, thirst, and muscular convulsions.

Baglivi agreed with Kircher that nobles could suffer from the affliction. He related a new case history in which a noble woman of Lecce in 1687 suffered a bite on her right femur. Soon afterwards she displayed the common physical symptoms, but her physician did not properly diagnose her condition. His lancing of the wound was ineffective. Her relatives, suspecting the tarantula's bite, brought in musicians and urged her to dance. When she heard one particular melody, she immediately lost all her modesty, jumped up from her bed, removed all her clothing, and danced vigorously for three days.[8]

When the French-born Neapolitan bookseller and man of letters, Antoine Bulifon, sought to disseminate information about tarantism at the end of the seventeenth century, he published in full an account sent him by a native Apulian and respected physician, Domenico Sangenito.[9] Sangenito too offered new case histories. He described the relapse of a miserable peasant, one Giovanni Giacomo Tesoro, who one summer day fell on the ground, had difficulty breathing, and began to turn black in the face and hands. Since all the local inhabitants knew of his previous affliction, they at once called for a guitar, and within a quarter of an hour the patient was dancing vigorously and breathing fully. Witnesses saw his natural color restored within an hour. Sangenito also related having seen five *tarantati* dance at the same time. Enlarging upon the social behavior of the afflicted, he related that four of these victims had been ploughmen and one a very beautiful country lass. Together each one of them had participated in a rather elaborate drama and had assumed a strange royal personality, all to the great pleasure and amusement of bystanders. At the end of three days of elaborate theatrics and dancing, the victims had no memory of their behavior. Sangenito remarked that since victims were generally practical men of meager means,[10] they commonly bypassed the physician by diagnosing the problem themselves and calling in the musician.

In 1706 another native of Apulia, the Benedictine monk Ludovico Valletta, produced a treatise on tarantism which again significantly enlarged the number of case histories available to physicians wishing to make theoretical statements about the disease. Valletta included in his treatise an account of an infant whose mother, a peasant "of the lowest type," had deposited the child in a bed of straw while she was gleaning wheat at harvest time. Bitten by a spider, the infant soon grew perilously sick. Unhappily, his mother did not recognize the cause of his distress and ineffectively administered herbal medicines. "Meanwhile by chance on that day a certain zither player had traveled there with his zither; and immediately when its music reached the ears of the small boy, he grew slightly better from the torment of his illness, and he agitated his hands and feet, although they were swaddled."[11] Since the child's face betrayed that anxiety characteristic of other *tarantati*, bystanding peasants were able to recognize the cause of his illness. Hence, the mother engaged the musician to play for the infant twice daily and "stupefying everyone assembled, including myself," the young boy jumped about and returned to full health.[12]

Which Spider?

In addition to a significant accretion of case histories over a period of 150 years, the accounts reveal that physicians and naturalists consistently came to pay greater attention to the tarantula itself. This reflects the widespread concern among early modern natural historians for precise and accurate descriptions of animal species subjected to the direct gaze of the observer. When Ferdinando wrote in 1621, he gave a textual description of the spider, noting that it was cold-blooded, solitary of habit, unable to tolerate cold, and variegated on its top but black on its bottom. In a description two decades later, Kircher illustrated the dorsal and ventral views of the animal, described its life cycle, and detailed the numerous varieties of the species in Apulia. Baglivi, at the end of the century, gave a full anatomical description, distinguished between the sexes, detailed the reproductive organs, and made references to the microscopical studies of insect and spider anatomy undertaken by Filippo Bonnani, Martin Lister, and Robert Hooke.[13] Also at century's close, the distinguished natural historian, Paolo Boccone, compared the tarantulas of Corsica with those of Apulia, discussing two species of spiders indigenous to Corsica and noting that the island variety had round black bodies with red variegations, lived in rocky places, and induced physical symptoms similar to those in Apulia. He also illustrated each species.[14]

Spiders have remained a taxonomical nightmare in the twentieth century, so it is not surprising that early modern classificatory attempts were confused. When the Dutch physician, anatomist, and chemist, Wilhelm Homberg, filed his report on spiders with the Parisian Académie des Sciences in 1707, he had used a microscope to examine them and included illustrations with his report. In his taxonomical scheme, the tarantula formed a sixth species, alongside domestic spiders (inhabiting human habitations), garden spiders (spinning webs in the air and hiding during the day), cave spiders, homeless spiders, and field spiders.[15] While Homberg's taxonomy does not strike a modern chord, his physical description of tarantulas was apt; they were larger and stronger than domestic spiders and had black underbellies, scaly heads, and hard, transparent shells. Given the relative instability of taxonomy in the seventeenth century, illustrations in medical texts of tarantulas were not always of the spiders we now recognize by that name. Such confusion of spider species was not lost on one seventeenth-century investigator of invertebrate anatomy, Martin Lister (1639- 1712). When Lister compiled his masterly volume on the spiders of England, he merely noted that exotic spiders were beyond his purview. He was quite aware of the bizarre medical phenomena attributed to the tarantula's bite, but he revealed his skepticism about these tales when he asked "whether this poisonous beastly little spider had ever been diligently examined according to my known standards."[16]

Transmission of Reports on Tarantism

Undoubtedly, a vastly improved communications network among learned physicians fueled the medical interest in tarantism in the early modern period. While the beginning of the 1600s found physicians reliant on printed books or medical texts for information about the disease, other forms of communication became available—and even came to predominate—by the end of the century. When Ferdinando published his discussion in 1621, it formed merely one of one hundred chapters, each devoted to a particular medical question or aspect of a specific disease. Such an arrangement of medical texts in one hundred unconnected chapters was standard; when the Dutch physician, Cornelis Stalpart van der Wiel, wrote about tarantism sixty years later, he used exactly the same format.[17] But discussions of tarantism also began to appear in books on natural history and natural philosophy, books not primarily aimed at an audience of physicians. Nevertheless, physicians revealed a remarkable familiarity with non-medical texts, indicating much overlap between the doctor and the naturalist in early modern Europe. Indeed, many of the most respected naturalists—including Conrad Gesner, Ulisse Aldrovandi, Francesco Redi, John Ray, Martin Lister, and Marcello Malpighi—had medical training and earned their livings as practicing physicians.

Although the medical text designed to cover all topics remained a staple in the diet of seventeenth-century medicine, short treatises devoted to a particular medical topic appeared increasingly. When the German physician, Hermann Grube (1637–1698), published his treatise on tarantism in Frankfurt in 1679, it filled a mere eighty pages. Wolferd Senguerd (1646–1724) had produced a text of equal brevity at Leyden on the same topic a few years earlier,[18] while Andreas Flachs had put forth a brief disputation in 1660, which amounted to no more than fourteen unpaginated sheets.[19]

While medical texts were undergoing marked changes in the seventeenth century, a far more significant stimulus to expanded and accelerated transmission of medical news came from scientific journals published under the aegis of newly formed scientific societies. Such formal scientific societies provided a new way for foreign scholars and physicians to discuss and to disseminate medical issues and to query learned men about intriguing medical matters. As early as 1642, the tarantula and its victims formed the topic of discussion in the informal scientific academy of the Bureau d'Adresse at Paris under the direction of Théophraste Renaudot.[20] Descriptions of both the spider and its victims also entered into the published accounts of the early scientific societies.

The physicians Martin Lister, Domenico Cirillo, and Tommaso Cornelio all published accounts in the *Philosophical Transactions of the Royal Society of London.*[21] The *Miscellanea curiosa sive Ephemeridum medico-physicarum germanicarum Academiae naturae curiosorum,* the official organ of one of the earliest German scientific societies contained a 1678 report offered by the German physician, Simon Aloysius Tudecius, who recounted witnessing the relapse of one *tarantati,* an Apulian sailor, on the docks of Marseille in 1666.[22] Likewise, the *Mémoires* of the French Académie des Sciences included a 1702 report by the physician, Étienne Geoffroy[23] and a 1707 report by another academician and active medical practitioner, Homberg.[24]

While none of the publications of these early scientific societies claimed to have vetted the reports submitted by members, the appearance of an account in the journal of a reputable scientific society lent it a certain respectability. And because the publications appeared several times a year, reports circulated far more quickly. When Ferdinando published his book in 1621, he had already observed *tarantati* for twenty years in his medical practice; when Geoffroy issued his report to the Académie, he had just returned from a visit to Rome, and his communication appeared almost immediately. Thus, the scientific journal offered a far quicker means of communication than the printed book, as well as a far wider circulation than the privately-circulated letter.

The less-than-private letter also became a mainstay of medical communication. The correspondence of Henry Oldenburg, secretary of the Royal Society of London, served to elicit communications from such foreigners as Cornelio, whose account Oldenburg later published in the *Philosophical Transactions.* Collections of letters sometimes formed the basis of books in natural history. Thus, Boccone's account of tarantulas in his *Museo di Fisica e di Esperienze* of 1697 was, in fact, merely his publication of an earlier letter written to the Polish Franck Bielinski, the Great Chamberlain of the Polish King in Warsaw. Similarly, Francesco Cancellieri's letter to David Ferdinand Koreff, professor of medicine at the University of Berlin, was scarcely conceived as a private communication. Replete with footnotes to every learned work which mentioned the word *tarantula,* the treatise joined another of his letters on the malignant air of Rome. He published both in an elegant volume in 1817.[25]

While scientific and medical communications had improved significantly by the end of the seventeenth century, a letter of inquiry about the tarantula did not necessarily produce an answer. In May 1694, Bulifon, the Neapolitan man of letters, wrote to Lorenzo Magliabechi asking him to solicit the famous Florentine physician, Francesco Redi, for his opinion on tarantism. Bulifon had planned to include Redi's letter in his volume on political, historical, and scholarly matters relating to the kingdom of Naples. Redi apparently never responded to Bulifon's request; his *Lettere memorabili, istoriche, politiche ed erudite* contained nothing by Redi, nor did any writing on the tarantula appear in Redi's posthumously edited correspondence. Instead, Bulifon was forced to rely on a far less famous–indeed virtually unknown—physician, Domenico Sangenito, to speak authoritatively on tarantulas and tarantism in his volume.

By the mid-eighteenth century, discussions of tarantism had even made their way into medical encyclopedias. The Swiss encyclopedist, Jean Jacques Manget, spent his life assembling massive reference works for the physician. In addition to enormous, multi-volume encyclopedias on anatomy and surgery, he also compiled a very successful, alphabetically arranged, four-volume work, entitled *Bibliotheca medico-practica sive rerum medicarum Thesaurus,* which appeared in Geneva in 1697. Here, listed under *T* the reader could find an abridged account of Baglivi's treatise on tarantism.[26]

In reading these accounts, it is quite apparent that communications about the disease among learned physicians spread with remarkable speed and thoroughness. They cited each others' case histories and texts incessantly and took issue with one another on the smallest detail. Physicians were surprisingly conversant with reports written by naturalists who did not necessarily belong to their own profession. Indeed, the very discussion on tarantism was wholly dependent on this exchange of information across the continent; how else could physicians in Germany, England, France, and Holland have generated so many printed pages discussing so learnedly a disease confined to one remote province of southern Italy?

Concerns about Witnesses and Experimental Verification

Over the course of the seventeenth and eighteenth centuries, learned physicians evinced an increasing desire to verify the baffling reports about tarantism. Discussions of the disease were contemporaneous with a flood of printed texts about marvelous medical phenomena found in the New World and the Far East, and natural philosophers and physicians became increasingly agitated over the issue of the credibility of witnesses.[27] As the editor of the report offered by Geoffroy to the Académie des Sciences in Paris noted, "with such extraordinary facts, it is not in the least surprising that a few fables get mixed up in the business."[28]

Because the disease had narrow spatial confines, physicians had to rely on local eyewitnesses. Sangenito reminded his editor, Bulifon, that he, "born and raised in Lucera, where the tarantulas are born" was giving "the testimony of my own eyes, and not the lies of others."[29] Neapolitan and Roman physicians were recognized not to have observed cases themselves. Thus, when the Apulian physician, Francesco Cancellieri, sent his account of tarantism to a physician at Berlin in 1817, he was careful to note that many others had requested his remarks since he was a native of the regions where tarantulas were poisonous. Cancellieri also endorsed the account of Sangenito because he, too, "was a learned and honest man and from the countryside of the tarantulas, namely from Lucera, or Nocera de'Saraceni, in the province of Capitanata, in the Kingdom of Naples."[30]

Since few had commercial reasons to travel to impoverished and remote Apulia, obtaining reports from travelers was especially difficult. Nevertheless, the well-connected English natural philosopher, Robert Boyle, did manage to find in London an unnamed witness who had traveled in the region.[31] Still, the observations of travelers could be remarkably brief and disappointing, as was that of John Ray, the distinguished English botanist. In a published account of his travels to Italy in 1663, Ray made short shrift of the tarantula and dismissed it merely as a "large sort of spider, the biting whereof is esteemed to be venemous."[32] The resourceful Henry Oldenburg tried, unsuccessfully, to get the French astronomer, Adrien Auzout, to send back reports on the tarantula's bite during his visit to Italy.[33]

Because the testimony of learned men was always deemed more credible than that of the unlearned, authors of the accounts immediately set forth their own credentials and those of their informants in order to earn the trust of their reader. Bulifon wrote that he had originally esteemed the reports about tarantism "not far from fictions" but that he had lost all his doubts once it had been confirmed to him by "persons worthy of faith" in whom "one could never have any suspicion of fictions."[34] Bulifon hinted that Sangenito was trustworthy because he was a "dottor Fisico."[35] And since, in early modern Europe, being learned implied membership in a certain social class, authors did not hesitate to describe the social background of their witnesses. Because the typical patient afflicted with the disease was from the lower classes, the "gente rozza ed inculta,"[36] it was important to convince the elite reader that the author or local informant was not of the same ilk as those under observation. When the English physician, Richard Mead

(1673–1754), wrote about tarantism, he acknowledged his indebtedness to both Baglivi and Valletta. Mead noted that some of the afflicted were "of great families, and so far from being dissemblers, that they would at any rate, to avoid shame, have concealed the misfortune which had befallen them."[37] Clearly, the social status of the early modern patient, as well as the early modern author, lent considerable credibility to the disease.

In addition to trustworthiness being accorded to noble and learned informants and patients, early modern authors tended to deem accounts of tarantism even more convincing if they corroborated earlier existing reports. Thus, Bulifon found Sangenito especially credible since he agreed with other famous authors, including Kircher.[38] Similarly, the editor of Geoffroy's presentation on tarantism to the Académie des Sciences noted that other learned men had corroborated Geoffroy's testimony. The editor revealed that the widely respected French Jesuit astronomer, Pierre Gouye, had earlier testified that a Jesuit father from Toulon had seen an Italian soldier dance on the docks for several days when he suffered an annual relapse of tarantism.[39]

Although the Neapolitan and Parisian editors took comfort in the corroboration of others, one author of a treatise on tarantism, Valletta, scorned all such testimony as adventitious and unnecessary. Valletta assured his reader that he had never even thumbed the pages of a single printed account of the disease and that he offered his own unaffected by the claims of others. Yet, since Valletta wished to challenge some details of Baglivi's narration, he was forced to concede that someone had thrust the learned physician's book into his hands just after he had entrusted his own manuscript to the printers. As a monk immune from the lures of fame and worldly esteem (which he clearly implied had motivated the learned Baglivi), Valletta boasted that the reader should grant his version greater credence.[40] But Valetta's disclaimer to be uninfluenced by earlier writers was clearly the exception to the general rule of wide and free citation—and facile acceptance—of second-hand reports in the early modern medical literature on tarantism.

In addition to efforts to ascertain whether a report emanating from Apulia came from a reliable source, authors also attempted to verify certain travelers' stories experimentally. Increasingly, physicians came to accept second-hand accounts only after making great efforts to observe directly the spiders and the strange effects of their poisons on the human body. Consistently, those who attempted first-hand observations and experiments found their efforts foiled. In the 1650s, Kircher had live tarantulas brought in glass phials to Rome in order to perform experiments with them, but they died in transit.[41] Similarly, almost five decades later, Baglivi had live specimens imported to his house at Naples. His hopes to observe directly the effects of the disease there were dashed, however. Although *his* spiders had survived the relative chill of the Neapolitan summer, he was disappointed to discover that they had lost their toxicity on humans.[42] Correspondence from the 1740s reveals that local officials in Apulia gathered live specimens to ship to Dresden for purposes of medical experimentation, but that the creatures again proved themselves to be remarkably fragile biological commodities.[43]

Although Kircher and Baglivi had been thwarted in their attempts to conduct experiments on live tarantulas in Rome and Naples, respectively, they lobbied their local sources for information derived from experiments conducted in Apulia. Reduced to relying on a handful of nobles, learned priests, and local physicians, they badgered their witnesses to come up with *new* observations. Kircher was delighted with the report of two fellow Jesuits about an experiment contrived to test whether or not the tarantulas themselves could be induced to dance by the same music played for human victims of their bite. Kircher's second-hand report of the experiment (which provided a positive answer) appeared as a third-hand report in the medical text of van der Wiel and many others.[44] In 1668, Henry Oldenburg sent an official inquiry to a Neapolitan correspondent of the Royal Society, Tommaso Cornelio, and asked him to verify the reports about the marvelous disease and its musical cure.[45]

In addition to experiments on the spiders themselves, some physicians were also eager to have experiments performed on lower animals. Although Baglivi may have been foiled in observing tarantulas afflict humans in Naples, he and a friend successfully subjected *rabbits* to live tarantulas and observed the process of the poisoning. Baglivi related that the two even called in a musician for the hapless animal, but to no avail. Moreover, the Neapolitan detailed for his reader the results of their post-mortem dissection of the rabbit: they found the brain slightly inflamed and spotted, and holding a large quantity of extravasated serum.[46]

Attempts to verify reports about tarantism from experiments continued throughout the eighteenth century. In 1756, no less a personage than the archbishop of Taranto busied himself with directing the capture and transport of tarantulas to foreign scientists. Although his efforts proved no more successful than Kircher's had more than a century earlier, the archbishop exhausted all the methods he could imagine to ensure the spiders' safe arrival. He dispatched some dead specimens preserved in brandy, he enclosed live specimens in an air-tight basket, and he placed yet other live tarantulas in a small perforated box "so they can breathe and have light."[47]

The fragility of the spiders directly impacted the fragility of the experimental evidence available to doctors interested in tarantism. Since foreign physicians could not themselves perform experiments with tarantulas or observe patients directly, they could only rely on second-hand reports. Although direct observation and the evidence of experiment were the rallying cries of medical reformers in the early modern period, in the case of tarantism each of these criteria of truth failed abysmally.

Theoretical Explanations of the Musical Cure

While experimental trials failed almost universally to meet expectations, foreign physicians and natural philosophers generally felt confident that the case histories provided particulars sufficient to deduce the action of the spider's toxin inside the human body. Drawing on these deductions—or speculations—they proceeded to construct an almost inexhaustible variety of explanations of the efficacy of the musical cure.

Ferdinando's case histories and observations of the climate and life-style of the inhabitants of Apulia were of special interest to his fellow learned physicians, and later writers ransacked his account for the smallest details. However, they did not find his *explanation* of the musical cure at all appealing. Indeed, Ferdinando's Aristotelian and Galenic explanation of the cold and dry character of the spider working on the hot temperament of the natives fell on deaf ears. By 1621, this seemed as dry as the Apulian plains in August. Willing to accept the narrative particulars of Ferdinando's account, his contemporaries cast aside his outmoded explanation of the second, third, and fourth degrees of dryness.

Instead, foreign physicians and natural philosophers preferred to devise alternative explanations of why music was an effective remedy. In so doing, they produced a variety of theories in keeping with their own understanding of how other diseases affected the human body. Broadly speaking, three new types of explanations—sympathetic, corpuscular, and chemical—appeared in the seventeenth century for the way in which music caused the body to expel the tarantula's poison.

Sympathetic explanations were the first to counter Ferdinando's Aristotelian discussion of dryness. A year before Ferdinando's medical treatise appeared in 1621, the Neapolitan natural philosopher, Tommaso Campanella (1568–1639), had spoken of tarantism in a treatise on sensation and natural magic. There, he pointed to an occult sympathy between the "friendly and appropriate sounds" of the music and the bodily spirits to explain the musical therapy.[48]

The German physician, Daniel Sennert (1572–1639), also relied on the doctrine of sympathy in his discussion of the disease. Unlike Campanella, who wrote about tarantism in a philosophical discussion on sensation, Sennert discussed it in a medical treatise on diseases arising from poisons external to the human body. Generally, Sennert endorsed the application of chemically prepared unguents in cases of poisoning induced by snakebite, insect bite, or rabid animals. But when he turned specifically to the poison injected by tarantulas, he conceded that chemical treatments failed. He denied the sufficiency of the Galenic explanation (the excessive heat of the spider's toxin) since there were many regions in the world hotter than Apulia. Instead, Sennert postulated the presence of an occult force placed in Apulia "either by celestial influx or divine decree,"[49] which caused the poisoned humors to respond to the music by twitching, convulsing, and dancing. His insistence on an occult sympathy between the music and the poison followed from his belief in a cosmos imbued with a divinely ordained order and regulated by musical harmony. Although Sennert considered man unable to penetrate fully all the details of this cosmic harmony, he deemed it within man's powers to observe the efficacious cure.

In 1639, the French physician, Lazare Meysonnier, invoked the harmony between the macrocosm and microcosm to explain how musical melodies restored balance to the body in victims of the tarantula's bite.[50] Going further than Sennert, Meysonnier believed that the stars and planets produced musical harmonies in their celestial motions which profoundly and directly affected the human body. Since man-made music could imitate the divinely created harmony of the celestial spheres, he claimed for music a power to cure all things contrary to the economy and harmony of the body. He considered the wandering musicians of Apulia modern counterparts of the young David who had cured King Saul's madness and melancholy by music.[51]

As noted above, the subject of tarantism also long enticed Kircher, and he inserted discussions of the disease in many of the books he wrote in the later decades of his life. Although he had broached the topic as early as the 1640s, descriptions of it appeared in his *Musurgia universalis* of 1650, his *Mundus subterraneus* of 1665, and his *Phonurgia nova* of 1673. While no physician himself, Kircher, like many early modern natural philosophers such as Boyle and Descartes, considered himself fully conversant on medical matters.[52]

Like Campanella and Sennert, Kircher was strongly influenced by Platonic and Hermetic beliefs in the power of music to cure physical and emotional disorders. Well versed in the classics, he knew that ancient Greek philosophers considered music allied with rhetoric, since each sought to move the emotions of the listener by employing rhythm, modulation of voice, and harmony. He invoked the Platonic and Pythagorean traditions to explain the connections between music, human emotions, and celestial harmonies and was especially mindful of Proclus's tenet that music imitated the harmony of the celestial spheres, as it attracted and allured celestial influxes.

Although deeply influenced by the Renaissance neo-Platonists and neo-Pythagoreans, Kircher elaborated his own peculiar theory to explain the efficacy of the musical cure relative to tarantism. He deemed it to work by an occult magnetism placed in the musical melody which mysteriously attracted the spider's toxin outside the body. While in essence his explanation was one of occult sympathy dressed up in the wrappings of magnetism, he also freely mixed Galenic humoral ideas, mechanistic ideas of percussion, and corpuscular theories into his eccentric explanation. From Galen, he adopted the belief that individual humoral disposition determined the efficacy of one particular melody or one particular musical instrument to jolt a semi-conscious victim into frenzied dancing. He described the case of a young girl with a particularly phlegmatic constitution who could only be roused to dance by the crashing of drums and the strident clanging of cymbals. Since Apulians were generally of a sanguine temperament, the playing of zithers and lyres worked most commonly. Kircher drew from corpuscular matter theory to explain how music impelled the movement of air particles outside the body, while he imposed his own iatromechanistic ideas to argue that agitated particles prompted movements inside the body when they entered the ear canal of the victim. Such agitated aerial particles then impressed their motions on the fine physical particles of the blood, which, in turn, affected the muscles where the spider's poison had lodged. The musically agitated particles of the blood so bombarded the muscle fibers that they caused them to twitch and vibrate, effects manifest in the victim. As the movement of the muscular fibers warmed the body, the pores of the skin opened, and the poison lodged in the agitated muscles vaporized and exited through the pores.

While Kircher's explanation of tarantism may seem incoherent, it appealed to many of his contemporaries. Kircher had no medical training, yet he had earned a certain respectability among physicians from his erudite works on natural history and from his earlier treatise on the plague. Like his English contemporary, Robert Boyle, he revealed in his published works a decided interest in many medical topics, including poisons, metallic medicines, kidney stones, and exotic botanical pharmaceuticals.

Other well respected physicians took up Kircher's explanation. The well-known Marco Aurelio Severino (1580–1656) stated that he had observed many cases of tarantism as an attending physician at the Ospedale degli Incurabili in Naples, where he later held the university's chair in surgery and anatomy. In a work of 1650 on the toxic effects of vipers, *Vipera pythia,* Severino cited both Kircher and Ferdinando and theorized that music accomplished its cure by provoking the spirits and by agitating the body. Severino's writings were markedly anti-Aristotelian, and he also employed Kircher's word *magnetismus* to describe the action between music and the spider's poison. Claiming that Kircher's doctrine of the magnetism of mystical nature was merely a new interpretation of the ancient teaching—like draws to like—he too drew on Platonic and Paracelsian ideas. Like Kircher, Severino believed the musical cure operated by an essential, yet occult similarity of the music with the poisonous substance injected by the spider into the bloodstream.[53]

Kircher's explanation was equally appealing to the German physician, Samuel Haffenreffer, who praised Kircher as "alter Argus vigilandissimus"[54] and reproduced exactly not only his text but also its illustrations and musical notations. He completely embraced Kircher's argument that occult magnetism or blind similitude of nature ultimately accounted for the musical cure.[55] Less surprising a disciple of Kircher was the German Jesuit, Gaspar Schott, who cited Kircher's case histories and explanatory mechanism verbatim in his discussion of natural magic and acoustics.[56]

Kircher's documented case histories of tarantism became widely incorporated in later discussions of the disease, even though many later physicians firmly eschewed all his references to occult sympathy and mysterious magnetic attractions. One group that opposed sympathetic notions was committed instead to corpuscular justifications of natural phenomena. These included the English physician, Walter Charleton (1620– 1707), who was greatly indebted to the atomistic natural philosophy espoused by the French natural philosopher, Pierre Gassendi (1592– 1655).[57] Drawing on Gassendi, who had written briefly about tarantism, Charleton elaborated a materialist explanation which denied the action of any magnetic or sympathetic attraction between the music and the venom. Although both Gassendi and Charleton accepted Kircher's explanation of aerial particles set into motion by music,

Charleton's physiology was fundamentally different from Kircher's. The Englishman believed that music set the spirits of the brain–rather than those of the blood–into agitated motion. Charleton invoked agitated spirits of the nervous system to explain the twitching and dancing of the muscles; in this way, he expanded the physiological processes involved in the disease. He held that poisons acted by thin, invisible particles which moved easily through the body, penetrating quickly even the thickest fibers or tissues of the organs. According to Charleton, music counteracted the poisonous corpuscles by impressing motion first on the surrounding air and then on the ear, from which they traveled to the brain and ultimately to the cerebral spirits. These cerebral spirits, in turn, transmitted their agitated motion to the poisonous particles causing them to vaporize, commingled with perspiration.

Such corpuscular theories appealed to Thomas Willis (1621-1675), the distinguished physician, anatomist, member of the Royal Society, and Sedleian professor of natural philosophy at Oxford. Interested in pathological conditions of the brain, Willis propounded a brief explanation of tarantism and its musical cure in a discussion of convulsive diseases in his *Pathologiae cerebri et nervosi generis Specimen* (1667). Like Charleton, he judged tarantism one of the many diseases which affected the "nervous liquor" rather than the blood. Yet Willis moved beyond mere corpuscular explanations and invoked the Helmontian concept of chemical ferments. He spoke of venomous miasms which spread through the body by the innate fermentative quality of the toxin. As the ferment acted on the nerves, it incited the disarray and internal disorganization of the "animal spirits," thus causing the painful muscular convulsions of the victim. Willis held that the ferment of the tarantula's toxin was potent enough to cause muscular spasms, but not to force the animal spirits into violent and destructive explosions. In his estimation, the disease was never fatal. He claimed that music reconvened the dissipated spirits to their normal places in the nerves, but provided no further explanation of this reassemblage. He was content to invoke the vague "gently soothing nature" of music.[58]

Following Charleton's and Willis's increasingly elaborate physiological explanations, the German, Wolferd Senguerd (1646-1724), also invoked the mechanical processes of the body to account for the physiological disturbances provoked by the tarantula's toxin. In his *Tractatus physicus de Tarantula* of 1668, Senguerd excluded almost no system of the body in his elaborate explanation of the poison's action. The venom's activity in the brain caused the symptoms of insanity evinced in so many victims; its motions in the voluntary muscles resulted in convulsions and twitching; it so agitated the heart that it produced the fevers so typical of the *tarantati.* Excessive heat in the heart, in turn, corrupted and putrefied the bile in the nearby vessels. The viscosity of the poison obstructed the pores and immobilized the spirits which otherwise traveled freely to the muscles, thereby explaining the languor and excessive sleep of some victims. Unlike Willis, Senguerd made no clear distinction between the spirits of the blood and the spirits of the nerves. While decidedly rejecting occult correspondences between music and the immaterial human soul, he tried to construct a chain of mechanically-induced physiological reactions which dispelled the toxin from the body.[59]

Though mechanical and corpuscular explanations gained increasing popularity among physicians and natural philosophers in the later part of the seventeenth century, they did not become canonical. Following an approach quite different from the iatromechanists was Johann Christiaan Frommann (1640–1730), provincial physician of Saxe-Coburg. In a larger work on fascination, Frommann included a treatise on tarantism. He considered Kircher's and Willis's accounts but rejected them as too grounded on trivial physiological detail. His own explanation emphasized the action of the spider's poison on the fantasy of the victim. Since he believed the fantasy governed the locomotive faculties, he posited the corruption of the native fantasy by the idea or governing design of the poison. Without specifying a mechanism by which the victim's native organizing principle was disrupted, Frommann envisioned a transmission of the idea from the spider to its victim. Thus, the *tarantati* took as much delight in music as the tarantulas themselves, whose fantasies and locomotive faculties were governed by musical melodies. That the dancing effected the cure by raising the body heat and allowing the poison to vaporize with the sweat was merely a coincidental and secondary result of the imprinting of the spider's *archeus* (or regulating principle) on whatever animal it may have bitten. In appealing to the disruption of the fantasy in this way, Frommann sought to view the disease within the broad Helmontian framework of disease and rejected the search of iatromechanists for physiological and mechanical explanations.[60]

Explanations of tarantism continued throughout the seventeenth and into the eighteenth century.[61] In them, one finds a striking eclecticism and a willingness to draw freely on Galenic temperaments, sympathetic harmonies, invisible material corpuscles, mechanical motions inside the body, chemical ferments, and Helmontian *archei.* No two explanations were identical; the disease and the medical theories invoked to explain it allowed for almost infinite personal, if not whimsical, permutations.

The intricacies of these explanations need not detain us further. It suffices to note that in all cases these learned physicians and philosophers invoked broader philosophical theories to explain the action of the music inside the poisoned body. And since, with the exception of Ferdinando, they all lived outside the confines of Apulia, they had no patients or toxic spiders on which to test their hypotheses. Indeed, most relied exclusively on second-hand published accounts and could not corroborate their theories by direct observation. In summary, in this disease, learned medicine remained learned; physicians created their bookish and abstract accounts for other physicians. In contrast with the growing spirit of empiricism so prominent in other parts of early modern natural philosophy and medicine, explanation, not treatment, remained their goal.

Withering Interest in Tarantism

No one set of experiments or toxicological discovery killed physicians' interest in tarantism in the early modern period. Rather, interest in the disease appears to have dissipated slowly, much as the symptoms of the *tarantati* gradually disappeared at the end of a week spent dancing. Nevertheless by the mid-eighteenth century there *was* a marked decrease in the number of medical treatises discussing tarantism. When Nicolai Caputi penned a dissertation on the tarantula in 1734, replete with illustrations, Italian sonnets, and letters of commendation, he could not find anyone willing to print it. His patron, the Neapolitan professor of medicine, Niccolò Cirillo, suggested he send it to the Royal Society of London. But despite its elegance, it found no takers in the learned body that had eagerly sought such information seven decades earlier.[62]

Medical skepticism fails to explain completely the waning interest in tarantism. A few physicians such as Cornelio and Baglivi had suggested that rustics might feign the disease, regarding a week's worth of dancing more appealing than working in the fields in the hottest part of summer. Cornelio had observed that most *tarantati* were "young wanton girls," who had persuaded themselves "according to the vulgar prejudices," to have been stung by a tarantula.[63] And Baglivi had noted that Apulian women might welcome a week's freedom from the social dictates of sexual repression. But despite their misgivings, belief in the disease as caused by the spider's poison and as cured by music survived intact in many of the most openly skeptical minds of the eighteenth century.[64]

If neither experimental evidence nor medical skepticism proved fatal to tarantism in the early modern period, what explains its disappearance? I suggest that given the constraints imposed on physicians by their inabilities to observe cases, the intellectual space in which learned physicians could maneuver their discourse was seriously restricted by the new emphasis on empirical experience. Earlier physicians had exhausted the theoretical discussions; iatromechanical and corpuscular explanations became as formulaic as the earlier Aristotelian and Galenic ones. Moreover, physicians had sought only to explain the efficacy of the musical remedy; they evinced not even the slightest interest in applying music as a therapy in other diseases. Early modern therapeutics remained remarkably conservative; purges, clisters, emetics, elaborate diets, and bleedings endured as standard medical therapies with markedly few changes. Despite the calls of many medical reformers—from Paracelsus to Robert Boyle—to advance medical knowledge by observation of folk remedies, these never seriously entered the canon of orthodox learned medicine. As this discussion of tarantism reveals, there was a sustained discontinuity between folk and learned medicine in the early modern period. Similarly, the hopes of some reformers to advance medical knowledge through experimental knowledge failed. Although interest in novel diseases and new ingredients for Western pharmacopeia ran high, and while new physiological and anatomical knowledge became acceptable, medical therapeutics remained conservative, and medical discourse remained elite. Once physicians had managed to constrain the bizarre features of a new disease within the imposing edifices of existing medical theories, interest in it evaporated.

Conclusion

Early modern medical discussions of tarantism reveal a keen and sustained interest on the part of learned physicians in a disease which most were never called upon to treat. Stimulated by more efficient forms of medical communication and goaded by more exacting observations of the spider and its Apulian victims, early modern doctors engaged in making authoritative statements about an intriguing disease most knew about only by written report. An historian of medicine is compelled to ask why tarantism did not simply remain *outside* the canon of learned medicine? I argue that entering the discussion offered the learned physician the welcome opportunity to advertise his chosen natural philosophy—be it sympathetic, corpuscular, mechanistic, chemical, or some eclectic mixture. By writing on tarantism, the physician could advocate his own philosophical allegiance and demonstrate its ability to account for new medical phenomena. Rather than being concerned with treating ailing patients, the early modern physician was eager instead to explain the forces of nature at work in the world.

Notes

1. Epiphanio Ferdinando, *Centum historiae seu Observationes et Casus medici* (Venice: apud T. Baglionum, 1621), p. 248.

2. Angelo Turchino dates the first account of the disease to 1590 in a letter from A. M. Manzoli, bishop of Gravina, to the famous naturalist of Bologna, Ulisse Aldrovandi; Turchino gives a printed transcription of the manuscript letter in his book, *Morso, Morbo, Morte, La Tarantola fra Cultura medica e Terapia popolare* (Milano: Franco Angeli Libri, 1987), pp. 201–203. Earlier references to the poisonous tarantula appeared throughout the sixteenth century in the literature on poisons but did not elaborate on the symptoms or the musical cure.

3. Modern incidence of the disease is found in the 1977 study of Miriam Castiglione, "Il Tarantismo oggi: Proposte per una Verifica," *La Critica sociologica* 11 (1977): 43–69. Today, victims who believe themselves possessed by the spider's poison congregate on the birthday of a local saint, S. Paolo di Galatina, in a chapel dedicated to him in the village of Galatina. A good overview of the recent literature on tarantism is found in Turchini, *Morso, Morbo, Morte;* here he discusses extensively the sociological and psychiatric explanations which scholars have offered of the contemporary phenomenon. All of this recent anthropological and sociological literature is heavily indebted to the field work of Ernesto De Martino, *La Terra del Rimorso: Contributo a una Storia religiosa del Sud,* 3d ed. (Milano: Il Saggiatore, 1976). On the historical connections between music and possession, see Gilbert Rouget, *La Musique et la Transe* (Paris: Gallimard, 1980); Marius Schneider, *Gli Animali simbolici e la loro Origine musicale nella Mitologia e nella Cultura antiche* (Milano: Ruscani, 1986); George Rosen, "Irrationality and Madness in Seventeenth- and Eighteenth-Century Europe," in *Madness in Society,* ed. Rosen (Chicago: University of Chicago Press, 1968); and Henry Sigerist, "The Story of Tarantism," in *Music and Medicine,* ed. Dorothy M. Schullian and Max Schoen (New York: H. Schuman, 1948), pp. 96–115.

4. Two Italian physicians are well known for their interest in toxicology—Francesco Redi (1626–1698) and Felice Fontana (1730–1805).

Surprisingly neither was drawn into the debate on tarantism. On toxicological experiments in the early modern period, see Peter Knoefel, *Francesco Redi on Vipers* (Leiden: Brill, 1988), and A. H. Maehle, *Johan Jakob Wepfer (1620–1695) als Toxikologe* (Aarau: Verlag Sauerlander, 1987).

5. Ferdinando, *Centum historiae,* pp. 248–268.

6. Athanasius Kircher, *Magnes sive de Arte magnetica,* 3d ed. (Rome: B. Deuersin & Z. Masotti, 1654), pp. 587–603.

7. This is according to Kircher's disciple, Gaspar Schott, who practically reproduced Kircher's account of tarantism verbatim in his own *Magiae Universalis naturae et artis, Pars II: Acoustica* (Würzburg: sumptibus Haereden Joannis Godefridi Schoenwetter, 1657), pp. 236–250. Although Schott had lived in Sicily for over a decade and had traveled frequently in southern Naples and Calabria, he did not not make good on his boast to add to Kircher's data. He merely repeated Kircher's cases.

8. Giorgio Baglivi, *The Practice of Physick* (London: Printed for Andr. Bell, Ral. Smith, Dan. Midwinter, Tho. Leigh, Will. Hawes, Will Davis, Geo. Stalon, Bern. Lintott, Ja. Round, & Jrff. Wale, 1704), pp. 345–409. On Baglivi's interest in tarantism, see Glauco de Bertolis and Carlo Agostoni, "Il Tarantolismo nell'Opera di Giorgio Baglivi," *Minerva medica* 58 (1967): 2620–2633.

9. Sangenito's letter is published in Antoine Bulifon, *Lettere memorabili, istoriche, politiche ed erudite,* 4 vols. (Naples: A. Bulifon, 1698), 2:107–113.

10. In Sangenito's words, the afflicted were "gente rozza ed inculta" such as harvesters and shepherds and "simili altri uomini camparecci." In Bulifon, *Lettere,* 2: 107.

11. Ludovico Valletta, *De Phalangio apulo* (Naples: ex typographia de Bonis, 1706), p. 164.

12. Ibid.

13. Baglivi, *Practice of Physick,* pp. 352–355.

14. Paolo Boccone, *Museo di Fisica e di Esperienze* (Venice: I. B. Zuccato, 1697), pp. 92–119.

15. Wilhelm Homberg, "Observations sur les Araignées", in *Suites des Mémoires de Mathématique et de Physique, tiréz des Registres de l'Académie royale des Sciences. De l'Année MDCCVII,* 4 vols. (Amsterdam: Pierre de Coup, 1746), 2:438–454. Homberg (1652–1715) received his M.D. from Wittemberg after extensive travels in Europe, practiced medicine for a time in Rome, received admission to the Académie des Sciences in 1691, and was named first physician to the Duc d'Orléans in 1704.

16. Martin Lister, *Historiae animalium angliae tres Tractatus* (London: apud Joh. Martyn, 1678), pp. 15–16.

17. Cornelis Stalpart van der Wiel, *Observationum rariorum medic. anatomic. chirurgicarum, Centuriae* (Leyden: Petrum van der Aa, 1687), pp. 424–448.

18. Hermann Grube, *De Ictu tarantulae & Vi musices in ejus Curatione, Conjecturae physico-medicae* (Francfurt: ex bibliopolio Hafniensi Danielis Paulli, 1679), and Wolferd Senguerd, *Tractatus physicus de Tarantula* (Leyden: apud Gaasbeeckios, 1668).

19. Andreas Flachs, *Aranea, Imprimis vero de Tarantulis* (Wittemberg: typ. J. Borckardi, 1660).

20. The proceedings of the Bureau d'Adresse were published as *Recueil genéral des Questions traitées es Conferences du Bureau d'Adresse, sur toute Sortes de Matière, par les plus beaux Esprits de ce Temps* (Paris: chez Jacques le Gras, 1665). For the account of the discussion on tarantism held on 19 May 1642, see pp. 169–172.

21. Martin Lister's account is found in the *Philosophical Transactions of the Royal Society of London* 6 (1671): 3002–3003; Cornelio's in 7 (1672): 4066–4067; and Cirillo's in 60 (1770): 233–238.

22. See *Miscellanea curiosa sive Ephemeridum medico-physicarum germanicarum Academiae naturae curiosorum* (1678): 281–282.

23. For Geoffroy's report, see "Diverses Observations de Physique genérale", in *Histoire de l'Académie royale des Sciences. Année MDCII* (Amsterdam, 1737), pp. 20–24. Geoffroy (1672–1731) accompanied the Abbé de Louvois to Italy as physician and friend. In 1704, he obtained an M.D.; in 1707, held the chair of chemistry at the Jardin du Roi, and then succeeded Tournefort at the Collège de France in the chair of medicine and pharmacy. In 1726, he was elected dean of the Faculté de Medecine at Paris. He was a frequent contributor to meetings of the Académie des Sciences.

24. Homberg, "Observations sur les Araignées."

25. Francesco Cancellieri, *Lettera al Ch Sig. Dottore Koreff sopra il Tarantismo, l'Aria di Roma e della sua Campagna* (Rome: n.p., 1817), pp. 1–14.

26. Jean Jacques Manget, *Bibliotheca medico-practica sive rerum medicarum Thesaurus,* 4 vols. (Geneva: sumptibus J. A. Chouet, 1697), 4:634.

27. On the broader issue of verifying experimental results in early modern science, see Steven Shapin and Simon Schaffer, *Leviathan and the Air-Pump: Hobbes, Boyle, and the Experimental Life* (Princeton: Princeton University Press, 1985); and Steven Shapin, *A Social History of Truth* (Chicago: University of Chicago Press, 1994).

28. Geoffroy, "Diverses Observations," p. 23.
29. Bulifon, *Lettere,* 2: 107-113.
30. Cancellieri, *Lettera*, p. 8.
31. Robert Boyle, "An Essay on the Great Effects of Even Languid and Unheeded Motion," (1685) in *The Works of the Hon. Robert Boyle,* ed. Thomas Birch, 5 vols. (London: printed for A. Millar, 1741), 2:26-0–261.
32. John Ray, *Observations Topographical, Moral and Physiological Made in a Journey through the Low-Countries, Germany, Italy and France* (London: printed for John Martin, 1673), p. 410.
33. *The Correspondence of Henry Oldenburg,* ed. A. Rupert Hall and Marie Boas Hall, 14 vols. (Madison: University of Wisconsin Press, 1965–1986), 5:300.
34. Bulifon, *Lettere,* 2:107.
35. Ibid.
36. Ibid.
37. Richard Mead, *The Medical Works of Richard Mead* (New York: AMS Press, 1978), pp. 48–56, in "A Mechanical Account of Poisons, in Several Essays," 1st ed. (1702).
38. Bulifon, *Lettere,* 2:107.
39. Geoffroy, "Diverses Observations," p. 23: "Voilà ce qui est attesté par des personnes dignes de foi & ce qui fut confirmé a l'Académie, non seulement par le soin que M. Geoffroy avoit eu de s'en informer en Italie, mais encore par des Lettres que lut le P. Gouye, où un P Jésuite de Toulon mandoit qu'il avoit vu danser plusieurs jours de suite un Soldat Italien mordu d'une Tarantule."
40. Valletta, *De Phalangio,* letter to the reader, p. 6.
41. Kircher, *Magnes,* pp. 587–603.
42. Baglivi, *Practice of Physick,* pp. 351, 369, and 385. When Boccone made field observations of the native tarantulas on Corsica in 1682, he noted that the Corsican variety did not produce the same effects as the Apulian ones and that music did not mitigate the unusual physical symptoms. Boccone, *Museo,* pp. 92–109. The lion's share of the reported cases of tarantism involved patients living in Apulia who responded to musical therapy. But one well respected Italian naturalist, Paolo Boccone (1633–1704), enlarged the medical literature by documenting cases of tarantism in Corsica. Boccone denied that Corsican tarantulas were the same species as Apulian ones, and he concurred with all other medical writers on the geographic specificity of tarantism when he noted that Corsican patients did not respond to musical therapy. He did present two fatal cases of Corsican tarantism, one afflicting a peasant bitten under the throat, and another afflicting a peasant on

the left breast. Not all the Corsican cases were mortal, however, for Boccone related that a local priest, Giovanni Domenico Cotti, was bitten one night in 1675 while sleeping. Muscle spasms in all of his joints followed the pain of the bite. After eight hours, heart palpitations immobilized Cotti. Lying unconscious for four hours, he later vomited four mouthsful of black blood, and his pain and muscle spasms returned. Treated with theriac, tepid wine, scarifications, cupping glasses and oil of scorpions, he passed two full days on the borders of life; finally, from the base of his fingernails there oozed painlessly a small amount of humor. Although his skin took on a livid hue for three continuous days, he ultimately returned to full health.

43. On the efforts of the Apulian archbishop to have specimens collected and transported to Dresden for scientific study, see his letters published by Franco Strazzullo, "Fragmenta historica," in *Settecento napoletano, Documenti,* ed. Franco Strazzullo (Napoli: Ligouri Editore, 1982), pp. 352–356.

44. Van der Wiel, *Observationum rariorum,* quotes Kircher's report of experiments conducted by his informants, pp. 438–439.

45. See *Correspondence of Oldenburg,* 9:45–46. Cornelio's reply, entitled "Extract of a Letter from Tommaso Cornelio, Neapolitan Philosopher and Physician Concerning Some Observations Made of Persons Pretending to be Stung by Tarantulas: English'd out of the Italian," appeared in the *Philosophical Transactions* 7 (1672): 4066–4067.

46. Baglivi, *Practice of Physick,* pp. 385–387.

47. Rossi's correspondence regarding the transport of the tarantulas is found in *Settecento napoletano, Documenti,* pp. 352–356. David Gentilcuore documents further efforts of seventeenth-century ecclesiastics to speak authoritatively about tarantism in his *From Bishop to Witch: The System of the Sacred in Early Modern Terra d'Otranto* (Manchester: Manchester University Press, 1992), pp. 149–161.

48. Tommaso Campanella, *Del Senso delle Cose e della Magia,* ed. and trans. A. Bruers (Bari: G Laterza & Figli, 1925), pp. 260–273. The first edition of the book appeared in 1620.

49. Daniel Sennert, "Medicinae praticae" in *Opera omnia,* 2 vols. (Venice: apud Franciscum Baba, 1641), 2:154. While concluding that the way is which music effected the cure was "obscurissimum," Sennert used the evidence of the disease presented by Ferdinando to insist that music attacked the poison by an "occult means." While offering the conclusion that music contained in it some special force, something divine, he considered tarantism one of the hidden wisdoms of the divine treasury which no man will easily unearth.

50. Lazare Meysonnier, *Pentagonum philosphicum-medicum* (Lyon: sumptibus Jacobi & Petri Prost, 1639), pp. 80–81.

51. Meysonnier, *Pentagonum,* pp. 80–81. See also his *Nova et arcana Doctrina febrium* (Lyon: sumptibus Petri Prost, 1641), p. 89, where he cites the universally accepted belief that music "can sweeten the ardors of fevers."

52. For twentieth-century evaluations of Kircher's contributions to medicine, see Harry Torrey, "Athanasius Kircher and the Progress of Medicine," *Osiris* 5 (1938): 246–275, and John Fletcher, "Medical Men and Medicine in the Correspondence of Athanasius Kircher," *Janus* 58 (1969): 259–277. Less reliable is Ralph Major, "Athanasius Kircher," *Annals of Medical History* 1 (1939): 105–120.

53. Marco Aurelio Severino, *Vipera pythia* (Padua: typis Pauli Frambotto, 1650), p. 336.

54. Samuel Haffenreffer, *Nosodochium, de Cutis affectibus* (Ulm: typis & expensis B. Kuhnen, 1660), p. 502.

55. Ibid., pp. 457–520.

56. Schott, *Magiae,* pp. 236-250.

57. Walter Charleton, *Physiologia epicuro-gassendo-charletonia* (London: printed by T. Newcomb, 1654), pp. 371–372. On Charleton's indebtedness to Epicurean atomism and his early attachment to Helmontian ideas, see Nina Rattner Gelbart, "The Intellectual Development of Walter Charleton," *Ambix* 18 (1971): 149–168; and Robert Kargon, "Walter Charleton, Robert Boyle, and the Acceptance of Epicurean Atomism in England," *Isis* 55 (1964): 184–192.

58. Thomas Willis, *The London Practice of Physick, being the Practical Part of Physick, Contain'd in the Works of the Famous Dr. Willis* (London: printed for T. Bassett, T. Dring, C. Harpers, W. Crook, 1692), pp 267–269. See Hansruedi Isler, *Thomas Willis, 1621–1675, Doctor and Scientist* (New York: Hafner Publishing Co., 1968).

59. Senguerd, *Tractatus.*

60. Johann Christian Frommann, *Tractatus de Fascinatione* (Nuremberg: sumptibus Wolfgangi Mauriitii Endteri & Johannis Andreae Endteri Haeredum, 1675), pp. 230–240.

61. One finds further explanations of tarantism in Cornelis Stalpart van der Wiel, Robert Boyle, Richard Meade, and Herman Grube.

62. Nicolai Caputi's "De tarantulae Anatome et Morsu" remained unpublished until 1741, when it issued forth from a press in the remote Lecce: Typis Dominici Viverito. The elegant manuscript version remains in the Oxford University Library.

63. For Cornelio's account, see *Philosophical Transactions* 7 (1672): 4067.

64. For instance, the author of the article "tarantisme" in the remarkably skeptical work of Denis Diderot and Jean d'Alembert, *Encyclopédie, ou dictionnaire raisonné des sciences, des arts et des metiers,* 17 vols. (Paris: Briasson, 1751–1765), 15:905-908, reviewed the evidence set forth by Geoffroy, Mead, Grube and Baglivi. Though suspicious that the bite might be imagined, he did not hesitate to assert that music truly produced efficacious cures; the author of the article "musique" was even more adamant that music cured the poisonous disease. On the largely unsuccessful attempts of medical skeptics to shake belief in the musical cure in late eighteenth-century Italy, see Vincenzo Ferrone, *I Profeti dell'Illuminismo, Le Metamorfosi della Ragione nel tardo Settecento italiano* (Bari: Editori Laterza, 1989), pp. 26–28.

NATURE AND CULTURE

IN THE *DISCOURSES* OF THE *VIRTUOSI* OF FRANCE

Kathleen Wellman

Nature, as the topics treated in this volume attest, embraces a wide range of issues. Even the most casual perusal of early modern texts suggests that nature functioned as the critical term defining the relationship of human beings to their world. But between the early modern view of nature and our own lies the intellectual chasm and historiographical minefield of the Scientific Revolution.[1] The work of Allen Debus,[2] in particular, has shaken our faith in positivistic accounts of this development; extensive exposure to the virtues of Paracelsus, van Helmont, Robert Fludd, John Dee, etc. has made clear both that the transition from early modern to mechanical views of nature was less than clear-cut and that there was extensive and empirically-grounded opposition to a mechanical view of nature throughout the early modern period.[3] Recent, more polemical studies have also attacked the Scientific Revolution as the source of an approach to nature which produced environmental disaster; that is to say, the mechanical philosophy is charged with condoning man's rapacious exploitation of the earth's natural resources.[4] This negative reassessment of mechanical science is grounded in a fundamental distinction between nature and culture, where nature is primary and good, and culture is secondary, and, if not bad, at least suspect as artificial.

When the *virtuosi* of France gathered to discuss a wide array of topics, they invoked nature as a fundamental term of authority or explanation. This paper will assess the contribution made by the conferences held at the *Bureau d'Adresse* from 1633–1642 to both the contemporary discussion of nature and to the subsequent historiographical debate. I have chosen to concentrate on two particular issues: 1) how nature itself is invoked in discussions of scientific issues, and 2) how discussions of nature are applied to culture and to particular social issues.

The *Discourses*

The *Bureau d'Adresse* was an innovative office founded in Paris by Théophraste Renaudot as part of this efforts as special *intendent des pauvres* to address the problem of poor relief in France. It was intended to serve as a clearing house for employment, housing, and information.

P. H. Theerman and K. Hunger Parshall (eds.), Experiencing Nature, 193–210.

Eventually, Renaudot expanded its services to include medical consultations for the poor on Tuesdays and to sponsor conferences on predetermined topics each Monday afternoon. The character of these conferences from 1633 to 1642 makes them particularly illuminating sources. They were intended to provide a new intellectual forum, deliberately undermining the typical rules of conduct of an academy; they were public, not private, and participants, unlike members of traditional French academies, were not exclusively nobles. Thus participants in these conferences included notable professional scholars,[5] and large numbers of professionals such as doctors and lawyers took part.[6] As contemporary critics often pointed out, the Bureau d'Adresse also attracted a number of dilettantes or *virtuosi,* as they called themselves, men of rank, leisure, and wealth.

The procedures governing the conferences were as unconventional as its membership. The doors of the Bureau were open to all; there was no limit set on the number of speakers and no moderator. All topics were to be admitted save religion and affairs of state. In essence, the conferences held at the Bureau d'Adresse were gatherings of people interested in new ideas who found the universities and academies stagnant, conservative, and unresponsive to popular concerns, especially those dealing with contemporary scientific issues.[7]

These unconventional conferences were held under the protection and auspices of Théophraste Renaudot, an influential bureaucrat under Louis XIII. Renaudot is best known for the creation of the *mont-de-piété,* a program under Louis XIII of government-sponsored low interest loans to the poor. To some degree, the nature and character of the Bureau d'Adresse were the reflection of Renaudot's interests. He was explicitly interested in the reform of pedagogical theory and practice; he argued, in a way that seems prescient of the Enlightenment, that an academy was a public trust and that the dissemination of information was vital for the well-being of the state and its citizens; and he established the conferences at the Bureau as a way to demonstrate the effect of a public academy.

The unorthodox nature of the *Discourses* themselves is perhaps also attributable in part to the interests of its sponsor. Renaudot's concerns no doubt made this group somewhat more likely to favor heterodox views. Renaudot was under the powerful protection of Richelieu, and, although the *Discourses* claimed to avoid the controversial areas of religion and politics, it seems likely that, so long as Richelieu lived, it would be possible for less conventional opinions to be expressed. Renaudot was a physician trained at the Montpellier school and therefore part of a more progressive and open scientific community than the

Faculty of Paris. Indeed, his espousal of Paracelsian medicine led to bitter disputes with that group. Finally, despite his expedient conversion to Catholicism, Renaudot would likely have remained sympathetic to unorthodox philosophical views.[8]

In addition, the *Discourses* have several features which make them a particularly intriguing source for the discussion of nature and culture. First of all, they deliberately present a wider array of opinion than the work of an individual. Each *Discourse* presents from four to twelve opinions on a given issue. The format was deliberately designed to allow for the presentation of many opinions, and no attempt was made to arrive at a consensus. Because the *Discourses* eschew the rigidity of the academic structure, they present a wider array of opinion by a more diverse group. For example, unlike the control exercised by the medical faculty on the issue of chemical remedies, the Bureau d'Adresse made no attempt to determine or impose a party line (even though Renaudot was a strong partisan on the question). Moreover, these *Discourses* also present a view of nature that is not circumscribed by the boundaries between the physical and the life sciences or even the sciences and the humanities. As a result, the conceptual schemes developed in them could be, and were, applied to a wide range of social issues.

The Views of Nature

The discussion of nature in the *Discourses* is shaped by premises sharply different from modern scientific attitudes. It is axiomatic that man is limited, that there are questions simply beyond his ability, and that he must acknowledge these limitations. For example, to try to understand the nature of light is a misbegotten enterprise because, as one speaker put it, "Light is a form, to seek the cause of light is to seek the reason of forms which is not known to us."[9] Another disparages prime matter by emphasizing the degree to which it differs from the divine, saying that "as God is incomprehensible by reason of his exceeding grandeur: so is this prime matter by reason of its baseness and lowness."[10] A speaker on sympathy and antipathy expresses what seems to be a prevalent attitude: "It is more fitting to admire the secret motions of nature than to seek the cause of them unprofitably," and "those who give reasons are not less ignorant but more vain."[11] One speaker, discussing how minerals grow, claims that all of nature's works are occult.[12]

But even if the human understanding of nature must be acknowledged as incomplete, participants insist that important things can be known because of the consistent ways nature operates. For example, nature, most speakers assume, acts according to the standards of propriety. In discussing earthquakes, a speaker says that, because the earth is the most ponderous body, earthquakes must be caused by fire, the noblest element.[13] Although nature supersedes human operations, there is a general assumption that those operations are analogous; for example, "as art can draw forth water by distillation and other ways taught by chemistry, so, by stronger reason, nature cannot want to do the same."[14] Note that the analogy works both ways: Nature's operations apply to man in a less grand fashion, but what man can do, nature must also be able to accomplish to a more admirable degree.

The standards of propriety work to man's advantage. Nature does not allow man to seek something which cannot be attained; thus, by implication, perpetual motion and the philosopher's stone both exist.[15] Nature has also given man what he needs; for example, "hearing is given to man to facilitate his natural desire to understand the thoughts of his species."[16] All facets of nature from the heavens to the simplest creatures are connected by the relationship between the macrocosm and the microcosm, a notion succinctly expressed as: "Superior bodies act on inferior, and all notions here below proceed from those of the celestial."[17] Thus one speaker contends that because there is perpetual motion in the heavens, perpetual motion is possible on earth, but it must be circular. Lightening and thunder are described as celestial activities analogous to the generation of metals in the earth. Divination is possible because there is such a close connection between superior and inferior.[18]

These attitudes might well be considered typical of the early modern period. However, the *Discourses* also demonstrate a striking skepticism; participants seem imbued with a sense that they are the advocates of a new standard of evidence. Several conferences explicitly contend that maxims, that is, popular sayings, are dangerous to health because they keep people from seeking medical care. Other speakers dismiss reports of monsters as apocryphal or the result of misguided credulity.[19] Participants assume a gap between the way the popular classes and elites approach these issues.

For example, one speaker explicitly cites the view that the anger of God against certain men causes their wives to bring forth monsters as the opinion of the vulgar.[20] Nonetheless, reports from exotic locations of unusual phenomena are sometimes credited. One participant reports that the water drawn from a river in Sicily cannot mingle with wine unless it is drawn by a chaste woman.[21] But references to monsters or irregularities of nature are only mentioned occasionally. The predominant sense in the *Discourses* is that any reported "freaks of nature" must be examined in terms of the regular, more significant operations of nature. In fact, all reports of monsters are to some degree undermined by speculation about the effects of the imagination on those individuals who claim to have seen them.

A conference on the subject of *incubi* and *succubae* seems to exemplify the attitude of conference participants on a wide range of subjects. "Two sorts of people err in this matter, the superstitious and ignorant vulgar who attribute everything to miracles and account the same done either by saints or devils, and the atheists and libertines who believe neither the one nor the other. Physicians take the middle way, distinguishing what is fit to be attributed to nature and her ordinary motions from what is supernatural."[22] Clearly, conference participants identified with the physicians as skeptical and critical students of natural phenomena.

Skepticism about unusual phenomena is reenforced by the fundamental belief which seems to characterize all discussion of nature in the *Discourses*, that is, that nature presents an ordered hierarchy which can be known from top to bottom. As one speaker put it, "God makes himself known to men by the marvellous effects of nature. The immaterial and incorruptible heavens communicate their virtues and influences here below: first, through the element of fire, which is most subtle; then through the air which is most pure in the upper region, more gross in the middle, and in the lower infected by the vapors and exhalations of the water and Earth; and through all compounds, in the production of which nature observes such an order that she always begins with the more simple and never passes from one extremity to another without a medium."[23]

Despite the general consensus that nature is ordered, the order itself is presumed to be based on a variety of principles. For example, nature, some speakers consider, keeps order by the opposition of contraries, a battle of forces in opposition. As one describes it, "nature subsists only by contrariety; life lasts only so long as the natural heat acts upon the radical moisture, when their combat is ended, man must necessarily die. His understanding has no better means to obtain truth, than by contrariety of opinions; for identity is as disagreeable to the mind as it is to nature."[24] Another understanding of nature presupposes that order is partially a natural process, an ordering of raw materials which is completed when God imposes a form.[25] Another speaker defines generation and growth as the fundamental processes of nature accomplished by the separation of humors by heat, the fundamental agent of nature.[26]

No paper presented in honor of Allen Debus could neglect the influence of a Paracelsian understanding of nature. Fortunately one does not have to go far to discover its influence on these *Discourses.* Renaudot himself was a Paracelsian who faced serious difficulties with the Faculty of Medicine, and it is not surprising that his rather unconventional gatherings would have attracted other Paracelsians. There are conferences dedicated to "Paracelsian topics" like the Fraternity of the Rosy Cross and the Weapon-salve controversy. Another conference analyzes dreams as signatures. Other participants claim that the senses tell us that most bodies are reducible to salt, sulphur, and mercury, and most speakers take for granted a macrocosm/microcosm relationship.[27]

Whatever principles are seen as fundamental, all of the conferences which directly address natural phenomena, and even those which simply discuss nature in passing, are preoccupied with the question of order in nature. But even a fundamental belief about order in nature does not reflect certainty about the operations of nature. As one speaker concludes, there are three kinds of events in nature, things which always happen, things which sometimes happen, and things which rarely happen.[28]

Discussions of nature also assume that nature is beneficent and acts in the best interests of the creatures of the earth. For example, nature has given animals attributes which suit their needs; as a result, predators have a particularly acute sense of smell.[29] The eye is so highly regarded by nature "that she has fortified it on all sides for its safety, with the bone of the forehead, eye-brows, eye-lids."[30] A creature grows "until the body has attained the proportion and stature requisite to its functions."[31] The instinct for self-preservation is taken to be one of the best indications of the care nature takes of her creatures. "Nature has given things two ways of preserving the being she has given them, namely to seek their good and flee their evil. Both of which animals do by attracting what is proper to their nature by right fibers and rejecting what is other by transverse fibers."[32] Belief in the beneficent teleology of nature is so pervasive that it is used to sustain arguments the modern reader finds peculiar. Even smallpox indicates nature's favor: "Yet those that are taken with this disease are usually the most healthy, and of a sanguine constitution, which is the most laudable. For this were to accuse nature either of imprudence or weakness; but she is good, wise, powerful and solicitous."[33]

Furthermore, nature is assumed to be beneficent regardless of one's view of the basic principles of nature. For example, even if nature is a battle of contraries, the contraries work in man's interest. "So great is the indulgence of nature that she has not thought it enough to bestow being and existence on the things she has produced, but she has also imprinted in them a strong inclination to preserve it, by fortifying them against the assault of their contraries, the presence of which sets them on such an edge that they become so much more active."[34] In fact, fundamental differences between the Galenic and the chemical views of medicaments are glossed over by a remark that whatever cures is a friend of nature, and all healing is done with such friends of nature.[35]

But in striking opposition to this view of the ordered beneficent nature, there is a dissenting view which sees nature as less scrutable, less predictable. The most extreme statement of this kind in the *Discourses* seems to assign to nature an almost malign character. "That there are certain secrets in nature of things dreadful to human reason, incredible, according to the principles of art and of our knowledge. That nature is the great *circe,* the grand sorceress, that the lodestone draws iron to it; that there is a certain stone called *pantarbe* which draws gold to it; that the hair or wool of a mangy beast being thrust in, for a certain time, under the bark of an aspen-tree, cures the beast of vermin; that the menstrua of women pollute springs and spoil looking-glasses." However the speaker hastens to qualify this statement, "it is not to be imagined that nature is so cruel a stepmother, but that there may be other means whereby they are preserved, and so the remedies may come from the same hand that caused the disease."[36] Even when proclaiming the circe-like quality of nature, participants assume that although nature is difficult to know she has nonetheless provided antidotes for any ill effects she may have produced. The very incomprehensibility of nature suggests beneficence. For example, one speaker claims that a comparison made between women and mules is really a compliment to women because nature, with clear intention, has hidden its highest mysteries in the lowest things. (The assumption is that, because mules do not propagate, nature is content with them as they are, and, as a result, any connection to these paragons of nature can only redound to the credit of women.[37]) The atom is another example of nature hiding her mysteries. As one speaker put it, "for an atom flying in the air is as true an essence as the whole earth." And although nature is capricious, even that quality seems to work for man's benefit, "for though nature loves change (of which she is the principle), yet it is only that of generation or of a lesser into a more noble substance; that of corruption and death she abhors."[38]

These *Discourses,* occurring just before the widespread acceptance of mechanism, ought to allow one to examine the transition from the pre-mechanical to the mechanical view of nature. They do not yet indicate any familiarity of the participants with the mechanical philosophy, and so one might expect them to reflect the pre-mechanical understanding of nature, with a female earth, revered as both nurturer and chaotic principle in an animistic universe. However, the picture is less conclusive than one might expect. The *Discourses* do not depict a feminine earth (although "nature" is a feminine word in French and thus the pronoun referent is "she"). Nor do they show a marked reverence for the earth. Those speakers who emphasize the mysteriousness of nature are more exasperated than reverent. Yet there is no reference to domination[39] or forcing nature's secrets from her and no analogies drawn between machines and natural processes. There is also no sense that the earth is alive, although generation is considered to be the most important natural process. Surprisingly, an animistic sensibility seems absent even in the discussion of Paracelsian topics. In fact, in the *Discourses,* discussion of Paracelsian topics seems considerably more abstract and less mysterious than earlier Paracelsian works. Although Paracelsians describe the operations of nature in terms of their own principles, nonetheless, they use the same methods of argumentation and the same standards of evidence as other participants. They, too, are predominantly concerned with the order and beneficence of nature. And all participants suggest that their arguments are empirically sound, based on direct observation.

What the *Discourses* suggest is that, although nature is not discussed in mechanical terms, it is nonetheless largely conceived in abstract terms. By virtue of the format of these discussions, that is to say, recorded minutes of meetings rather than full debates, individual speakers, at least on some issues, seem to present "position papers," articulating the Paracelsian, Aristotelian, or religiously orthodox views. (Other opinions are less clearly categorizable.) However, the final impression conveyed by any specific *Discourse* is of an open-minded eclecticism. Individuals present their views in quest of a systematic or sustainable foundation for scientific knowledge. There seems to be an openness towards a philosophical foundation for understanding nature without an overt desire to impose a particular philosophical view.

Ultimately, the question the participants grapple with is not whether to submit to unknown natural processes or to subjugate nature by force but rather to assess the degree to which nature can be known. In essence, it is a discussion of how occult or opaque nature's methods are. The discussion of nature in the *Discourses* suggests that an organic view of nature as vitalistic, mysterious, and revered had broken down well before the seventeenth century. Nature instead is shaped to human reason and reduced, or at least invoked, as an abstract ordering principle. But it seems reasonable to suggest that an understanding of nature as abstract but ordered and good might well have laid the foundation for the success of the mechanical philosophy. In other words, perhaps the erosion through abstraction of a personified nature prepared the way for a mechanical world view. And, as part of this process, nature as a source of nurture and chaos had been rationalized into a principle of order and beneficence.

Nature and Culture

But if nature had become an abstraction whose fundamental operations were discussed under a wide variety of terms, it was no less effective as a principle appropriate to comment on social and cultural concerns. Indeed, in the *Discourses,* nature is invoked in a decisive fashion on virtually every topic, whether scientific or more generally rhetorical. The term "nature" is frequently and explicitly invoked to comment on society. The participants see, or at least express, no dissonance between nature and culture (at least culture understood as social organization). Nature is assumed both to provide a fundamental principle for society and the state and to be directly concerned in making sure the affairs of state function well. Nature is directly invoked to insure that her gifts are apportioned to further the interests of society. For example, health is defined as "a disposition, according to nature, which renders a man capable of performing the offices of life aright."[140] Nature, recognizing the fundamentally social nature of human beings, gave man speech because society is "no way imaginable without it."[141] In fact, nature suggests a rationale for public life; as a speaker on the vacuum confidently proclaimed, "since nature suffers violence to herself to prevent inanity, and all things quit their particular interest for that of the public, undoubtedly, there is no such thing as vacuum in nature."[142] Once again, just as analogies from macrocosm to microcosm and microcosm to macrocosm are assumed to be equally compelling, so too analogies from nature to society and society to nature are both considered persuasive.

Not only has nature sanctioned society, but, conference participants also assume confidently, nature's principles pervade society. For example, a conference on harmony discusses the many ways in which the natural principle of harmony extends into human life. "Musical harmony aims at the instruction of men; that of man's body is the admirable artifice of the formative faculty, which Galen calls divine; but the harmony of the world almost puts our curiosity to a non-plus . . . . The cause of this chain and connection is an universal virtue comprehended in the extent of each being . . . nature must have allotted them a power which may guide them to that end."[43] Because nature and society are governed by the same principles, "he who lives long can learn nothing new in the world which is but a revolution and repetition of the same effects always produced by the same causes; not only in nature, whose course and changes may be seen in the revolution of the four seasons of the year, but even in affairs of state and private matters, about which nothing is said or done, but that which has been practiced before."[44]

The great flexibility of nature as applied to society is indicated in a conference on number where one speaker claims that natures operates according to the number two. "The binary is the first of all numbers. Nature is so delighted with it that she has exactly observed it in the structure of man, whose sense organs and almost all his members are double, and therefore it is also so carefully observed by architects." Another argues for the number three, "But the ternary . . . is competent to the deity and his works; the world is distinguished into three ages; there are three sorts of souls in nature, three faculties in man and three principal parts in his body."[45] In fact, arguments for the numerical basis for the operations of nature go up to seven.

Nature, as benevolent and ordered, also provides a particularly useful analogy for discussing both the hierarchical arrangement of society and for sanctioning particular social policies. For example, social roles suggest natural analogies. "As gold is the master-piece of nature so is glass of art, which cannot produce anything more noble. Hence in France, the making of it is permitted only to the nobles."[46] More importantly, nature explicitly sanctions social hierarchy. For example, as a speaker on the topic "Why men love to command more than to obey" put it, "nature has established this law, that inferior things ought to obey superior, the less worthy the more worthy, so that obedience and command are the different consonances which compose the harmony of the world."[47] There is also a physiological basis for different social roles; "those who lead are called by the philosopher, lords and masters of nature, having an heroic spirit, and capable of governing not only themselves but others as well; their bodies being usually weak and delicate, hair fine, skin smooth and thin. Others are servants by nature, being strong and sturdy, fit to carry burdens, to undergo labor, and such inconvenience as attend those who are subject to another's will."[48] Even slavery is sanctioned by nature. "Servitude may be termed natural, being founded by the inequality of the sufficiencies and abilities of men; some being born with organs so nimble and pliant, that their mind acts almost divinely, others are so dull, that the soul seems mired in soil."[49]

Even the hidden operations of nature can sanction the actions of the state. For example, one speaker justifies the lies governments tell the people by saying, "this nature teaches us, as she reveals only the surface of the earth, but has hidden all its treasures, . . . that which hinders the effects of state-counsels, where secrecy is the soil, is letting of them be discovered, which is treason. That which takes away credit from all the arts and professions is making them common."[50]

If speakers in these conferences acknowledge a fissure in their world view, it is not between nature and society but rather between nature and art. This question is most explicitly addressed in a conference on "Whether a piece of iron laid upon the cask prevents thunder from marring wine contained within it, and why?" when a speaker contends that

> the operations of nature are not like those of art, her ways and contrivances are more obscure and the causes of things are occult; as for example, the lodestone draws iron even though no body of air or smoke can be perceived issuing out of the lodestone. And magnetic balsam, or weapon-salve, cures a wounded person, though at a great distance, having only some part of his clothes . . . so secret and silent is nature in her operations. On the contrary the designs and contrivance of art may be as easily discovered as those of a clock . . . but the reason of this diversity of operations between art and nature, is that art goes to work publicly and before the senses, and nature does her business within doors and secretly.[51]

At least as far as this conference is concerned, mechanism would have more in common with mere art. It would be predicatable and obvious, and nature would be considered more difficult to know than simple mechanical processes.

Their understanding of nature endowed these conference participants with such extraordinary confidence that, in their deliberations on the question of the ancients versus the moderns, they invariably argue that nature supports the moderns or, at the very least, equally favored the ancients and the moderns. "For nature being still as wise and powerful as any previous time, and the universal causes being the same, their operations must also be as perfect, and their effects as excellent in these days as they have been in any. As for our minds, they are so far from being impaired, that they improve more and more in accurateness and being of the same nature with those of the ancients, have such an advantage beyond them, as a pigmy has upon the shoulders of a giant."[52]

Conclusion

The view of nature which the *Discourses* suggest varies widely in terms of particular principles but is strikingly similar in the general expectations about the character of nature and the legitimacy that comparisons to nature confer on social institutions. Confidence in the order and beneficence of nature carried over into expressions of confidence in the order and legitimacy of social structures, institutions, and policies. It is sometimes suggested that the absolutist state of seventeenth-century France favored or even required the manipulative power of a mechanical understanding of nature.[53] However, these *Discourses* suggest that the existing science was readily adapted to support the authority of the absolutist state and that the contemporary understanding of nature endowed an unassailable legitimacy on existing social structures. Thus, the rationalization of seventeenth-century institutions did not require a mechanical view of nature. Instead, the *Discourses* demonstrate that early modern science, with its myriad descriptions of natural processes, proved sufficiently flexible for participants in these conferences to assess with confidence the social and cultural reality they confronted.

In discussions of both nature and culture, the *Discourses* suggest a kind of midpoint; that is to say, nature is understood neither as a machine nor as an animistic world spirit. As far as the relationship between nature and culture is concerned, participants in these conferences suggest that they are mutually sustaining. Participants do not consider their understanding of nature deficient or inadequate as a means to discuss cultural institutions, nor do they seem compelled to look at nature in a new way to deal with changes in their social reality. In discussions of nature in general and its more specific relationship to the culture of early modern France, the *Discourses* are almost chaotically eclectic, but this very eclecticism bespeaks openness and flexibility.

Notes

1. For an interesting collection of essays discussing some of the historiographical implications of the Scientific Revolution, see David

Lindberg and Robert Westman, eds., *Reappraisals of the Scientific Revolution* (Cambridge: Cambridge University Press, 1990).

2. See, in particular, by Allen G. Debus: *The French Paracelsians* (Cambridge: Cambridge University Press, 1991); *Man and Nature in the Renaissance* (Cambridge: Cambridge University Press, 1978); and *The Chemical Philosophy: Paracelsian Science and Medicine in the Sixteenth and Seventeenth Centuries,* 2 vols. (New York: Science History Publications, 1977).

3. Important additional sources on Paracelsus and van Helmont are by Walter Pagel: *Paracelsus: An Introduction to Philosophical Medicine in the Era of the Renaissance* (Basel: S. Karger, 1958); "The Religious and Philosophical Aspects of van Helmont's Science," *Bulletin of the History of Medicine,* supplement no. 2 (Baltimore: Johns Hopkins University Press, 1944); *Joan Baptista van Helmont: Reformer of Science and Medicine* (Cambridge: Cambridge University Press, 1982); "Recent Paracelsian Studies," *History of Science* 12 (1974): 200–211. By Allen G. Debus: *The English Paracelsians* (London: Oldbourne Press, 1965); "The Chemical Philosophers: Chemical Medicine from Paracelsus to Van Helmont," *History of Science* 12 (1974): 235–259.

4. Carolyn Merchant argues that before the Scientific Revolution, female earth was central to an organic cosmology which was undermined by the Scientific Revolution. As she put it: "Nature was contrasted with art (*techne*) and with artificially created things. It was personified as a female-being, e.g. Dame nature; she was alternately a prudent lady, an empress, a mother, etc. The course of nature and the laws of nature were the actualization of her force. The state of nature was the state of mankind prior to social organization and prior to the state of grace. Nature spirits, nature deities, virgin nymphs, and elemental forces were thought to reside in or be associated with natural objects." *The Death of Nature: Women, Ecology and the Scientific Revolution* (New York: Harper and Row, 1980) p. xxiii.

5. J. B. Morin, Tommaso Campanella, and Etienne de Claves are some of the important contemporary intellectuals who are known to have participated.

6. Because of Théophraste Renaudot's connection with the medical faculty of Montpellier, it is assumed that large numbers of physicians, especially those excluded from Parisian medical practice, offered medical services at Renaudot's clinic and participated in the conferences. Some sources estimate that there were one hundred participants, but provide no evidence for such an estimate. There are a total of 643 topics addressed. The first sixty meetings had paired topics, but subsequent meetings treated one topic. On Renaudot's difficulties with the

Faculty of Medicine, see Howard Solomon, *Public Welfare, Science and Propaganda in Seventeenth Century France: The Innovations of Théophraste Renaudot* (Princeton: Princeton University Press, 1972), pp. 162–200.

7. Reports of these conferences were published each week from 1634 to 1641 by Théophraste Renaudot; the conferences of 1642 were published by his son in 1655. These reports comprise five volumes of approximately five hundred pages each.

8. Although Renaudot's career and biography have been the subject of scholarly interest and attention, the *Discourses* themselves, strangely enough, have been neglected. The best source on Renaudot's career is Solomon, *Public Welfare, Science and Propaganda in Seventeenth Century France.* For other biographies of Renaudot, see Gaston Bonnefont, *Un Docteur d'autrefois* (Limoges: E. Ardant, 1893); *Un Oublie, Théophraste Renaudot: Créateur de la Press, de la Publicité, des Dispensaires, de Monts-de-piété* (Limoges: E. Ardant, 1899); Georges Gilles de la Tourette, *Théophraste Renaudot, d'aprés des Documents inédits* (Paris: E. Plon, 1884); *La Vie et les Oeuvres de Théophraste Renaudot* (Paris: Editions du Comité, 1892); and Albin Rousselet, *Théophraste Renaudot, Fondateur de Policliniques* (Paris: Payot, 1819).

9. Conference 59, part 1, "De la Lumière," *Seconde Centurie des Questions traitées ez Conférences de Bureau d'Adresse, depuis le 3. Jour de Novembre 1634. jusques à l'11 Février 1636. Dédiée à Monseigneur le Chancellier. Avec une Table de Matières.* (Paris: Bureau d'Adresse, 1636) pp. 74–75. Except where indicated, translations from French to English are my own.

10. Conference 2, part 1, "Des Principes et de la Fin de toutes Choses," *Recueil général des Questions traitées es Conférences du Bureau d'Adresse sur toutes Sortes de Materières, par les plus beaux Esprits de ce Temps.* (Paris: Bureau d'Adresse, 1634) pp. 9–17.

11. Conference 9, part 1, "De la Terre," *Recueil général,* pp. 71.

12. Conference 44, part 1, "Comment croissent les Mineraux?," *Recueil général,* pp. 354.

13. Conference 73, "Du Tremblement de Terre," *Seconde Centurie,* pp. 185.

14. Conference 32, part 1, "De la Sympathie & Antipathie," *Recueil général,* p. 258.

15. Conference 43, part 1, "De la Pierre philosophale," *Recueil général,* p. 347.

16. Conference 57, part 1, "De l'Ouye," *Seconde Centurie,* p.58.

17. Carolyn Merchant provides a particularly useful description of the Renaissance cosmos: "the Renaissance shared certain presupposi-

tions about nature. The Renaissance cosmos was a living unit, of which all parts were interconnected in a tightly organized system. The orthodox view inherited from the medieval interpretation of Aristotle was an earth-centered hierarchical cosmos extending upward from the four inanimate elements which were mixed together to form the minerals, vegetables, and animals found in the sublunar region of change, the other unchanging ether-filled spheres of the seven planets, with their associated hierarchies of angels, above the moon. Beyond the planets was the sphere of the primum mobile, source of the daily rotation of the heavens, then the sphere of the fixed stars and zodiacal constellations, and finally the Empyrean heaven of god. Together they comprised a living chain of being, each member a step in a stable, ordered, spherically-enclosed world, each member sharing some particular feature with the steps below and above, yet excelling in some unique characteristic. Man was linked to the animal world below, with which he shared sensation, and to the angels above, with whom he shared rationality. Each part of his body was governed by one of the zodiacal signs, so that as a microcosm, he was a miniature replica of the celestial spheres, or macrocosm." *Death of Nature,* pp. 100–101.

18. Conference 80, part 2, "S'il y a quelque Art de deviner," *Seconde Centurie,* p. 247.

19. For example, the first speaker notes: "So that lying and consequently fables, and all the species thereof, have the same proportion to our understanding as monsters have to nature; our understanding cannot endure them, whereas truth is its sustenance." Conference 283, "Des Fables," *Quatriesme Centurie des Questions traitées aux Conférences du Bureau d'Adresse, depuis le 24 Janvier 1639. Jusques au 10 Juin 1641.* (Paris: Bureau d'Adresse, 1641) p. 411.

20. Conference 10, part 2, "Des deux Frères monstrueux, vivans en un mesme Corps, que se voyent en cette Ville," *Recueil général,* pp. 68–72.

21. Conference 11, part 1, "De la petite Fille velue qui se voit en cette Ville," *Recueil général,* p. 85.

22. Conference 128, "Des Incubes & Succubes & Files Demons peuvent engendrer," *Troisiesme Centurie des Questions traictées aux Conférences du Bureau d' Adresse, depuis le 18 Fevrier 1636. jusques au 17 Janvier 1639* (Paris: Bureau d'Adresse, 1642), p. 87.

23. Conference 48, part 1, "Si tout ce qui nourrit d'Animal doit avoir eu Vie," *Recueil général,* p. 385.

24. Conference 111, part 2, "Quel est le plus communicatif du Bien ou du Mal," *Troisiesme Centurie,* p. 90

25. The possible explanations for the introduction of the form are theological or philosophical. "For, such is the order of nature, that when a subject possesses all of the dispositions requisite for introduction of a form, the Author of Nature, or (according to Plato) the Idea, or that soul of the world (which Avicenna held to be an Intelligence designated to the generation of substantial forms) concurs in the production of the Form." Conference 122, "De l'Origine des Formes," *Troisiesme Centurie,* p. 163.

26. The separation of heat is an explanation for birth marks. "As the degrees of life have dominion over the First Qualities, so they have authority one over another, each in his order. The vegetative life in man makes use of the elementary qualities at pleasure, even to the prejudice of their own nature. So, heat congregates things of the same and separates those of different nature; but our vegetative soul makes it do the contrary; namely, unite the four humours in the veins though different in nature, instead of segregating things: for in this case, heat acts not with full authority but as the soul's officer, following her intentions. And the reason is, because these four humours being ingredients to the nativity of man, they must necessarily pass into his nourishment; which they cannot do without being mingled together . . . so the mother's spirits keep the same course and rule towards the embryo, so that those which served to the mother's touch, go to find that same place in the child's body, there they mark the image which they brought from the brain; nature find ways for her intention where none appear." Conference 121, "D'ou viennent les Marques que les Enfans apportent du Ventre de leur Mère," *Troisiesme Centurie,* p. 159.

27. For a more extensive discussion of Paracelsianism in the *Discourses,* see Debus, *French Paracelsians,* pp. 84–95.

28. Conference 135, "De l'Heur & du Mal-heur, & si les Hommes sont heureux ou malheureux," *Troisiesme Centurie,* p. 214.

29. Conference 56, part 1, "De l'Odorat," *Seconde Centurie,* p. 51.

30. Conference 58, part 2, "De la Peinture," *Seconde Centurie,* p. 71.

31. Conference 131, "Comment se fait l'Accroissement" [second speaker], *Troisiesme Centurie,* p. 199. The fourth speaker also applies this principle to human beings: "wise nature has set herself such bounds as she judged convenient, beyond which most do not grow; which are, for men, between six and seven foot" (p. 200).

32. Conference 118, "Des Maux de la Mer," *Troisiesme Centurie,* p. 143.

33. Conference 126, "Des Causes de la petite Vérole," *Troisiesme Centurie,* p. 179.

34. Conference 351, "Antiperistaze," *Quatriesme Centurie,* p. 351.

35. The second speaker most strongly makes this point: "That those two principles which seem contrary one to the other, are not so, if rightly understood; for when the chemists say, the *similar curantur similibus,* they speak not of diseases, as the Galenists do, with whom they agree, that the same are augmented by the use of resembling things; but of the part diseased which (being the seat of affections against nature) can alone be said to be cured, and not the disease which is only a privation, errour, or disorder of the body . . . which the dogmatists, as well as the chymists, cure by remedies like in substance to the nature of those parts which they strengthen; For what ever is a friend to nature, called by Hippocrates, *marborun medicatrix,* is also an enemy to that which is against nature." Conference 49, part 1, "S'il y a des Remèdes spécifiques à chaque Maladie," *Recueil général,* pp. 393–394. This translation is from the English translation of the seventeenth century.

36. Conference 289, "Si le Fer appliqué sur un Tonneau empesche le Tonnere de corrompre le Vin. Et pourquoi," *Quatriesme Centurie,* p. 436.

37. As the fourth speaker proclaimed in making the comparison, "Nature shows that she is not satisfied with her other productions when she makes other animals propagate by generation; but when she has made a mule, she stops there, as having found what she sought." Conference 46, part 2, "Du Caprice des Femmes," *Recueil général,* p. 375.

38. Conference 139, "Quel est le plus désirable de vivre peu ou longuement," *Troisiesme Centurie,* p. 230.

39. Conference 48, part 1, "Si tout ce qui nourrit d'Animal doit avoir eu Vie," *Recueil général,* pp. 385–390. This is the one discourse which explicitly raises the question of whether nature exists for man's benefit. It provokes vehement opposition. Note for example the objections of the fifth speaker (p. 388): "the life of man cost nature dear, if it must be maintained at the expence of so many other animals' lives. If you say, that being made for man, the greatest happiness that can befall them is to serve him in something be it by the loss of their lives. But it is a fair excuse to cover our cruelty and luxury; to say that Animals are no more proper than Plants to nourish man . . . the similitude of substance is of little consideration; for animals live not of their like, and the Cannibals are ordinarily all Leprous. That a thing may be food, it is sufficient that it have an humidity or a substance proportionate to ours, in whatever order of things soever it be found. And nature has had no less care of nourishing an animal . . . ." This translation is taken from the seventeenth-century English translation by G. Havers.

40. Conference 123, "Si les Maigres sont plus sains, & de plus longue Vie que les Gras," *Troisiesme Centurie,* p. 167.

41. Conference 156, "Si les Hommes se formeroient un Langage n'en ayans point appris d'Autre," *Troisiesme Centurie,* p. 398.

42. Conference 46, part 1, "Du Vuide," *Recueil général,* p. 317.

43. Conference 57, part 2, "De la Harmonie," *Seconde Centurie,* p. 63.

44. Conference 2, "Des Principes & de la Fin de toutes Choses," *Receuil général,* p. 10.

45. Conference 57, part 1, "De l'Harmonie," *Seconde Centurie,* p. 68.

46. Conference 103, part 1, "Du Verre," *Troisiesme Centurie,* p. 26.

47. Conference 40, part 2, "Pourquoi tous aiment-ils mieux commander qu'obéir?," *Recueil général,* p. 321.

48. A more orthodox opinion on this question is expressed by the sixth speaker who claims that man was originally created to lead and thus ambition is just a sign that he is trying to revert to his nature "that man having been created by God for command, as Holy Writ attests, he always retains the remembrance of his origin and would be master everywhere." Conference 40, part 2, "Pourquoi tous aiment-ils mieux commander qu'obéir?," *Recueil général,* p. 322.

49. In Conference 7, part 2, "S'il est meilleur à un Estat d'avoir des Esclaves," p. 54. Participants in this conference express several interesting opinions. For example, one speaker contends that it is less un-Christian to make criminals slaves than to kill them. Another speaker claims that the French should lead in the crusade against slavery because "if there is any place which ought to carry the cause, it is France where a slave from other countries is freed when he sets foot in France." *Recueil général,* pp. 54–57.

50. Conference 37, part 2, "S'il faut toujours dire la Vérité," *Recueil général,* p. 301.

51. Conference 289, "Si le Fer appliqué sur un Tonneau empesche le Tonnere de corrompre le Vin. Et pourquoi," *Quatriesme Centurie,* p. 436.

52. The fourth speaker uses God as the basis of his argument that the moderns cannot be less favored than the ancients. "For God does not create souls now with any less advantages and grace than formerly; he is as liberal of his favours as ever, especially in the ages of grace". Conference 144, "S'il y eu de plus grands Hommes un quelque'un des Siècles précédens qu'en cettui-ci," *Troisiesme Centurie,* pp. 350–351.

53. See Margaret C. Jacob, *The Cultural Meaning of the Scientific Revolution* (New York: Knopf, 1988); and Merchant, *Death of Nature.*

DIONYSIUS LARDNER'S AMERICAN TOUR:

A CASE STUDY IN ANTEBELLUM AMERICAN INTEREST IN SCIENCE, TECHNOLOGY, AND NATURE

Paul H. Theerman

Dionysius Lardner, a British popularizer of science already well-known in Europe, arrived in New York City from France on 29 September 1840. A little over a year later, on 18 November 1841, he started a lecture tour through the United States that would occupy him for the next five years. Starting in New York City, he then lectured in the major East Coast cities of Boston and Philadelphia, and subsequently set out on the southern and western circuits. His tour included Charleston, South Carolina; Mobile, Alabama; New Orleans; Vicksburg and Jackson, Mississippi; St. Louis, Cincinnati, Pittsburgh, "and other Western and Central cities." He concluded by revisiting the East Coast cities where he started.

Lardner's tour was unprecedented and wildly successful. In science lecturing, the country had witnessed nothing remotely similar in the number of cities visited, in audience, and in proceeds. He came close to filling the halls in which he spoke. His last appearance in Philadelphia garnered an audience of 2,000—and he reportedly earned £40,000 for his effort.[1] This was a phenomenal sum, roughly equivalent in American terms to $200,000, in an era when a princely income was $25,000 a year. Lardner capped his five-year tour with a two-volume work, *Popular Lectures on Science and Art,* written and published in 1846.[2] After his American experience, he returned to Paris by way of a similar lecture tour in Cuba.[3] In his remaining years he continued his popular writing; he died in Paris in 1859.[4]

Several motivations lay behind Lardner's tour of the United States. Most immediately, Lardner was fleeing the consequences of an extramarital affair of the late 1830s. The scandal broke in 1840 and forced Lardner and his lover to move to Paris. The removal did not stop the offended man, a cavalry officer, from finding them there and beating Lardner. The captain sued for divorce and won; Lardner was fined £8,000 and found that he could not return to Britain—his friends there censured his actions, thinking them "folly and crime."[5] As a consequence, he and his lover (who eventually became his new wife) set out for the new world to recoup his fortunes.

How did Lardner come to find himself in this position? Born in 1793 in Dublin, he was an 1817 graduate of Trinity College of that city. He took orders as an Anglican priest and then settled in London. He secured a position as professor of Natural Philosophy and Astronomy

P. H. Theerman and K. Hunger Parshall (eds.), Experiencing Nature, 211–236.

at University College in 1827 and occupied this post until 1831. Most likely he owed this position to the aid of Henry Brougham, the powerful Whig politician and supporter of the university. During this same time, Lardner contributed to Brougham's Library of Useful Knowledge, then in the full flower of its production.

While yet a college professor, Lardner had found his true calling in popularization: lecturing on physical science in the different popular lecturing institutions of London. In 1829 he started the Cabinet Cyclopædia, a venture to publish small treatises covering the whole of knowledge. He enlisted eminent men for the project: Sir Walter Scott wrote on the history of Scotland, James MacIntosh on the history of England, Robert Southey on British naval heroes, and Connop Thirlwall on ancient Greece. Lardner wrote or co-wrote many of the volumes in the physical and mathematical sciences: hydrostatics, pneumatics, arithmetic, geometry, electricity, magnetism, and meteorology. Several of these short books became justly famous. J. F. W. Herschel first published his books *Preliminary Discourse on the Study of Natural Philosophy* and *Treatise on Astronomy* as parts of Lardner's Cabinet Cyclopædia series; as did John Stevens Henslow his *Botany.* The series was completed at 133 volumes in 1849.[6] On the basis of the encyclopedia, and of his separate treatise, *The Steam Engine,* Lardner had an international reputation before he reached American shores.[7]

Lardner's American tour provides a way of examining the popularization of science in antebellum America. This paper investigates some of the causes of his popularity; looks at how Lardner may have affected the antebellum American audience; and considers the effect of the tour both on Lardner and on the rise of science popularization generally.

Generating a Lecturing Reputation

The beginning of Lardner's lecturing, in November 1841, showed the immediate success with which he met. His reputation gave him a necessary boost.[8] As reported in the *New York Tribune:*

> Despite the attractions of the Tabernacle [a rival hall] last evening, the Lecture Room of Clinton Hall was filled to overflowing, aisles, galleries, and every other place where footing could be obtained to listen to the Lecture, introductory to his proposed course, delivered by DR. LARDNER. The subject was Astronomy–and it was treated in a manner worthy the high reputation of the lecturer . . . . While the audience was collect-

> ing he exhibited a beautiful mode of the Steam engine, which he said at some future time, if his lectures were deemed worthy of attention, he should explain at length.[9]

A close relationship developed between the lecturer and the newspaper, newly founded and desperate at this time to secure a place in New York's competitive newspaper market. The *Tribune*'s story slipped imperceptibly from the third person of the reportage to the first person of the lecturer, either through a "quick writing" scheme of the reporter, or collaboration with Lardner, or both. Very rapidly, the paper knew it was onto something good. By the third lecture, the newspaper story had moved from page two to page one; the sixth lecture also takes up the front page, filling more than five of the six columns![10] At both the fifth and seventh lectures, the *Tribune* reported that the audience had grown from its previous size.[11]

Lardner worked to increase his showmanship and his draw. Visual aids, in the form of a "moving diorama in which all the stars visible in the firmament should be laid down with their proper magnitude and in their proper positions, together with the double stars, nebulae, &c," diagrams of the moon, and the steam-engine model, were added to the lectures.[12] Towards the middle of the series, Lardner "introduced an innovation" and resolved "to divide his lecture into two parts, or, said he 'as I am upon the stage, I suppose I may say into two acts, devoting them to the consideration of different sections of knowledge.'" The beginning of the evening continued the astronomy series; the second part was on the steam engine, and in between, he showed his models and dioramas.[13] By the twelfth and final night of the series, Lardner—and the *Tribune*—were triumphant. Lardner announced a second series—the two general topics were astronomy and electricity—which ran on eight successive days, excepting Sunday and Christmas, from 21 December. The large audience that concluded his first series, the *Tribune* reported, "have taken tickets for the whole course" of the second.[14] The newspaper promised: "The successive Lectures of the course will be regularly presented to our readers and extra copies of the papers containing them may be obtained at the office."[15] Lardner had enthralled New York, and continued his success in Philadelphia and Boston. By the end of a year's time Lardner was able to start his extended tour of the South and the West.

The Audience

Who heard (and read) Lardner is a difficult question to answer. Certainly the best one can say is that the number was large; for science lecturing, it was precedent-setting. Lardner himself cites 1,200 to 2,000 in the audience per session in the large cities of Philadelphia, Boston, and New York. This is close to the best that one can draw; Lardner approached the stars of antebellum lecturers. By way of comparison, Benjamin Silliman, perhaps the most famous American scientist at the time, gave popular lectures on chemistry in Boston that drew up to 1,500.[16] In 1853, the largest lecturing hall in New York City held 3,000, and only a handful of speakers could fill it.[17] "[S]tars like [Henry Ward] Beecher, [Bayard] Taylor, and [Horace] Greeley" drew 1,500 to 3,000."[18]

The aggregate number to which Lardner lectured may be estimated; it was a significant percentage of the urban population of the country. Lardner lectured several nights or even weeks in succession in the major cities, over the course of a number of nights in the smaller, and probably only a night or two in the intervening towns. His charge is estimated at a quarter per session, as Joseph Henry cites the ticket price as "6d to a shilling sterling per head," which at the prevailing exchange was approximately 25 cents to half of that. Certainly he would not charge less than Peale's Museum, whose contemporary general admission ticket was a quarter.[19] Given the claim that Lardner earned $200,000 over his four years, he then had approximately three-quarters million admissions.[20] Even assuming multiple attendances, it is unlikely that the mean number of sessions per attendee was more than five. One might then estimate that about 150,000 people heard Lardner in the four years that he was on the circuit. This amounted to up to three-quarters of one percent of the estimated United States population during the period of the tour, which was between eighteen and twenty million. However, it comprises upwards of five percent of the city and town population, which stood at about 2.7 million in the mid-1840s.[21] Lardner spoke to a significant number of the Americans.

Lardner's audience was increased by the printing and distribution of the lectures, due in large part to Horace Greeley and the *New York Tribune.* Greeley latched onto Lardner for a variety of reasons. Lardner's emphasis on the beneficent aspects of science and engineering, the spread of public enlightenment and truth through science lectures, fit in well with the newspaperman's reforming spirit; Greeley was a man who flirted seriously with Fourierism in the 1840s, after all. Just beginning the *New York Tribune* in the spring of 1841, Greeley no doubt thought that publishing Lardner's lectures helped to boost circulation in the greatly competitive New York market. The *Tribune* reported verbatim apparently all his lectures in New York City; it had a regular daily circulation of more than 10,000 by the end of 1841, and would grow throughout the '40s.[22] As noted above, Lardner's lectures were the occasions for printing extra numbers of the paper, for sale at the editorial office. Reports of the lectures printed in the *Tribune* were likely picked up and reprinted in other papers around the country, as this was standard practice among newspapers; the *Tribune*'s weekly edition was a national paper.[23] Finally, Greeley and his business partner Thomas McElrath published the collected lectures as *Popular Lectures on Science and Art,* a phenomenally successful book. The first edition of 1846 reached a fourteenth edition by 1852, and a fifteen in 1859.[24] Lardner's sizeable personal audience was thus further increased through print.

Getting a closer look at the audience is difficult; contemporary newspaper accounts do not describe it. One hostile commenter on the tour, Joseph Henry, ranked Lardner's tour with the most popular entertainments, implying that the audience was the most common. In his draft of a letter to English botanist John Stevens Henslow, Henry wrote,

> It is surprising how much noterity . . . the compiler of . . . English popular works gets in this country. Dr. Lardner was before he came to this country a much greater man than Herschell. . . .
>
> I do not [?exult] in the misfortune he has brought on himself but regard him rather as an ob[j]ect of pit[y] than resentment. He has met with no encouragement from scientific men of any standing in our country. He has adopted the position of an itineran[t] lecturer and during the past winter has managed to draw tollerably large audiences in the theaters of New York by occup[y]ing the stage on alternate nights with jugle[r]s and public dancers at from 6d to a shilling sterling per head, and with this as an indication of the state of morals among us I am pleased. He is certainly a very interesting writer and a man of considerable talent but he has sadly been wanting in an essential element of a scientific character[,] a sacred regard to truth. I hold that no person can be trusted as the historian of sci[ence] who could be guilty of the crime of which he is charged.[25]

Henry's stern sense of propriety, offended by the circumstances that brought Lardner to America—and probably by his theatricality and scanty research credentials—led him to disparage Lardner. But though

Henry meant to equate Lardner with jugglers and dancers—and to belittle him as an itinerant lecturer, not one who was invited to give his talk—the simple truth was that he just lectured in public theaters. Lardner's venues were the Clinton Hall, Niblo's Gardens, and Palma's Theaters in New York City; the Chestnut Street Theater in Philadelphia: the Tremont and Melodeon Theaters in Boston; and the St. Charles Theater in New Orleans. In other, smaller cities, Lardner occupied the city theater, invariably called after the city itself; for smaller towns we have no information. The only site which was not a public stage was Peale's Museum in Philadelphia. But, as the chronicler of that institution notes, "Museums were turning into theaters."[26] While the purchase of the Museum building by the American Philosophical Society in August 1841 had restricted its new thousand-seat auditorium to more scientific activities, the previous three years had seen it extensively used for "music and light drama," and "entertainment that combines instruction with amusement."[27] There was a continuum between theater and science in Lardner's tour.

There were few moral qualms about the setting of the lectures, as there were few other appropriate sites available.[28] A major part of the Smithsonian's first public activities comprised public lecturing. But the first lectures of this venerable institution were given at Carusi's Saloon, in Washington, for lack of a better place. The other major scientific lecturing institution of the 1840s, the Lowell Institute at Boston, was specifically prohibited from constructing a building, so it relied on rented quarters, that is, on hiring the Odeon theater at $20 per night.[29]

Horace Greeley, in many ways Lardner's patron, was also a popular lecturer (though he practiced that art to spread his message, not to earn money, as he liked to point out).[30] His audience was likely to have been similar to Lardner's: men and women interested in "progress," attracted by a name, and concerned with keeping current. A contemporary description of Greeley's New York audience survives, colored by the tones of an adulatory biographer, but accurate enough:

> The audience, we observe, has decidedly the air of a country audience. Fine ladies and fine gentlemen there are none. Of farmers who look as if they took the Weekly Tribune and are in town tonight by accident, there are hundreds. City mechanics are present in considerable numbers. An ardent-looking young man, with a spacious forehead and a turn-over shirt-collar, may be seen here and there. A few ladies in Bloomer costume of surpassing ugliness—the costume, not the ladies—come down the steep aisles now and then, with a well-preserved air of uncon-

> sciousness. In *that* assembly no one laughs at them. The audience is sturdy, solid-looking, appreciative, and opinionative, ready for broad views and broad humor, and hard hits. Every third man is reading a newspaper, for they are men of progress, and must make haste to keep up with the times, and the times are fast. Men are going about offering books for sale—perhaps Uncle Tom, perhaps a treatise on Water-Cure, and perhaps Horace Greeley's Hints toward Reforms; but certainly something which belongs to the Nineteenth Century. A good many free and independent citizens keep their hats on, and some 'speak right out in meeting,' as they converse with their neighbors.[31]

If the lectures were not lower-class entertainment, neither did they attract only the upper strata. Lardner's claim that he addressed primarily a professional audience must be dismissed as flattering his listeners. The sheer numbers of people he addressed, plus his venue, made that impossible. That the audience was *more* professionally oriented than the run of population is probably true. That it was overwhelmingly white and middle class was no doubt the case. And though established members of the professions were probably prominent, the audience was more probably younger and more unsettled and unformed professionally; indeed antebellum lectures seem to have been attended by those *wishing* to attain to a professional or quasi-professional status, at a time when the traditional professions were considered "overcrowded" and when science and engineering beckoned as opportunities.[32]

Once the initial hesitation about Lardner's personal life had eased, a good proportion of professional people were probably in attendance. Professional men generally had some background and interest in natural philosophy, the almost exclusive topics of Lardner's lectures. Most professionals were college graduates, and hence most had been through the year-long natural philosophy course that was part of the college curriculum. The connection of medical doctors with science was of course a natural one, but many lawyers also betrayed a strong interest in science, especially natural philosophy. Many of Joseph Henry's students, later members of the bar, wrote that his presentation of the subject, with Henry's emphasis that one only *understood* a subject when one subsumed it under general laws, appealed to and strengthened their understanding of civil as well as natural law. Finally, like lawyers, many clergy were college-educated and the strength of the natural theology tradition was yet sufficiently strong that through interest and background they could be expected to form part of an audience.

From what little is known of the public's attendance at lectures, we do know that even the most popularly oriented science exhibitions and lectures attracted people whose occupations, while not in the traditional professions, granted them some status in life. Robert Collyer's demonstration of the Armstrong hydroelectric machine attracted one of Henry's students, Edward Drayton, as well as Drayton's father, Joseph, an esteemed scientific illustrator and member of the Wilkes expedition scientific corps. Joseph Faber's demonstration of his speaking automaton was attended by a committee of the American Philosophical Society, distinguished citizens all—who nonetheless had a private, not public showing of the machine.

Yet, on the basis of the projected attendance figures, they could not have supplied the bulk of the lecture-goers. That audience was no doubt far broader and reached much farther down into the strata of society. Recent work on an itinerant natural philosophy lecturer in upstate New York, Charles Came, has shown that straightforward electrical experiments, leavened with astronomical lectures and magic lantern shows, could still earn a person a modest but sufficient living well into the 1840s.[33] Even Benjamin Silliman's 1842 Lowell lectures on chemistry were initially disrupted by rowdies in the back balcony. "The fourth gallery—at first filled—is now empty. Those who have fallen off are the lighter sort that attempted to get up a disturbance & having been arrested &c followed up by the police officers many of them have fallen off."[34]

Popularity

Why was Lardner so popular? The question does not admit of an easy answer. Obviously there was a "fit" between the substance and style of his presentations and what the public sought. A partial consideration of the points of contact indicates three factors: the rise of national lecturing; the doctrine of progress; and, partially combining the two, the constructive role of Horace Greeley and the *New York Tribune.*

Most historians of nineteenth-century public lecturing point to the 1840s as the turning point in the organization of the institution. Prior to that time, lectures had been predominantly local affairs, certainly locally booked, but also locally manned and of local influence. The first such were private talks given to select groups: the literary and philosophical society, or a church mission. The Lyceum movement, linked to Josiah Holbrook's founding of the town lyceum in 1826, broadened the offering by increasing the scope of topics and making the lectures public.

This movement grew throughout the 1830s to include library societies, mechanics' institutes, and charitable groups. Though aimed at the general public, they most often relied on local lecturers. The lectures were usually free, the lecturers most often unpaid.[35]

The pattern began to change in the 1840s, and by the middle of that decade, lecturing had changed from being predominantly local to being regional, and in the case of some top speakers, national. Advertisement, sponsorship of a lyceum or other estimable committee, and possible printing of the lecture as a pamphlet, all pointed to the high intrinsic value and moral worth of the public lecture. The lecturing system was, in effect, "a quasi-civil institution."[36] Between 1840 and 1860 it grew to be of enormous size. "By the early 1840s there probably were between 3,500 and 4,000 communities that contained a society sponsoring public lectures. . . . Indeed, by the mid-1840s few northern towns of 1,000 or more people lacked at least one association sponsoring lectures."[37] Lardner came to America just at the point when the lecturing system was broadening to national scope, and he offered himself as a speaker with literally an international reputation. He was eagerly embraced.

What Lardner had to say in his lectures, and how he delivered them, further increased his popularity. Lardner paid careful attention to his presentation, to his theatricality. He had been a part of the London theatrical society in the 1830s, and his English popular lecturing experience had given him definite views on the value of good presentation:[38]

> [T]he lectures as orally given, though similar in substance . . . , differed considerably in form and expression [from the written]. This must necessarily be the case. The oral discourses were strictly extemporaneous, in the only sense in which didactic discourses ever are so. They were delivered from the stage of the theater without reference to any written notes or memoranda. The general outline of the subject, the leading arguments, and the most important illustrations and examples, alone were previously registered in the memory of the speaker. The language in which these were clothed, and the more minute details of the subjects, arguments, examples, and illustrations, were left to the suggestion and inspiration of the moment. Nor was this course adopted merely to save labor, or from any necessity arising from the over-pressure of business. It was pursued because it was found, from long practical experience in public lectures, to be the best.[39]

Lardner thought this the best practice, because it allowed for a "more animated" delivery, best suited for the "character, capacity, and acquirements" of the audience.[40] In his estimation of what constituted good lectures and good lecturing, Lardner discounted the benefits of high intellectual ability alone. Though not "very exalted intellectual gifts," the qualities of familiarity—that is, not exclusive mastery—of the subject, a long experience in teaching, fluency of language, clarity, order, a good voice, and a "manner and appearance . . . exempt from any peculiarities calculated to excite repugnance"[41] were the keys to a good lecture. In fact, he wrote of delivering in one night, successive lectures on "two, three, and sometimes four subjects, . . . having no connexion whatever with each other. Thus astronomy, electricity, and the steam-engine, would be successively noticed"; in this, Lardner likened his practice to "performances in a dramatic theater."[42]

Lardner started from the assumption that the capacities of audiences differed; truth may be one, but its form could be manifold. Lardner gauged his audience and adjusted his presentation accordingly. Though his written lectures were certainly more formal than his apparently lively oral ones, Lardner's style remained clear and direct, in sharp contrast both to the convoluted style that often passed for rhetoric in the early nineteenth century, and to the writing dense with facts and material that often characterized earlier attempts at popularization, such as the tracts of the Society for the Diffusion of Useful Knowledge. In Lardner's lectures, both oral and written, mathematical demonstrations hardly ever made their presence felt.[43] Neither were the lectures *precise.* Lardner always used round numbers—for example using a hundred million miles as the distance of the earth to the sun—saying they "have the advantage of being easily impressed on the memory; and for the purposes of the readers for whose use these volumes are intended, they have all the necessary utility."[44]

Beyond their presentation, the content of his lectures appealed to American audiences. The basis of his English reputation had been "the wonders of science and technology."[45] Lardner believed in progress, certainly scientific progress as expressed in his essays. According to him, progress was inevitable in astronomy (an unremarkable claim at the time), but he also asserted continual advance in the study of electricity. Lardner believed and maintained that progress in knowledge implied technological and moral progress as well. These he presented with an emphasis on self-education and moral improvement that arose from his early days as a writer for the Society of the Diffusion of Useful Knowledge, and as a lecturer for the London Institution, the London Mechanics' Institution, and elsewhere. But "[t]he tide of self-improvement and self-education that was so strong in the 1820s and 1830s was already ebbing, . . . and so Lardner was left high and dry" by the '40s.[46] Lardner's move to America, prompted by circumstance, turned out to be fortuitous. To continue the metaphor, that tide of scientific and technological wonder, was cresting in the United States in the 1840s. As Joseph Henry wrote in 1849: "From Buffalo to Albany I passed over a line I had not traversed before since 1826 and then the travel occupied me 15 days in a canal boat. I now passed over the same distance in 15 hours. This is the go a head age—"[47] The electromagnetic telegraph, the steam locomotive, and other evidences of improvement prompted such oratory. Further, science possessed a particular cachet for the 1840s American: it was essentially *useful* knowledge that lecture-goers could employ in their attempts to "give them the hold on life that their aspirations seemed to require."[48] In ways that had passed in Britain, science lecturing promised benefits of intellectual and moral growth that could lead to useful work—and perhaps to increased social position. It was small wonder that social reformer Robert Dale Owen pushed to use the bequest of James Smithson to support lectures on useful topics, and to publish them subsequently as "tracts" (although his design was never brought to fruition.) Lardner's lectures, tailored to the common man, met this need magnificently.

Lardner started his American career with surefire topics: astronomy, the steam engine, and electricity. But in the full repertoire of his lectures, his topics mostly followed the basic natural philosophy course as taught throughout Britain and the United States in the early nineteenth century. Lardner spoke on matter and its properties; on mechanics (including hydrostatics, hydrodynamics, and pneumatics); on physics—light, heat, electricity, and magnetism; on astronomy; on meteorology; and on the technical arts—though with the exception of the discussion of some simple machines such as the lever and the pulley, he limited himself to the steam engine, which he presented in great detail. Nonetheless, Lardner's audience—and more so his readers—could justly expect to have the basics of physical science explained to them.

Beyond the natural philosophy course, which comprised the subjects of about half of the sixty printed lectures, Lardner also lectured on other, supplementary topics of a more particular nature. In this he discussed nature as it related most directly to his audience: the weather and the heavens. Lardner spoke on almanacs, the relationship of the moon and weather and other purported "lunar influences," popular fallacies in science and nature, the return of Halley's comet, and the ongoing discussion on the plurality of worlds, that is, whether there had been other creations throughout the universe, and, if so, what this implied for Christian doctrine. Much of Lardner's work was new on the American scene. For example, he apparently was the first to present Karl Mädler's phenomenal lunar map, published in 1837, to American audiences. His essay on Halley's comet came from an *Edinburgh Review* article of 1833, the year of the comet's return, but the topic was not too stale by the early 1840s. Lardner's discussion of the plurality of worlds came in the wake of the celebrated dispute in the 1830s between William Whewell and David Brewster on that topic. In appearance, if not always in fact, Lardner gave his audience the sense that he was presenting both the full and basic principles of physical science, and its latest discoveries and conclusions.[49]

Lardner's appeal, then, came from his presentation and from his subject matter, but also from what might be called style, or the overtones of his presentation. First of all, Lardner's lectures allowed the audience not only to learn physics, but to identify with the respected figure of the natural philosopher. And as both privileged interpreter of nature and source of material and moral progress, the scientist possessed inherent authority, which Lardner insisted on claiming.

Lardner presented himself as an authentic man of science. On the title page of *Popular Lectures on Science and Art,* Lardner's name was followed by four lines of small type listing his academic honor and affiliations. Lardner portrayed himself as one who had expert competent training in the physical sciences and experience in lecturing on them. He contrasted himself to American lecturers, who he said were selected for their public *social*—not intellectual—standing, and who composed their lectures not from their experience, but rather from their libraries, with scissors and paste-pot:

> The real fitness of the individuals by their talents, acquirements, or habits, to fulfil the duty of a public instructor is little regarded. But the title of the Honorable A. B., senator from the state of C. D., or, if senators cannot be found, the Honorable E. F., member of the house of representatives, is regarded as a qualification of the first order. In any case an honorable is most important. . . . When the important evening, in the fulness of time, arrives, the lecturer is ushered in solemn form by the members of the committee to the pulpit, where a decanter of water, a glass goblet, and a pair of wax candles, are duly provided, and the members of the society are entertained for an hour and one half with selections from Boswell's Life of Johnson, in the formation of which the use of the scissors bears an unconscionable ratio to that of the pen.[50]

Lardner took pains to promote the natural philosopher as a figure of considerable authority and standing, based on personal moral qualities. Again and again Lardner contrasted the irrational desires of the populace with the rational and calm calculations and considerations of the scientist. This came out strongly in his treatments of weather almanacs and other popular fallacies. Lardner's discussion showed that the predictive power of a prominent almanac by a Mr. P. Murphy was worse than chance alone would have provided.[51] Yet, he complained, "This is the production for which the demand was so urgent, and for which the public impatience was so irrepressible, that the shop of the bookseller, like those of bakers in a famine, was obliged to be protected by the police, so violent was the demand of the thousands who flocked to obtain it."[52] In connection with these phenomena, Lardner mentioned four other instances when an incredulous public almost rioted over the interpretation of nature. In 1832 arose the scare that Biela's comet was going to hit the earth; Arago had to calm the fears in France.[53] A London "water panic" scared the inhabitants of the city that their water was no longer healthy. "It was in vain that Brande analyzed the water at the Royal Institute, and Faraday attempted to reason London into its senses."[54] Further fears of adulteration successively fixed on white bread and on the air.[55] For Lardner, the world was virtually turned upside down. "Knowledge ceased to be power; philosophy lost its authority."[56] He called on the public to trust the scientist:

> [The weather] is a problem to which, hitherto, scarcely an approximation has been made, even by the most eminent natural philosophers; and, as it is one of the details of which the public in general cannot be expected to understand, they can only regulate the confidence which they will place in its pretended solutions by the reputation and authority of those who propound them.
>
> Who, then, it may be asked, are the persons that put forth those predictions; and on what grounds do they ask the faith of the public? Among these prognosticators, is any name found holding a respectable rank in the community of science? What have the labors and researches of these persons contributed to the actual advancement of our knowledge of nature? What are the works on which their reputations are founded? Do these weather-prophets possess any of the recognized qualifications, founded on education and previous attainments which would fit them for encountering such a problem? What learned societies in Europe have these pretenders enriched by their labors?

> Where are the transactions in which their investigations and discoveries have appeared? These questions would be answered by a mere enumeration of their names—names utterly unknown in philosophy or letters. It would be answered that among them there is found not one individual whose presence would be tolerated in any scientific reunion in Europe. Such are the class of persons to whom the public, in the contemptuous silence of the great leaders and guides of science in every part of the world, surrendered their faith.[57]

In contrast, he invoked a scientific methodology that said that science is rational calculation, no different in kind from the types of calculation that guided—or should guide—daily life. For example, when talking about the plurality of worlds, Lardner supported the likelihood of multiple creations of life with an appeal to methodology. Science, he said, proceeds by means of the probabilistic extension of fruitful analogies: "Modern discovery has collected together a mass of facts . . . which forms a body of analogies bearing on this inquiry even more cogent and convincing than the proofs on the strength of which we daily dispose of the property and lives of our fellow-citizens, and hazard our own."[58] In thus glorifying and idealizing the scientist, Lardner provided an opportunity for his audience to identify with him. Americans coming to the lectures for self-improvement could see the social authority that accrued to the man of knowledge, and which could begin to accrue to themselves.

Yet there was an inherent disjuncture between this public face of science—stressing sobriety, morality, care, and the primacy of reason—and its presentation. The latter relied on drama, fortune, amazement, and astonishment. In Lardner's lectures, this dichotomy came together through the themes of genius, chance, providence, and the marvelous.

"Lardner's writings expressed his child-like delight in scientific wonders."[59] This came out especially with respect to electricity and with steam, the two new motive powers of the 1830s.[60] Lardner here came close to some very real constraints against the charge of impiety. There was a centuries-old concern that a fascination with the creation would tend to lead the Christian to lose sight of the Creator.[61] Lardner evaded this in a characteristic way. The natural theology tradition, that the rational consideration of the creation would lead one to knowledge of the Creator, played only a small part in his work. Instead, Lardner stressed the emotional response that nature could provide. His discussion of Laplace's nebular hypothesis, which had a reputation of leading to atheism, stressed that the ideas did not challenge the supremacy of God, but rather made one marvel at the wonderful secondary causes by which He carried out His creative commands.

Wonder and amazement pervaded the scientific process as well. Rational calculation could only work on facts, but *discovery* was presented as a dramatic event, in which *happenstance* governed. What would have been the result, Lardner queried his audience, if Franklin, so gifted in many areas, *had not* directed his talents to the study of electricity; if Galvani *had not* arranged the frog leg on the metal rail; if Davy *had not* had access to the powerful electro-chemical battery of the Royal Institution? Clearly the understanding, even the perception, of fundamental powers of nature would be lacking, as well as their attendant technological benefits. Without Franklin's having turned his attention to electricity, for example, the world would not know about lightning protection; without Davy's fortuitous work on electrochemistry, there would be no electroplating.

Happenstance, wonder, and amazement provided a dramatic frame for Lardner's presentation. Yet they posed problems also. Not only were they in sharp contrast to the calm rationality otherwise ascribed to legitimate scientists, but making science rely on chance could undermine the belief in progress. Steady improvement, one of the staples of the lectures, seemed at odds with the workings of genius. Lardner resolved this problem by claiming that chance was not really chance at all, but the working of Providence, "the prompting of HIM, whose will is that intellectual Progression shall be incessant."[62] Providence guaranteed that one could reasonably be sure that science would continue to uncover fundamental entities and powers and fruitfully exploit them. Science still operated by means of men of genius, a commonplace of his time, but Providence insured that genius, itself the imperfect reflection of the creative act of God, would be manifested at the proper time.

> If the supply of capital facts be occasionally due to chance *or to the Being by whom what is miscalled chance is directed,* it is to the operation of the faculties of exalted minds that the development of the laws of nature is due: if rude lumps of the natural ore of science be now and then thrown under the feet of philosophy, the discovery of the vein itself, its depth and direction, its quality and value, the separation of the precious metal it contains from its baser elements, the demonstration of its connexion with the phenomena of nature, and its adaptation to the uses of life, are all and severally the work of that noble faculty of intellect, that image of his own essence, which the

> Creator of the universe has impressed upon man, and which is never more worthily exercised than in the investigation of those laws of the material work, in all of which, whether they affect the vast bodies of the universe, or the imperceptible molecules of those around us, *there is ever conspicuous a provident care for the wellbeing of his creatures.*[63]

In connection with this, wonder at the spectacle of nature is changed into pleasure at its contemplation, thus elevating it and making it acceptable. Lardner evidently wanted to evoke the affective possibilities of science in the quotation from Cicero which prefaced his volumes. But rather than evoking stoic equanimity, Lardner alluded to Christian love. Science provided deep intellectual pleasure, not dispassionate quiet: "It is in the perception of truth itself—in the contemplation of connected propositions leading by the mere operation of the intellectual faculties, exercized on individual physical facts, to the development of those great general laws by which the universe is maintained—exalted pleasure, compared with which the mere attainment of convenience and utility in the economy of life is poor and mean."[64] Amazement, wonder, genius, chance—the contingencies of life which provided the drama that Lardner's setting prompted and his success well-nigh demanded—these were used, channeled, and tamed by subsuming them under the characters of Providence and Christian love. In doing so, science retained the moral and intellectual authority that it would remain the foundation of progress, both moral and material. By being founded on such Christian concepts, science could not be called into question without showing an appalling lack of faith in God's goodness.

The Outcome of the Tour

Lardner's new career defined the rest of his life's work. Certainly this career was foreshadowed in his earlier editorial work on the Cabinet Cyclopædia (which indeed continued during his American years) and in his lectures for the London Institution and elsewhere. However, the American tour focussed Lardner's efforts into science writing alone, away from the wider bounds of the Cabinet Cyclopædia, and away from his role as editorial impresario. More specifically, the two-volume *Lectures on Science and the Arts* became—spectacularly—the basis for his (initially anonymously presented) twelve-volume set, *The Museum of Science and Art.*[65] Large sections from one appear almost verbatim in the other, expanded and updated. In this work, Lardner also went on to other topics in science: the telegraph, animal instinct, and geology. Lardner printed large parts of the *Museum* under his own name, and the whole agglomeration of books became the form in which his work was most widely distributed to the British.

Revision, republication, and appearance as the textbook were the final incarnations of the literary remains of Lardner's American tour.[66] Lardner produced the *Handbook of Natural Philosophy and Astronomy,* first in five volumes between 1851 and 1853, and then independently as all possible subsets of the material.[67] Then came *Natural Philosophy for Schools, Animal Physiology for Schools* (based on a more general popular book of the subject), and *Chemistry for Schools.*[68] For the rest of his life, Lardner wrote for a living, interspersing a few scientific papers and a fine review of the Crystal Palace exposition of 1851, among repetition upon repetition of scientific subjects.[69]

As a model for the career of scientific journalist, though, Lardner had only a modest immediate influence. On the other hand, he was a leader in exploiting the American lecturing scene and showing the way for others. After Lardner, the number of well-known European scientists lecturing in America increased markedly.

Europeans had often lectured in the United States, but prior to Lardner, no one of note had come. Instead, most had been British itinerants, or instrument makers and apothecaries, who offered lectures on the side to supplement their income. One prominent one was George Chilton (1797–1836) of New York City, who emigrated to the United States in 1797 at age thirty. His chemical, astronomical, and natural philosophical lectures in New York and Newark, New Jersey, in 1803 to 1806 made his reputation and secured his instrument-making and paint and pigment business.[70] Early nineteenth-century science education for women, who were generally debarred from a college education, also provided an audience for itinerant lecturers.[71]

The itinerant foreign lecturer continued through the 1840s. Joseph Faber, a German who had developed a speaking automaton, came to ruin not once, but twice on American lecture tours, in 1844 and 1845-–1846, with rescue only coming when P. T. Barnum purchased his machine.[72] In 1845 Robert Collyer, an English lecturer, demonstrated in America William Armstrong's "hydroelectric machine," a device that generated static charges by steam friction. Collyer's lack of success eventually forced him to ask Joseph Henry for help, proposing that Henry offer lectures on Collyer's behalf and in return receive access to the hydroelectric machine.[73]

After Lardner's tour, however, a stream of prominent European scientists came to America to lecture. Charles Lyell toured and lectured twice (1841–1842 and 1845–1846), as did William Scoresby, the arctic explorer (1844 and 1848). J. W. F. Johnston, the agricultural chemist, and John Pringle Nichol, the Glasgow astronomer and popular lecturer, each undertook the tour once (in 1848 and 1849, respectively).[74] Lyell and Johnston delivered the lucrative Lowell lectures, as did the Swiss scientists Louis Agassiz and Arnold Guyot. Agassiz followed his 1846 outing with a successful grand tour through New York, Charleston, and Washington in the succeeding years.[75] Guyot, the meteorologist and geographer, saw his lectures translated from French by Harvard classicist C. C. Felton, published in the Boston paper, and afterwards edited into a popular volume.[76]

For naturalists and geologists such as Lyell, lecturing was a supplement, perhaps more for honor than for money, on a trip to observe and collect in the new world. But the pecuniary reason was the determining factor in Lardner's tour, and also in others'. Agassiz and Guyot used lecturing to establish themselves in America before a regular appointment secured their estates. Nichol took up the lecture tour in an effort to regain solvency, after impoverishing himself equipping his Glasgow observatory.[77]

Lardner's tour showed that fame and money were possible on the lecture circuit. Certainly the Lowell Institute lectures in themselves were a powerful incentive, and having contacts in America helped. Yet the pattern that Lardner had established was already there, and few European scientists in America failed to offer popular lectures, if necessary in parallel with the specialized ones to select audiences. This, for example, was Agassiz's pattern. Evidence that the popular European lecturer was prevalent and desired in America by the end of the decade comes from Joseph Henry. He always tried to promote the native scientist, considering many Europeans better rhetoricians than scientists—and no doubt had Lardner in mind. Yet in his private memorandum on setting up the Smithsonian's scientific lecture series, which began in 1849, he nonetheless acknowledged the Europeans' prestige and draw: "I shall mainly depend upon the lecturers from abroad to give the popular exhibitions. . . . [I will call] to Washington [American] men distinguished for their original researches not professed lecturers. Though by this course we may not suceede in amusing yet shall we not fail to instruct our audience."[78] The mystique of the European lecturer—formed in great part by the notorious Dionysius Lardner—was strong in antebellum America.

Finally, what longer term effects might the lecture tour have had? As just noted, the tour led to a spate of books by Lardner, and a whole series of American lecture tours by Europeans. While he and his tour were not the only cause, they nonetheless strongly contributed to the formation of a new career, the scientific man of letters.

Though forced into this career by circumstance, Lardner nonetheless showed that it was possible to make a living, even a handsome living, by lecturing and writing for the public on topics of science alone.[79] Beyond this, Lardner managed to keep a level of status and respectability in doing so. Lardner's career was an innovation, and contrasts with those of contemporary scientific writers for the populace. He was not independently wealthy, as were J. F. W. Herschel and Charles Babbage. He had no remunerative position on which he might rely, as did Faraday, Arago, Liebig, Nichol, and other men who wrote or lectured for a popular audience. In other words, his public career was defined neither by wealth or position. He was not a woman, like the famous popularizers of science, Jane Marcet and Mary Somerville, with the attendant loss of status that that gender implied, but which in turn sanctioned popularization as a suitable occupation. Similarly, Lardner did not come out of the traditions of the instrument maker or the itinerant physical demonstrator. Both these occupations had well-established identities by the 1830s, going back to the late seventeenth century. Men in both these professions lectured and, in some cases, notably John Desaguliers and Benjamin Martin, published their lectures. But by the early nineteenth century, the man of science had become distinct from the itinerant and the instrument maker, and the trade and manual connotations of these professions did not stick to Lardner. While there appeared to be some hesitation about Lardner's career soon after his scandal—Joseph Henry referred to him as an itinerant lecturer, as noted above—Lardner's innovation was in being able to regain his status, and to make popularization an honorable, full-time occupation for a educated man.

The contemporary figure that comes closest to this ideal of the scientific man of letters was David Brewster. Yet the contrast with Lardner is telling. Brewster also lived for many years as a scientific writer and editor. His reputation as a popular science writer was considerable. Through his books on the kaleidoscope and the stereoscope, he gained an worldwide following; his biographies of Galileo, Tycho, and Kepler found an international printing, while he was the acknowledged master biographer of Newton in the nineteenth century. Brewster was a prolific and widely read reviewer in Victorian periodicals.[80] But in contrast to Lardner, Brewster *desired* an academic career, he did not give one up. He stayed an active optical researcher and professional organizer, and did not define himself and his career by his writing and editing. Furthermore, by all accounts, he was a miserable lecturer![81]

Brewster was the master of one kind of popularization, that aimed at the highly educated audience that wished to keep current with the ideas and controversies, both scientific and professional, of the scientific community. Lardner, by contrast, was the master of a quite different sort of audience, that of the middling sort of people, who thought that familiarity with the basic concepts of science was something worth acquiring.[82] In this way, Lardner broke with the tradition of popularization that included—as well as Brewster—John Herschel, Charles Babbage, and François Arago; rather he initiated a style of popularization that would come to include Thomas Henry Huxley, Herbert Spencer, and the American editor and popularizer Edward Youmans.

A new professional role and a new type of audience were the social aspects of the change in popularization. The form or genre of popularization also changed. The essence of this new style of popularization was the creation of the popular scientific essay, that is, the adapting of the high-style essay form for a popular audience. Prior to Lardner, written popularization for the moderately educated usually took the form of a dialogue, meant to be read aloud, or a textbook, to be studied silently. The essay form in science was used for the review literature; it was aimed (as mentioned above with respect to David Brewster) at the more highly educated. Lardner bridged the divide. To a general audience, Lardner presented lectures that consisted of short essays on the basic principles of physical science, and of interesting, informed accounts of the physical world, such as on Halley's comet or life on other planets. But while essayistic in form, they were conversational in style: he based his writing on his lecturing. In this way he helped to form a new style of science presentation that takes us from the early nineteenth century up to Thomas Henry Huxley and his "secular sermons" on science for the populace.

Thus Dionysius Lardner's American tour had repercussions at many different levels. His wide-spread lecturing and subsequent publication reached throughout the country, satisfying a desire to learn more about science. The lecture tour became an exemplar for similar efforts; it virtually opened up the "American market" to European lecturers of note. And the career that Lardner forged out of his American experience, that of scientific man of letters, resulted in an new genre, the popular scientific essay, and a new mode of interaction between scientist and populace, the professional popularizer.

Notes

1. *Dictionary of National Biography,* s.v. "Lardner, Dionysius."

2. Dionysius Lardner, *Popular Lectures on Science and Art,* 2 vols. (New York: Greeley & McElrath, 1846). This was preceded by revised offprints of newpaper accounts, for which see note 15, and followed by many subsequent editions, for which see note 24. Unless otherwise indicated, all references are to the first edition.

3. Ibid.; the article does say "reportedly."

4. Ibid.; J. N. Hays, "The Rise and Fall of Dionysius Lardner," *Annals of Science* 38 (1981): 527–542.

5. William Charles Macready, *The Diaries of William Charles Macready, 1833–1851,* 2 vols. (1912; reprint ed., New York and London: Benjamin Blom, 1969), 2:53; entry for March 21, 1840.

6. *Dictionary of National Biography,* s.v. "Lardner, Dionysius"; Morse Peckham, "Dr. Lardner's Cabinet Cyclopædia," *Papers of the Bibliographical Society of America* 45 (1951): 37–58.

7. Hays, "Rise and Fall," p. 258. The *British Library Catalogue of Printed Books* lists several editions, under varying names, from 1828 through 1851.

8. One contemporary comment on what makes a successful lecturer amid the avid lecturing climate of the early 1850s was "Reputation, it has been discovered, will '*draw*'. Reputation *alone* will draw. That airy nothing is, through the instrumentality of the new institution [of lecturing], convertible into solid cash, into a large pile of solid cash." J. Parton, *The Life of Horace Greeley, Editor of the New York Tribune* (New York: Mason Brothers, 1855), p. 327.

9. *New York Daily Tribune,* 19 November 1841, p. 2, reporting lecture of 18 November 1841.

10. *New York Daily Tribune,* 6 and 14 December 1841.

11. *New York Daily Tribune,* 10 December 1841, p. 4; 15 December 1841, p. 2.

12. *New York Daily Tribune,* 15 December 1841, p. 2.

13. *New York Daily Tribune,* 6 December 1841, p. 2.

14. *New York Daily Tribune,* 21 December 1841, p. 2.

15. *New York Daily Tribune,* 21 December 1841, p. 2. Both courses of lectures were also presented in pamphlet form by the *Tribune:* Dionysius Lardner, *Course of Lectures* (New York: Greeley and McElrath, 1842), and idem, *The Second Course of Dr. Lardner's Lectures on the Sun, Comets, the Fixed Stars, Electricity, Light and Sound, Steam Navigation, &c., &c., rev. & corr.* (New York: The Tribune Office, 1842).

16. Margaret W. Rossiter, "Benjamin Silliman and the Lowell Institute: The Popularization of Science in Nineteenth-Century America," *The New England Quarterly* 44 (1971): 602–626, p. 617.

17. Parton, *Life of Greeley,* p. 329.

18. Donald Scott, "The Popular Lecture and the Creation of a Public in Mid-Nineteenth Century America," *The Journal of American History* 66 (1980): 791–809, p. 800.

19. Charles Coleman Sellers, *Mr. Peale's Museum: Charles Willson Peale and the First Popular Museum of Natural Science and Art* (New York: W. W. Norton & Company, Inc., 1980), p. 274 ill.

20. Though Lardner's earning are large, they do not seem to be impossible or inflated. An English actor and friend of Lardner's, William Charles Macready, tallied that he had earned £5,500 "clear of expenses" during exactly one year on the stage in America in 1843–-1844. This was approximately $27,500. Macready, *Diaries,* 2:275, entry for September 25, 1844. Compare the contemporary statement about author William Makepeace Thackeray, who "crosses the Atlantic, and, in one short season, . . . gains thirteen thousand [dollars], and could have gained twice as much if he had been half as much a man of business as he is a man of genius." Parton, *Life of Greeley,* p. 327, probably referring to Thackeray's American tour of 1855.

21. U.S. population for 1840, 17,069,453: U.S. Bureau of the Census, *Historical Statistics of the United States: Colonial Times to 1970,* 2 parts (Washington: Government Printing Office, 1975), 1:8. The estimate for 1842–1845 is 18 to 20 million; ibid., p. 8. *Urban* (as opposed to rural) population was about 2.7 million in mid-1840s; ibid., p. 12.

22. Glydon G. Van Deusen, *Horace Greeley: Nineteenth-Century Crusader* (Philadelphia: University of Pennsylvania Press, 1953), p. 54. The *Weekly Tribune* had a circulation of 15,000 by the end of 1841 (ibid.) and, according to Scott, a press run of 175,000 by the 1850s. Scott, "Popular Lecture," pp. 798–799.

23. Scott, "Popular Lecture," p. 799 n. 30, states that "The role of the *New York Tribune* and other metropolitan dailies . . . in the emergence of a national communications system was enormous." The *Tribune* had a special section, called the "Sketches of Lectures," that reported New York City lectures. This was often reprinted in newspapers all around the country. Ibid., p. 798.

24. Dionysius Lardner, *Popular lectures on Science and Art,* 14th ed., 2 vols. (New York: Greeley and McElrath, 1852); 15th ed. (New York: Blakeman and Mason, 1859).

25. Joseph Henry to John Stevens Henslow, 5 July 1842, draft, Henry Papers, Smithsonian Archives, printed in *The Papers of Joseph Henry,* ed. Nathan Reingold, Marc Rothenberg, *et al.,* 6 vols. to date (Washington, D.C.: Smithsonian Institution Press, 1972–), 5:244–245.

Part of Henry's disdain arose from personal antipathy. At a meeting of the British Association for the Advancement of Science in 1837,

Lardner challenged Henry's recounting of the speeds of American steam boats, steam navigation being a subject on which Lardner claimed a special knowledge. The encounter created a minor political disturbance at the meeting; Henry never forgot the personal and national slight. *Joseph Henry Papers,* 3:508; 4:100–101; 5:158, 244–245.

26. Sellers, *Mr. Peale's Museum,* p. 281.

27. Sellers, *Mr. Peale's Museum,* pp. 281, 280, 288.

28. Lawrence W. Levine also notes that "America in the first half of the nineteenth century did experience greater cultural sharing [between high- and lowbrow] in the sense that cultural lines were more fluid, cultural spaces less rigidly subdivided than they were to become. "William Shakespeare and the American People: A Study in Cultural Transformation," *American Historical Review* 89 (1984): 34–66, quotation on p. 65. See also the same author's *Highbrow/Lowbrow: The Emergence of Cultural Hierarchy in America* (Cambridge, Mass.: Harvard University Press, 1988).

29. Rossiter, "Silliman and the Lowell Institute," p. 206 n. 24.

30. Henry Luther Stoddard, *Horace Greeley: Printer, Editor, Crusader* (New York: G. P. Putnam's Sons, 1946), pp. 137–138.

31. Parton, *Life of Greeley,* p. 330.

32. Consider Joseph Henry's advice to an unknown young man seeking career counseling in engineering: "I think you have made a good choise and provided you can persever in the course you are now in there is every probability that you will become an important and useful man. The professions of Law and Medicine are so much crowded that provided a person has the proper talents for the business of practical engineering I think his chance of success is greater in this line than in either of the others." 15 March 1845, retained copy, Henry Papers, Smithsonian Archives, printed in *Joseph Henry Papers,* 6:246–250, quotation on pp. 246–247. See also Scott, "Popular Lecture," pp. 795–798, for his assessment of the audience, on which my account is partially based.

33. Roger Sherman, "Charles Came, Itinerant Science Lecturer, and His 'Splendid Apparatus,'" *Rittenhouse* 5 (1991): 118–128.

34. Silliman's "Personal Notices" for 19 March 1842, quoted in Rossiter, "Silliman and the Lowell Institute," p. 612.

35. Scott, "Popular Lecture," p. 792.

36. Ibid., p. 793.

37. Ibid., p. 792.

38. Hays, "Rise and Fall," pp. 528–529.

39. Lardner, *Popular Lectures,* 1:14.

40. Ibid., 1:16.

41. Ibid., 1:15–16.
42. Ibid., 1:17.
43. In another work, *A Manual of Electricity, Magnetism, and Meteorology,* compiled and edited by Charles V. Walker for his Cabinet Cyclopædia, when he introduced analytic mathematics at one point, he apologized to his reader. See Hays, "Rise and Fall," p. 531.
44. Lardner, *Popular Lectures,* 1:20.
45. Hays, "Rise and Fall," pp. 533ff.
46. Ibid., p. 541.
47. Henry to Alexander Dallas Bache, 23 October 1849, draft, Bache Papers, Smithsonian Archives.
48. Scott, "Popular Lecture," pp. 797, 801–802, quotation on p. 801.
49. However, much of Lardner's other material was dated. Though he claimed that the written essays were mostly prepared fresh for the occasion, (*Popular Lectures,* 1:18), he certainly based them on older material. Some of the lectures seem to be derived from lectures from the 1820s, no doubt from Lardner's days as a professor at University College, London. When discussing topics in physics proper, he often took an historical approach. His discussion of electricity never mentioned any material dating past 1800; for galvanism, none past 1820; and for electromagnetism, not past 1830. Though the lectures were collected and first published in 1846, they contain nothing on the telegraph, neither S. B. F. Morse's recent successes, nor Charles Wheatstone's system of the previous decade.
50. Lardner, *Popular Lectures,* 1:10.
51. Ibid., 1:160–165.
52. Ibid., 1:164.
53. Ibid., 1:159.
54. Ibid., 1:160.
55. Ibid., 1:160.
56. Ibid., 1:160.
57. Ibid., 1:165–166.
58. Ibid., 1:52.
59. Hays, "Rise and Fall," p. 533.
60. Ibid.
61. For an introduction to this theme, see Richard Olson, *Science Deified and Science Defied: The Historical Significance of Science in Western Culture* (Berkeley and Los Angeles: University of California Press, 1982), pp. 160–164; and A. C. Crombie, *Medieval and Early Modern Science,* 2 vols. (Garden City, New York: Doubleday Anchor Books, Doubleday & Company, Inc, 1959), 1:15–16.
62. Lardner, *Popular Lectures,* 1:103–104.

63. Ibid., 1:363, italics supplied.

64. Ibid., 1:119.

65. Dionysius Lardner, *The Museum of Science and Art,* 12 vols. as 6 (London: Walton and Maberly, 1855–1859).

66. Compare the fate of Jane Marcet's works as textbooks, M. Susan Lindee, "The American Career of Jane Marcet's *Conversations on Chemistry,* 1806–1853," *Isis* 82 (1991): 8–23.

67. Dionysius Lardner, *The Handbook of Natural Philosophy and Astronomy* (London: [Walton and Maberly,] 1851–1853).

68. Dionysius Lardner, *Natural Philosophy for Schools* (London, 1857); *Animal Physiology for Schools* (London: Walton and Maberly, 1858), based on *Animal Physics: or, The Body and Its Functions Familiarly Explained* (London, 1857); *Chemistry for Schools* (London, 1859).

69. Dionysius Lardner, *The Great Exhibition and London in 1851* (London, 1852). Although I have made a case for Lardner as a scientific man of letters, the *Dictionary of National Biography* reports that he was also reputed to be the Paris correspondent of the London *Daily News.*

70. "Obituary—the late Mr. George Chilton," Silliman's *American Journal of Arts and Sciences* 31 (1836–1837): 421–424; Rita Gottesman, ed., *The Arts and Crafts in New York* (1938; reprint ed., New York: Da Capo Press, 1970), pp. 384–385.

71. See Deborah Jean Warner, *Graceanna Lewis: Scientist and Humanitarian* (Washington, D.C.: Smithsonian Institution Press, 1979), pp. 10–11, and her "Science Education for Women in Nineteenth-Century America," *Isis* 69 (1978): 58–67. I am grateful to Ms. Warner for pointing out this material.

72. For an account, see the *Joseph Henry Papers,* 6:361–363 and *Proceedings of the American Philosophical Society* 4 (1843–1847): 83–84, 422.

73. *Joseph Henry Papers,* 6:320, 324.

74. For recent work on Nichol and his popular lecturing, see Simon Schaffer, "The Nebular Hypothesis and the Science of Progress," in *History, Humanity and Evolution: Essays for John C. Greene,* ed. James R. Moore (Cambridge: Cambridge University Press, 1989), pp. 131–164, especially p. 150.

75. Rossiter, "Silliman and the Lowell Institute," p. 609.

76. Arnold H. Guyot, *The Earth and Man,* trans. C. C. Felton (Boston: Gould and Lincoln; and New York: Sheldon Blackenor and Co., 1856).

77. Schaffer, "Nebular Hypothesis," p. 150.

78. Joseph Henry Desk Diary entry for 6 May 1849, Henry Papers, Smithsonian Archives.

Much as the Smithsonian bequest was considered an endowment to support American science publishing, so it was considered an endowment for American science lecturing. According to the plans of Robert Dale Owen, who proposed much of the early legislation and was an early active regent of the institution, the Smithsonian was to sponsor lectures on science and useful subjects, and to provide for their publication as tracts. There was an active lecture program prior to the Civil War, but the lectures paid relatively little. Furthermore, only some of the lectures were published, and those as part of the annual report of the Smithsonian to Congress. The program declined after the war. For Owen's plans, see the bills he sponsored in the House of Representatives in William Jones Rhees, comp. and ed., *The Smithsonian Institution: Documents Relative to Its Origin and History,* 2 vols. continuously paginated, Smithsonian Miscellaneous Collections, vols. 42 and 43 (Washington: Government Printing Office, 1901), pp. 322–330;, and the report he wrote as regent on the initial organization of the institution, *First Report of the Committee on Organization* (Washington, D.C.: [Smithsonian Institution,] 1847).

79. Macready also saw Lardner speak on great men of the day, but this was clearly a sideline.

80. Consider the remarks of a young lawyer in New York City: "I have read with great interest Sir David Brewsters article in the last North British Review [reviewing Humboldt's *Cosmos*]. I think he has the faculty of writing on scientific subjects *for the public* in a greater degree than any one I ever read. Almost every one else takes too much for granted—(I mean, that his readers know more than they really do)—while he adopts the lawyers rule for a jury and 'explains every thing.'" Henry M. Alexander to Joseph Henry, 9 January 1846, Henry Papers, Smithsonian Archives, printed in *Joseph Henry Papers,* 6:364–-368, quotation on p. 365.

81. According to the biographical account in the *Dictionary of National Biography,* Brewster gave up the ministry because of poor speaking ability.

82. On earlier attempts to reach this audience, and especially their children, and the influence of the Lockean tradition of general education in science, see James A. Secord, "Newton in the Nursery: Tom Telescope and the Philosophy of Tops and Balls, 1761–1838," *History of Science* 23 (1985): 127–151.

FROM THE SCIENCES TO HISTORY:

A PERSONAL AND INTELLECTUAL JOURNEY

Allen G. Debus

The history of science and history of medicine are fortunate fields in that some of the pioneers are still active and highly visible. I. Bernard Cohen, a student of George Sarton, Marshall Clagett, a student of Lynn Thorndike, and Ilza Veith, a student of Henry Sigerist, remain major figures in their respective areas of research. They are surely not the only ones I might have mentioned, but just to name these three underscores the fact that as fields the history of science and medicine are recent arrivals on the academic scene. Those who attend the annual meetings of the History of Science Society or the American Association of the History of Medicine today, where there are many hundreds present, would scarcely recognize the meetings of thirty years ago where almost everyone knew each other. Many of those who were active in the field were primarily scientists or philosophers rather than historians, and there were very few graduate programs in existence. When I first began to look for a graduate program in the history of science in the early 1950s, there seemed to be only three possibilities: Harvard, Wisconsin, and Cornell.

But I am already ahead of myself. How did one find the history of science at a time when it rarely formed part of the normal exposure of students majoring either in the sciences or in history? I am sure that each person to whom this question is asked will have his or her own answer, and I can only present my own.

From Chicago to Chicago

Unlike the families of many colleagues I have known, mine, of Protestant-German American descent, had no academics in it.[1] The men of my family were businessmen.[2] My parents and my grandparents were very conservative. The isolationist world of the twenties was still part of the world I lived in as a boy a decade later. And although the family businesses were not large ones, Franklin Roosevelt and the Democratic Party were looked on as enemies of American values. One of my early memories is of election night, 1932, one of sadness in our family. Herbert Hoover had been our candidate.

My father had hoped to become an electrical engineer, but in the early decades of this century, formal education was not considered as

P. H. Theerman and K. Hunger Parshall (eds.), Experiencing Nature, 237–280.

essential as it is now. On the other hand, it was deemed proper to help contribute to the family expenses. Accordingly, my father left high school at the end of his sophomore year to take a job continuing his education in night school. During the first World War, he enlisted and was accepted in the Signal Corps. As luck would have it, he was scheduled to leave for Camp on 11 November 1918. He reported for duty at the railroad station, but because of the Armistice, he was told to return home. We still have his government check for a single day's pay. My mother had attended Northwestern University for two years, and my parents met each other in a youth group at their church. They were married in 1924 and moved into a new house on a secluded street in north Evanston the following year. I was born in Chicago on 16 August 1926 and educated in the Evanston public school system, which at that time was considered one of the best in the nation.

My interest in history began at an early age and was encouraged by my mother who drove me almost weekly to the great Chicago museums. I had a special love of the great hall of prehistoric life at the Field Museum with its magnificent paintings by Charles R. Knight. This interest was intensified by the full-scale models of dinosaurs at the Sinclair exhibit and by the "World a Million Years Ago" both at the Chicago World's Fair in 1933. I was also fascinated by the Oriental Institute at the University of Chicago which–along with the Egyptian halls at the Field Museum–sparked an interest in ancient Near Eastern history. (I could probably still give a credible account of Egyptian dynastic history on the spur of the moment.) This interest continued for many years, and I recall going to the University to meet John A. Wilson and Henri Frankfort both of the Oriental Institute when I was in high school. My interest in history also grew from reading–first general accounts for children such as that written by Hendryk van Loon, then surveys such as H. G. Well's *Outline of History* and college textbooks which I found in bookstores, and finally a number of classics. Gibbon's *Decline and Fall of the Roman Empire* and James Henry Breasted's *A History of Egypt* were two of my favorites.[3]

My interest in history developed also from a more unlikely source. My great aunts had purchased an imposing Victrola in 1908, which they had put into storage in an outside hall with the advent of radio. It was filled with records they had bought between 1908 and 1923. In a sense, I was introduced to the early decades of our century through its music of dialect songs and topical material that reflected a world different from the 1930s. I have always remained interested in this subject. The experience of listening to and studying this music gave me an opportunity early on to place past events in their historical context, and the subject has continued to captivate me to the present day.

The future was cloudy for many students of my generation. The attack on Pearl Harbor in December 1941 affected everyone; it certainly affected those of us who were in school. From a widespread "America First" isolationist attitude, we suddenly found ourselves plunged into a righteous war against enemies frequently pictured as little more than barbarians. At school, we organized savings bond and stamp drives and were involved in collections for the wounded in hospitals as well as the servicemen overseas. Before graduation, we had all learned of the death of friends both in the South Pacific and in Europe. It was not a normal period for any of us, and we all expected to be inducted into the army or the navy within months of our graduation in June of 1944.

I was still seventeen and had a scholarship to attend a small engineering college in southern Indiana in the fall. However, because of the war I applied to the Technological Institute at Northwestern University in chemical engineering for the summer session in order to await my induction notice at home. In fact, I never did serve in the armed forces. Although at various times I was classified 1-A, I was never called. I did continue an accelerated program, graduating at Northwestern in the summer of 1947. The engineering program there was tightly and rigidly organized. In chemical engineering, I took the basic courses in chemistry and physics as well as advanced courses in chemistry and chemical engineering. Practical courses in machine shop, pattern making, and mechanical drawing were all part of the program as well as work experience in professional laboratories. In my own case, I worked several quarters at the laboratories of Montgomery Ward, where I was among the first to make samples of DDT for testing for civilian purposes. Because of the program's rigid requirements, I was not able to take an elective course until the end of my second year. At that time, only one history course fit my schedule—one on Renaissance history taught by a young Assistant Professor, John J. Murray, who remained a close friend until his recent death. I enjoyed his course so much that I soon decided to transfer from engineering to liberal arts and put my chemistry courses toward the quick completion of a major in that field. As a result, history courses largely filled my final quarters at Northwestern: more Early Modern history from Murray, British history from Leland Carlson, American Intellectual History from Ray Billington, and Russian history from Leften Stavros Stavrianos. By the time I graduated with a Bachelor of Science degree in chemistry, I had almost completed a second major in history.

John Murray left Northwestern to take a position at Indiana University in 1946, and he offered me an assistantship there if I wanted to continue on to graduate work in history. I accepted his offer and left for Bloomington in September 1947. Indiana was truly a rural state then. Driving south one would frequently overtake farmers in Model T Fords. And Saturday nights in Bloomington witnessed a huge influx of farmers around the Court House. At Indiana, my work was entirely in history, and as Murray was primarily teaching early modern English history, I took his courses and seminars. It was in his seminars at Indiana University that I had my first real contact with graduate research. Murray had excellent students, and he was a master at guiding them toward worthwhile research topics. He influenced me greatly by pointing out that I should be working on some aspect of the history of science because of my degree in chemistry. Accordingly, I wrote papers on Francis Bacon, William Gilbert, and Robert Boyle. When it came time to begin work on my master's thesis, I continued my work on Boyle with "Robert Boyle and Chemistry in England 1660–1700," which was accepted in time for graduation in June 1949. Murray also arranged for *The Indiana Journal for Bookmen* to publish a derivative paper (1949–1).[4] This was my first publication in the history of science.

During my two years in Bloomington, I had read widely in *Isis*, and I had been struck by many of the essays of its editor, George Sarton. In one of these, he had written that the proper training of an historian of science should involve the completion of two master's degrees, one in history and the other in a science, prior to proceeding on to graduate work in the history of science. I now had the first degree in history, but I did not have enough money to continue on to the second one. Accordingly, I returned home to work as a salesman for my father at Modern Boxes. However, I knew that I did not want to devote my life to business. At the time, I was convinced that I had to obtain the second master's degree, and with support from my parents I reapplied for admission at Indiana University, this time in chemistry. I began classes there in the autumn of 1950. The chair of the department of chemistry was Frank T. Gucker. A distinguished physical chemist, Gucker had formerly taught at Northwestern University where he had occasionally offered a course in the history of chemistry. In addition to my regular classes, I began a research project under his direction on the electrical conductivity of salts at very low concentrations. The equipment I used was the same he had employed for his own doctoral research some thirty years earlier. My coursework included classes in physical chemistry from Jerry Schmidt, who also taught a one-semester course on the history of chemistry that I greatly enjoyed.

While at Indiana, I met another graduate student in chemistry, Brunilda López Rodríguez. We fell in love and decided to marry. By June 1951, I had completed all but one course required for my Master of Science degree in chemistry, and I have always regretted the fact that I neither took that last course nor completed my research. Still, this past summer Bruni and I celebrated our fortieth anniversary, and the uncompleted M.S. is a small loss in comparison with a long and happy marriage.

Bruni's family lived in Puerto Rico and mine in Chicago. We compromised as to the marriage site with both families meeting in Miami Beach. We had very little money, and we drove back to Chicago hoping to find a job in chemistry. Sometimes I wonder at the confidence and optimism of youth! Luckily, fate was kind to us. Several positions in chemistry were available and we finally accepted jobs at Abbott Laboratories in North Chicago, primarily because they wanted to hire both of us.

I have never been sorry for the five years we spent at Abbott Laboratories. Bruni worked there as a research microanalytical chemist, and I worked in both the research and the development laboratories, where I had projects as varied as the search for a tasteless aspirin (aluminum aspirin) and for a more economical synthesis for the local anaesthetic, Procaine (novocaine). My work led to five patents[5] which I like to think of, in a sense, as the equivalent of the chemical research I never completed under Gucker at Indiana.

In the end, our stay at Abbott Laboratories confirmed our desire to return to school. To prepare myself in the necessary languages, I contacted the father of a grammar school friend of mine, A. Haire Forster. A retired professor of New Testament Greek at the Seabury-Western Theological Seminary in Evanston, Forster had guided my studies of classical Greek some years before, and I now approached him again in regard to language preparation. For several years, Bruni and I drove to Evanston from Waukegan once a week where we read Latin, French, and German with him. This training ultimately proved invaluable, since it enabled me to pass my examinations immediately after entering Harvard.

My work in the chemical laboratories provided me with a second advantage for the future. As an organic chemist, I had to set up reactions that often took long hours, even days, to complete. In such cases I was able to read. Although I kept up with the chemical literature in my field, I also focused on works in the history of science. I avidly read general accounts of the history of science as well as more specialized works in the history of chemistry. The general surveys clearly indicated that the study of early science centered on the physics of motion and on the acceptance of Copernican astronomy. Although I had taken a number of courses in physics at Northwestern, I had never studied astronomy and so began to make up for this.

As early as 1953, we began to look into the possibility of entering the graduate program in the history of science at the University of Wisconsin, and we visited Marshall Clagett, Robert Stauffer, and Aaron J. Ihde there at that time. I was admitted to the program with a small fellowship, but because of our limited funds we decided at the last minute to stay at Abbott Laboratories. Over the next two years, we became ever more convinced that I should return to graduate school. We now settled on Harvard, and I made several trips to Cambridge both to talk to I. Bernard Cohen and to see if I could find a part-time job. At one time, I interviewed at Lincoln Laboratories, where I would have worked on an enormous primitive computer that filled several rooms. At another time, I interviewed at the Polaroid Company, where they were looking for a chemist to do research on an instant color film. And at another time, I interviewed for a position as a chemist with a Boston paint company. In short, I was willing to take on almost any job, but the fact that I wanted a part-time position so that I could get a Ph.D. in the history of science rather than in a science presented a problem in every case. It was only in 1955—after receiving a bequest from my grandfather's estate—that we felt safe in making definite plans. We moved to Cambridge late in the summer of 1956, with our year-old son and first child, Allen. After getting settled, I stopped in at the Widener Library's Room 189, which served as Bernard Cohen's office as well as that of the graduate program in the history of science. It was an impressive book-filled suite that had been the office of the recently deceased George Sarton. There were only a few students in the field at Harvard, and there were only three of us—George Basalla, William Coleman, and myself—entering the program in the fall of 1956. My courses for the next few years centered on the offerings of Charles H. Taylor, Wilbur K. Jordan, and Cohen. Taylor gave impressive lectures, never failing to make the intellectual debates of the Middle Ages interesting. Jordan aimed in his seminars on Tudor and Stuart England to teach us how to pursue a research topic. Cohen's seminars taught us how to read Galilean texts carefully and to relate them to their medieval background. At one of the Cohens' teas, I met Leonard K. Nash, a professor of chemistry with an interest in the history of chemistry. I became one of his teaching assistants the following year in Natural Sciences 4, an introduction, largely through chemistry, to the sciences for non-science majors. This course, first developed by James B. Conant in his effort to reform the teaching of science in the United States, revolved about a series of case histories. My feeling was then— and remains today—that these courses were at best partially successful; they were not historical enough for the historians and too historical for the scientists.

The actual teaching of the history of science at Harvard fell almost exclusively to Cohen.[6] Trained originally in mathematics and physics, Cohen emphasized the development of the physical sciences in the period from Copernicus through Newton. He was then and now remains one of the most eminent interpreters of this period. As students, we read very little of the work of George Sarton, but much of Alexandre Koyré, the Russian historian and philosopher of science. Koyré had written of the "glorious progress" of the evolution of scientific ideas, and he interpreted the Scientific Revolution as the replacement of the Aristotelian cosmos with the Copernican world view. For him, this shift was clearly allied with the mathematization of the sciences in the sixteenth and seventeenth centuries. Koyré's approach was internalistic, one that centered on the development of the physical sciences from Copernicus through Newton. He spent little time on other sciences and did not consider the relationship of science to society.

In addition to coursework, students in the Harvard program of the late fifties had the opportunity to listen to numerous visitors. I particularly recall lectures by Marshall Clagett, Herbert Butterfield, and Alexander Koyré, but personal commitments at times prevented my attendance at these extracurricular events. As a married man with a family, mine was a different life style than that of most other students. My years in pharmaceutical research had conditioned me to rise early in the morning for reading and course preparation; evenings were reserved for my family. Furthermore, because my research interests centered on early modern chemistry, a subject of relatively little interest to other students, I rarely spoke to others in detail about my work, and I had little time for afternoon coffee or tea at the hangouts on Massachusetts Avenue. The late afternoon often found me shopping with my wife at the local grocery stores. From the start, I felt that my job was to get my degree as quickly as possible.

The reading I had done before going to Harvard had only strengthened my interest in the history of early chemistry. The mystical and allegorical writings of the alchemists seemed too difficult to decipher with certainty, but the works of Paracelsus and his followers were different. To be sure, on occasion, their writings sank just as deeply into obscurity as those of the earlier alchemists, but not always: they had written with both reform and a real desire to be understood in mind. It first occurred to me to write a dissertation on the work of the Paracelsians and the rise of chemical medicine in the sixteenth and seventeenth centuries. My fellow students in the graduate program warned me against this, pointing to the fact that the study of the Scientific Revolution was firmly anchored in the physics of local motion and the acceptance of the heliocentric system. Any study of chemistry would necessarily be of lesser interest. Moreover, no one at Harvard had a specific interest in the area I wanted to study. Jordan's seminar on Tudor England actually brought this problem to a successful resolution.

Although he was not particularly interested in the subject, Jordan permitted me to write my seminar paper on the English followers of Paracelsus. At that time, relatively little had been done on sixteenth-century English chemical medicine, but it has since proven to be a fertile field rich in new topics for research. Fortunately, the Harvard libraries contained many source materials for this research, but it soon became apparent that a number of essential works should be studied in England. Accordingly, I applied for a Fulbright award which was granted and augmented with a fellowship from the Social Science Research Council. We embarked on the S. S. *United States* for Southampton and then London in September 1959.

Not long before our departure, I had read Walter Pagel's six-part essay, "Religious Motives in the Medical Biology of the XVIIth Century" in the 1935 volume of the *Bulletin of the Institute of the History of Medicine.* His magisterial *Paracelsus: An Introduction to Philosophical Medicine in the Era of the Renaissance* had appeared in 1958. Pagel's approach seemed closer to my own than that of any other author I had encountered. Before leaving for England, I wrote to ask him if we could meet sometime during my year in London. He replied that he would look forward to meeting me, and I first visited him at his home in Mill Hill in the fall of 1959. This first meeting soon led to weekly get-togethers. After tea at 3:00 P.M. with Mrs. Pagel, we would retire to his library for several hours of discussion. There, a remarkable library of sixteenth- and seventeenth-century volumes surrounded us, and we referred to it frequently as our conversation moved from one subject to another. Pagel's love of early editions surely rubbed off on me. Since rare books on science and medicine were much less expensive then, even a graduate student on a limited budget could buy some sixteenth- and seventeenth-century editions. In addition to Pagel's vast knowledge of Paracelsus and his followers, I benefited greatly from his contextual approach to these texts, an approach far different from the positivism of George Sarton. We became friends and remained close until his death in 1983. Indeed the informal weekly "seminars" we began during my student days resumed whenever I was in England in succeeding years.[7]

In London, I worked primarily at the British Museum and the Wellcome Historical Medical Library. During the course of my year abroad, I managed both to complete my dissertation research and to make considerable headway in my writing. I had requested to be associated with the department of the history and philosophy of science at University College London, in part because of the libraries in London, and in part because of the presence there of Douglas McKie, a renowned scholar of the chemical revolution of the eighteenth century. I made a point of listening to the course lectures offered at the department; they were held in the evenings because nearly all of the students taught science in secondary schools during the day. These students primarily sought master's degrees as a means to advancement in their school posts. I met no one who planned to go on to a Ph.D. in the field or to do serious research.

McKie gave the lectures on the history of chemistry, and he was assisted by a recent appointee, William A. Smeaton. Angus Armitage lectured on the history of astronomy, while E. Ashworth Underwood spoke on the history of medicine, and J. S. Wilkie lectured on the history of biology. All of these lecture series presented factual surveys, and to me they seemed to reflect the compartmentalized histories of specific sciences like those found in Whewell's nineteenth-century *History of Inductive Sciences*. Thus, although they departed from the Sarton tradition, which had emphasized the unity of the sciences, they were nonetheless useful.

During our time in London, we got to know Bill Smeaton very well. We frequently had dinner and went to the theater with him as we still do on our visits to England. We also became close friends with P. M. Rattansi, whom I first met at the home of Walter Pagel. He was then a graduate student at the London School of Economics, and he is now (1992) the Head of the Department of the History and Philosophy of Science at University College.

We returned to the United States late in August 1960 on the Cunard Liner, the *Ivernia*. My last year at Harvard was devoted to the completion of my dissertation and to looking for a job. I was the only student in the program who planned to get his degree in June 1961. Indeed, I may have been the only one in the history of science in the country. Although the job situation was not much better then than it had been in 1956, both Ohio State and the University of Chicago were looking for an historian of science. Chicago appealed to me not only because of its stature but also because a job there would allow me to be close to my parents who then lived in Northbrook.

In December 1960, the annual meeting of the History of Science Society took place in New York City. At that time, the meetings were relatively small, and my paper, "Solution Analyses Prior to Robert Boyle," was slated for a full hour following that of the well-known historian of astronomy, Doris Hellman. I had never spoken before outside of a seminar, and Cohen "auditioned" me at his home before the meeting. During my lecture at the meeting itself, he sat in the front row with Cyril Stanley Smith from Chicago. It was a kindness I will never forget. After the meeting, an offer from Ohio State preceded any word from Chicago. When my father fell ill unexpectedly, however, I made a short trip home to see him and took the opportunity to contact the department of history at the University of Chicago. They arranged for a luncheon meeting while I was there, and although I did not give a paper, the final outcome was a job offer which I accepted.

Back in Cambridge, we were preparing for the birth of our third child. Our second son, Richard, had been born in 1957, and Karl was born toward the end of May 1961, only weeks before my graduation. My parents came to Cambridge for the event, and they took the two older boys back to Chicago with them so that we could prepare for the move. During the summer we arranged to buy a house in Park Forest, and we enjoyed our final weeks in Cambridge. The Boston area seemed like home after our five pleasant years there, but, with our things finally packed and our furniture loaded and shipped, we left Massachusetts to return to Illinois at the end of August.

The University of Chicago

In the fall of 1961, the University of Chicago entered a new phase of its history under a new Chancellor, George Beadle. At the same time, the department of history had a new chair, William McNeill. These two appointments brought significant changes: Beadle placed a stronger emphasis on the sciences within the University as a whole, and McNeill radically redirected the department of history. Walter Johnson, McNeill's predecessor, had by all accounts been a good and well-liked chair who had been reappointed a number of times. However, he believed that the foundation of any program in history must be American history. McNeill, on the other hand, has always been an historical polymath with an interest in global problems that touch every area of historical inquiry. In regard to the history of science, he was highly critical of the traditional, internalist studies that seemed to separate the subject from the broader historical enterprise. In line with his own breadth of interests, he was to encourage the development of programs in areas that had not previously been represented within the department. The history of science was one such area to benefit from his support.

My own appointment had been arranged in the closing months of Johnson's term as chair. It had evidently been felt that my two-year experience assisting in Natural Sciences 4 at Harvard could prove useful to the College at Chicago, the legacy of President Robert Maynard Hutchins, which still divided the faculty into an undergraduate teaching group and a research-oriented faculty associated with the "Upper Divisions." My own appointment straddled the two. Two-thirds of my appointment involved teaching sections in the undergraduate physical sciences sequence, the rest introducing a course in the history of science to graduate students through the department of history.

I found the course in the physical sciences a more difficult undertaking than I had anticipated. The Chicago course, unlike the Harvard counterpart I had taught, centered on physics and the topics differed from the pre-war tradition I had been exposed to at Northwestern more than fifteen years earlier. Furthermore, Chicago emphasized sections as opposed to the general lectures favored at Harvard, and it seemed evident to me that the result was even less successful. Although many theoretical arguments may be advanced to support these courses, they seem to have largely disappeared from the academic curricula in more recent years.

Although mine was the first appointment made by the department of history in the history of science, the University already supported others in related fields. Ilza Veith had taught the history of medicine at Chicago for a number of years prior to my arrival. Her office and the center of her teaching activities were in Billings Hospital, consonant with her training in the German tradition under Henry Sigerist. I have heard former students of hers praise her frequent discussion luncheons. However, her students were almost all medical students. Ellis Avenue separates the main campus from the medical complex, and although the distance is a small one, it seems to act as a barrier separating these two parts of the University. Reflective of this, although Veith had strong connections with students and faculty in medicine, she seems to have had rather limited contact with the members with the department of history, and she had little impact on the development of a program in the history of medicine there.

Another person with whom I came in contact shortly after my arrival was Robert Palter. A member of the department of philosophy, Palter was my colleague in the physical sciences sequence. He had already started a small course in the history and philosophy of science for undergraduates in the Collegiate Physical Sciences Division (1959), but there was no room for me to participate in it. Two other philosophers, Sylvain Bromberger and Dudley Shapere, also had a special interest in the history of science. With the exception of Dudley Shapere, however, these philosophers of science—and Ilza Veith as well—all soon left Chicago for other positions.

No less important for the development of the program in the history of science was Robert Rosenthal, who was Chicago's rare book librarian and the Curator of its Department of Special Collections. Rosenthal made a point to meet members of the faculty and to find out their interests. He actively bought books in fields in which current research was underway, and the history of science was no exception. Until his death in 1989 on a book-buying trip in Scotland, Rosenthal actively sought to purchase early chemical books that he knew I would use. At the same time, he worked to acquire other major collections in the history of science such as the Schaffner Collection and the Fishbein Collection in the history of medicine. He also played an important role in the transfer of the Crerar Library in the sciences and technology from the campus of the Illinois Institute of Technology to the University of Chicago (1981; date of merger agreement).

With all of these positive factors, the University of Chicago seemed an ideal place for the establishment of a major program in the history of science. The scientists I had met were encouraging, while others in the social sciences and the humanities showed an active interest in the field. I have always felt that the history of science is properly a field of history and that students should receive historical training beyond their courses in the history of science. Accordingly, I was particularly gratified that William McNeill looked favorably on the history of science. He actively encouraged the new members of the department to propose major fields in their specialties. This was not accomplished without some opposition, and I can recall James Cate, our medievalist, arguing that the department of history at Chicago should hold to its strength in the more traditional fields of American, European, medieval, and ancient history rather than turning to some of the then more fashionable fields like African history. Nevertheless, at an evening meeting of the department in 1963, I proposed the creation of a major field in the history of science. This was approved, and at the same meeting a field in southeast Asian history was also established.

To Alan Simpson, Dean of the College, I further argued for a separate major undergraduate major in the history of science. This also received approval. With the official acceptance of the field on both the undergraduate and graduate levels, I applied to the National Science Foundation for a grant (NDEA–Title 4) to develop the program. The members of the Oriental Institute at the University had an interest in appointing David Pingree, whose study of ancient texts involved an astonishing range of languages. The Institute's Director, Robert C. McCormick Adams, discussed with me the establishment of a program in the history of science that would be limited from antiquity to about 1600. I rejected this proposal outright for it seemed to me then—and it does now—that no program in the field could ignore the Scientific Revolution and all that it has led to in our world. Although I disagreed with Adams on this point, David Pingree was appointed in 1963, and he remained a valued member of our program until his departure for Brown University some years later.

At the end of my first year at Chicago, a chance meeting in Ithaca, New York, with Michael Hoskin at the International Congress of the History and Philosophy of Science ultimately led to our return to England. Hoskin had only recently inaugurated a new series in the history of science through the Oldbourne Press, and invited me to submit my dissertation to him. Eventually, this chance meeting led to the publication of *The English Paracelsians* (1965–1).

In the course of our correspondence, and at later meetings, Hoskin informed me that his college at Cambridge, Churchill, had instituted a new category of fellowships for foreign scholars. He suggested that I apply for one, and in 1964 I did. I was granted one of the early Overseas Fellowships at Churchill College for the year 1965–1966 only to find that my colleague, Shapere, also had a research leave for that year. I felt that it would be a disaster for our programs if both of us were away from Chicago at the same time. Since I had obtained another fellowship for the year that made it possible to remain at home for my research, I stayed in Chicago to watch over our program. One of our graduate students in history, Barton Hacker, gave my courses in the history of science during this leave and Rattansi came to Chicago as a visiting professor that summer.

Happily, the Overseas Fellowship at Churchill College was delayed to 1966–1967, but this necessitated finding a replacement to teach the courses in the history of science through the department of history. After a search, we appointed Edward J. Collins as Instructor. Collins had entered the Harvard program during my final year there, and he came to Chicago with glowing letters in spite of his as yet unfinished dissertation on Descartes. The program also expanded into the history of medicine with the appointment of Lester King as Professorial Lecturer. His research on seventeenth- and eighteenth-century medicine was already well-known, and although he could not devote all of his time to teaching in our program due to his work as a Senior Editor at the American Medical Association, his appointment did much to broaden its scope.

I had been promoted to Associate Professor in 1965 with the publication of my first book, *The English Paracelsians*, and I planned to devote the coming year in England to the completion of a volume on the chemical philosophy of the sixteenth and seventeenth centuries. We left home greatly concerned about my mother, who had suffered a stroke on a vacation in Florida in March. Still, the fellowship could not be postponed another year, and we took the train on my birthday from Chicago to Montreal, where we embarked on the *Empress of Canada* the next day.

The crossing was a smooth one, and after settling into our flat at Churchill College, we adjusted once again to life in England. Sir John and Lady Cockcroft helped in this regard, serving as gracious hosts to all of the Fellows and welcoming them with great kindness. As we had on our last stay in England, we bought a car which allowed me to resume my afternoon "seminars" with Walter Pagel. I found also that there were many invitations to speak. At the University of Cambridge, I lectured in a series on "Science and Technology Before the Copernican Revolution" (21 November 1966). Among the other speakers in the series were G. E. R. Lloyd and Joseph Needham. I also gave a required Overseas Lecture at Churchill College (2 June 1967). Other lectures were given at the University of Durham, University College London, and the University of Leeds. Professor Hugh Trevor-Roper invited me to speak in his seminar on Renaissance Intellectual History at Oxford, and I also talked at the History of Science Seminar at Cambridge. At that time, it was customary to devote a spring meeting of the British Society for the History of Science to a lecture by a foreign visitor, and I was chosen for the meeting held at the Science Museum in South Kensington on 8 June. There, I took "The Paracelsians and the Scientific Renaissance" as my topic. All of these invitations were flattering, but they cut deeply into my research time. Although I did make progress, I did not manage to complete my work on the chemical philosophy.[8] I did not realize then that it would take years to find enough time to complete the project.

Today, I wish that I had been more involved in the history of science program at Cambridge. Large and active, it drew scholars from many areas of the academic community. I was invited to do some tutorial work, but I declined. With the time I spent on my research for the book, the preparation of the lectures I had agreed to give, and my weekly visits to Mill Hill to see Walter Pagel, there was little time left for anything other than my family.

Historians and Philosophers

One of the blessings of being away from home is that the problems we face there seem far removed. I certainly did not realize on our return to Chicago that the new academic year 1967–1968 would be so difficult. Indeed, it witnessed the beginning of a two-year debate with the philosophers of science that eventually reached the Council of the University Senate.

During my year abroad, Dudley Shapere had proposed the establishment of a Committee on the History and Philosophy of Science in a document submitted to the dean of the Division of the Humanities, Robert Streeter. This had been tabled at the time because I was out of the country. With my return, Shapere called a joint meeting of historians of science and philosophers of science on 17 August 1967, where he argued for a separate program in the history and philosophy of science. For him, the departmental rules in philosophy did not seem to permit the proper training in this field. He also held that historians of science needed the philosophy of science. While unable to question his first point, we did not agree that the history of science depended on the philosophy of science. In a written reply, we responded that the history of science was already a professionalized field and that the historians of science in the department of history had been trained professionally. Furthermore, because the history department reflected a broad concept of history with a strong emphasis on intellectual history, this was the proper place for such a program. We pointed out that the historians of science at Chicago had a broad approach to history and that, with the flexibility of the departmental rules, they could accommodate a wide variety of students with greatly differing backgrounds. We had had no trouble in arranging programs for individuals in the science faculties for graduate students interested in pursuing research in technical and internalist subjects. In this written statement, I concluded that "we feel that students who have an interest in relating science to society—to religion—to literature—to economics—or to a host of other topics should be able to find a home in any proper history of science program. Ours—from the first—has been designed to accommodate both types of students." Moreover, in a survey of our students we counted four first-year students, another thirteen advanced students, and two recent Ph.D.s. In short, we believed that we had already developed a successful program within the department of history and that nothing was to be gained by forming a degree-granting committee with narrower goals than those we had established.

Within the next few months we learned more of the proposed new program. By this time, John Hope Franklin had succeeded William McNeill as chair of the department of history, and I noted in my calendar numerous meetings with him during the fall quarter for discussions of this problem. Dudley Shapere had certainly not lost his enthusiasm, and it became evident that he intended to proceed with his plans no matter what the historians thought. In November 1967, he presented us with a "Proposed Committee on the Foundations and Implications of Science." There he argued again that "under present arrangements, according to which students must study the philosophy of science within the Philosophy Department, there is a considerable amount of material which they must master which, if not completely irrelevant to their central concerns, is at least peripheral thereto." But as the philosophers in his program needed more science and history of science, so too, he argued, the historians of science needed the philosophy of science so that they might properly deal with "the relative roles of evidence and theory." Therefore, he sought a committee with participants representing the philosophy of science, the history of science, and the sciences. He suggested a program leading toward the Ph.D. that involved four areas of examination: the philosophy of science, a field of the history of science, a field of science, and "an appropriate field concerned with the primary interest of the student's program."

In reply to this proposal, the historians of science (David Pingree, Edward J. Collins, Lester S. King, and myself) submitted an eleven-page document which pointed out that the program developed within the department of history could easily duplicate that proposed by the philosophers of science. We argued that the history of science is basically part of intellectual history and that historians of science should be in contact with a wide range of specialists—not only in the sciences, but in the history of religion, the history of social thought, the history of literature, the history of philosophy, and other pertinent areas.

> For us the philosophy of science is one of many fields in this category and we see no reason why it should have a closer relationship with the history of science than with any of the others. Indeed, for many historians of science a special relation-

> ship with the history of philosophy or the history of religion would prove more valuable. However, individual research problems require different–and often unexpected–fields of expertise. The most important point in any such program should be its flexibility so that the interests of individual students can be properly accommodated within it. Thus, if a History Department is oriented toward intellectual history in its broadest sense–and if such a flexible program is permitted by departmental regulations–we believe that the study of the history of science should be centered there. One flaw found in older programs in the history of science–separated from other departments–has been the insularity which has fostered an exclusively internalist-technical approach to the subject at the expense of its broader ties with other fields of history. We doubt whether the real future of the history of science is to be found in this tradition or in the formation of dipolar History-Philosophy combinations.

We pointed out that our program emphasized original research and that a number of papers written in our seminars had appeared in major journals in the history of science and medicine. Furthermore, we noted that the University had supported our program through a series of appointments beginning with that of David Pingree in 1963. We already had an established series of lectures, and we had been granted six federally-funded NDEA–Title IV fellowships for our students.

In answering specific charges made against the departmental program, we felt it necessary to point again to the fact that we encouraged our students to take fields outside of history and that most of these fields had been taken with faculty members in the physical and biological sciences. When the philosophers suggested that their proposed committee might prove attractive to those interested in various societal problems, we rightly noted that our students already had this option available to them through history. Our program was widely publicized within the University of Chicago, and our descriptive flyers were distributed nationally and internationally. We argued once more that the program proposed by the philosophers could be exactly duplicated within the history department. From our point of view, the plan seemed at best a carbon copy of what we already had, and at worst retrogressive because it lacked the flexibility of our department rules. Clearly, at an impasse, we suggested either the philosophers of science (1) ally themselves with the program in the Department of History or (2) they form a committee of the philosophy of science rather than one on the history and philosophy of science. We added that "if such a clearly defined Committee on the Philosophy of Science were set up, the members of our program would gladly cooperate with it."

The November "Proposal" of the philosophers of science had been prepared for a visiting committee invited by them and the Division of the Humanities with no input from the historians. It was composed of two philosophers, Adolf Grünbaum and Carl G. Hempel, one historian of recent physics, Martin J. Klein, and one historian of medicine, C. Donald O'Malley; it was to be joined by Saunders MacLane of our Department of Mathematics. After the visitors met first with Shapere privately, the full Committee joined the philosophers of science and the historians of science for a dinner meeting on 29 November. Private meetings and discussions followed the next morning.

Although a few history students were summoned to give "evidence" at these morning sessions, no faculty members from history were invited to support their program or to clarify statements made at that time. Worse, although the department of history program had been operating since 1963, the chair of that department, John Hope Franklin, was not asked to appear either. This was all the more regrettable since Franklin had been unable to attend the dinner the evening before. Given these facts, it is hardly surprising that the members of the Committee wrote a report favoring the philosopher's report. Although dated 3 January 1968, this report was not made available to the historians until much later. In presenting its verdict, the Committee wrote that, after the evening dinner at which the historians were present, "our *initial* sympathy . . . was on the side of the historians, who charge the proposed new Committee with being a mere instance of Parkinson's Law." Only after their morning discussions had they changed their minds. They decided that (1) the demand for teachers in the field existed in departments of philosophy and science rather than in departments of history, (2) the historians at Chicago were only interested in science to 1700 and had no interest in adding specialists in more modern periods, (3) students in history do not have time to acquire a sufficient training in the sciences, and (4) fellowship money for such students would be very limited in a large department. In fact, all of these statements were unfounded. Derek J. de Solla Price's then current "A Guide to Graduate Study and Research in the History of Science and Medicine" showed that 89% of the teaching positions in the history of science in the United States were in departments of history or in independent departments of the history of science and/or medicine rather than in history and philosophy of science programs.[9] Nor was it correct that the historians of science were only interested in science prior to 1700. Unsuccessful offers had gone out earlier to two specialists in nineteenth-century science, and there was no plan to abandon the search for historians of recent science.

As for the claim that the history program did not allow enough time for students to prepare a field of science in order to do research in a recent topic, we replied that there was no time limit involved in the completion of a doctorate in history. Students were expected to have sufficient preparation to carry out research in their chosen field: those who required training in the sciences, just like those who needed unusual languages, would have to take whatever time it required to achieve proficiency. We further noted, that advanced scientific training was not necessary for all areas of research in the field since some valid subjects did not require a detailed understanding of a modern science.

Nor was fellowship money a problem. Applicants to history in the history of science had consistently ranked highly in regard to applicants in other fields of history. We easily showed that our current students had all received substantial financial aid and that those who had been accepted for the following year had secured full scholarships.

Our statement rejecting the conclusions of the visiting committee was sent to Dean Streeter with a cover letter from John Hope Franklin on 24 May, already the end of the academic year. In short, the year since my return from England had been a difficult one, but we had made progress. We had joined with other fields within the department of history to support the appointments of Keith Baker, whose work in eighteenth-century French intellectual history touched on the history of science; Noel Swerdlow, who specialized in mathematics and astronomy from antiquity through the Renaissance; and George Stocking, who was already well-known for his work in the history of anthropology. During this trying year I had also received an attractive offer to leave Chicago, which I probably would have accepted had it not meant leaving my father alone. One result of this offer was my promotion to Full Professor.[10]

With the advent of a new academic year, Shapere began work on a new plan for an "Interdivisional Committee in the Foundations and Development of Science," which we first saw in January 1969. When this plan also proved unacceptable to the historians, Richard Lewontin sought to bridge the gap between the historians and the philosophers by proposing a "Committee on the History and Philosophy of Science" (10 March 1969). To the historians, his proposal seemed to ignore much of the content of the previous year's debate. On 19 March the department of history discussed Shapere's proposal along with other proposals that had been made to that time for the establishment of a separate committee incorporating the historians and the philosophers of science. In a letter to the Provost, John Wilson, and to Deans Robert Streeter and D. Gale Johnson, John Hope Franklin wrote that

> At its meeting on 19 March 1969, the Department of History discussed the several proposals which have been made to create a joint committee on the history and philosophy of science. Special attention was paid to the "Proposal for the Creation of an Interdivisional Committee in the Foundations and Development of Science." At that meeting the Department voted unanimously that it could not support either this or any other such proposal. If approved, such a committee would serve as a second graduate program in the history of science outside the Department of History. The graduate program in the history of science has flourished in the Department of History since its inception in 1963, and it seems unrealistic to establish a duplicate program . . . .
>
> At the same meeting on 19 March the Department of History voted in favor of supporting the formation of a committee on the philosophy of science within the Division of Humanities. At the same time the historians of science and the Department as a whole reiterated their willingness to cooperate with either a subdepartmental or an independent committee on the philosophy of science.

The next step was to carry the proposal to the Council of the University Senate, which was to meet on 28 May 1969. The historians were not invited to that meeting, but enough questions were raised to delay action. The next meeting of the Council was scheduled for 10 June with historians present. Prior to that meeting, Franklin had sent a letter to all members of the Council outlining the opposing position of the history department. I summarized the position of the historians at the meeting on 10 June. The minutes of that Council meeting reflect the heated debate that took place, but in the end it was decided to establish a subcommittee to examine the entire problem.

With the beginning of summer, we left for England where we took up house through early September once again at Churchill College. Back in Chicago, Collins kept track of new developments and on 22 July he wrote to tell me of the naming of the president's committee to investigate the history and philosophy of science; William H. Kruskal, then chair of the department of statistics, served as its chair, with Kenneth J. Northcott, chair of the department of Germanic languages, and John R. Bormuth, Associate Professor of Education, as its members. This Committee had hoped initially to complete its report by the end of the summer, but this proved impossible. I find in my files numerous letters to Kruskal and to others relating to this problem. I closed a letter to George Stocking dated 29 September with the statement that "I am furious with the whole affair. I spend all of my time writing statements on our position. I have absolutely no time for research or for reading or for preparing new lectures."

The Committee completed its report on 20 October 1969 and discussed it with the concerned parties in the Provost's office on 5 November. A revised report was submitted five days later. Here it was noted that the program in the history of science operated easily within the department of history, but that the philosophers could not develop the kind of program they desired within the department of philosophy. The recommendation was to establish a degree-granting committee on the conceptual foundations of science within the Division of Humanities, but also with the connections to the sciences, and perhaps to other fields. The members of the new committee as well as the historians always made clear the differences between the two programs. Above all, it was important from the historians' point of view that "the Committee, as part of its announcements, make plain its concern with the study of the philosophy of science in large part through the analysis of specific scientific activities of particular eras."

The revised report was presented to the Council of the University Senate on 11 November 1969 and passed without dissent. More than two years had passed since Shapere had initiated the plan to unite the history and philosophy of science. In the end, the historians had succeeded in maintaining our program within the department of history, and we had managed to prevent the philosophers of science from calling their program one in the history of science. On the other hand, the philosophers of science had obtained a degree-granting committee with appointive powers. The debate had really centered on both the definition of the history of science and the still current problem of internalist vs. externalist interpretations. For the historians, these two interpretations were equally important, while for the philosophers internalist–and most frequently positivist–interpretations were preeminent. We certainly hoped that now that this matter had been settled we would be able to return to our research.

The Morris Fishbein Center for the Study of the History of Science and Medicine

As the relationship of the new committee and the department of history program was being determined, a new development occurred. In July 1969, Morris Fishbein celebrated his eightieth birthday. The editor of the *Journal of the American Medical Association* for a quarter of a century and the author of many books on medicine, Fishbein was surely one of the best-known American physicians. Furthermore, he had always had a deep love of medical history. His study of medical quackery, *The Medical Follies* (1925), had been one of the best-selling books of the twenties, and his *Doctors at War* (1945) remains a basic study of the medical profession during the second World War. These and his many historical articles aimed primarily to inform physicians of their heritage. Morris Fishbein had been one of the founding fathers of the Society of Medical History of Chicago in 1909, and later he had been instrumental in establishing its journal, the first such journal of medical history in this country. The Fishbein Collection of early medical books already formed an important part of the Division of Special Collections at the University of Chicago, and on his eightieth birthday, his friends sought to honor him further.

The celebration itself was in the form of a gala dinner at the Ambassador East Hotel. Morris Fishbein recalled the event in the annual booklet, "Season's Greetings" which he sent to his friends at the close of the year.

> In the evening (July 22nd) a great multitude came to the Guildhall, where Rabbi Binstock gave the invocation, Howard Rusk toastmastered, and Elmer Bobst and Clement Stone gave toasts. Next, President Edward Levi, University of Chicago, reminisced wittily. Mayor Richard Daley presented old Pepys with Chicago's Medal of Merit. Then Nanna Svartz, Komei Nakayama, and Christian Barnard spoke and Secretary Finch read a greeting from President Nixon. Herbert Mayes gave a short-short biographical sketch of old Pepys' 80 years via telephone from London while his remarks were cleverly illustrated with pictures from the family album and from old Pepys' travels. To all this old Pepys replied full of emotion and Cardinal Cody gave a beautiful benediction. All old Pepys' publishers, associates and friends delighted in the special program prepare by Gaylord Donnelly. Thus the eightieth birthday was gorgeously celebrated.

The Chicago *Maroon* for 24 July 1969 emphasized a different point about the celebration:

> A testimonial dinner for Dr. Morris Fishbein, world famous medical author and editor, has netted over a quarter of a million dollars for a planned University Center for the study of the history of science and medicine.
>
> The center, to be named in honor of Dr. Fishbein, will be a interdisciplinary undertaking, drawing from the faculty and facilities of the biological and social sciences to examine science and medicine in the broad context of the cultural history of mankind.

There was at first some indecision as to whether the money should endow a center or a Fishbein-named professorship in the history of medicine. A high-level search committee was organized to secure the full-time appointment of a senior figure in medical history. A number of candidates were invited to visit, and one German scholar even spent a quarter in Chicago. Still, no one seemed quite right for the position, and eventually we appointed a promising young scholar, Jerome Bylebyl, after the departure of Edward Collins.[11]

Hardly less important was the discussion as to where to locate the new Center academically and how to structure it administratively. Primarily due to the efforts of William McNeill and H. Stanley Bennett, dean of the Division of Biological Sciences, the Fishbein funds were associated with the Department of History through the Division of Social Sciences. The Center itself was formally inaugurated with a four-week lecture series beginning on 11 May 1970. I. Bernard Cohen of Harvard University, who at that time also served as the President of the International Union of the History and Philosophy of Science, spoke first on "The Individual and the World of Numbers–The Rise of a Statistical Point of View with Respect to Man, Science, and Society." The following week James A. Shannon, a former Director of the National Institutes of Health, spoke on "Biomedical Science in Medical Education–Some Contemporary Problems." On 25 May, Mirko D. Grmek of the International Academy of the History of Science came from Paris to talk on "Hippocrates and the Enigma of Life." C. Donald O'Malley of UCLA., the President of the International Academy of the History of Medicine, had prepared to give the final lecture but had died unexpectedly a few weeks before the lecture series began. I read his paper on "Desiderius Erasmus: The Spirit and the Flesh" (1 June). The series pleased Morris and Anna Fishbein, and although they left for

Europe almost immediately afterward, Fishbein wrote me from Rome in July that "we shall never forget the wonderful lecture series which were of such a high quality and distinction."

The structure of the Center resulted directly from the discussion and debate of the preceding three years. At a meeting on 9 December 1970 presided over by John Wilson, Provost of the University, it was resolved that the center would be an association of faculty devoted to research in the history of the sciences and medicine and that its membership would include professionally trained historians of science and medicine as well as historians, scientists, and other interested parties. The first brochure of the Fishbein Center (1971–1972) listed the history department faculty as David Pingree, Professor of the History of Science; George W. Stocking, Jr., Associate Professor of History and Anthropology; Keith M. Baker, Assistant Professor of European History; Noel M. Swerdlow, Assistant Professor of Medieval History; Edward J. Collins, Lecturer in the History of Science; Lester S. King, Professorial Lecturer in the History of Medicine; and myself.

The Center was neither to have appointive powers nor offer courses nor grant degrees. Instead a Policy Committee would advise the chair of the department of history of the Division of Social Sciences in regard to new members, faculty needs, fellowships, and other matters of concern. This short statement cemented the close relationship of the history of science program with the department of history. In 1972, the document was rewritten in slightly more detail, but its general provisions remained the same.

I was appointed the first director of the Fishbein Center in 1971, and three years later, I was reappointed for a second three-year term. I was particularly fortunate in having Mrs. Helen Little as a secretary during these years. Not only did she act independently when it was necessary, she took a genuine interest in the students and their problems.

These six years witnessed a period of real growth and achievement. Although David Pingree left for Brown University and Edward Collins was replaced by Jerome Bylebyl, the teaching staff remained largely constant until Bylebyl left for The John Hopkins University. Surely by the mid-1970s, we at Chicago had one of the strongest programs internationally for the period through the Scientific Revolution of the seventeenth and eighteenth centuries. In addition to this core, the Center included a number of other members from various departments within the University. The 1976–1978 brochure lists Leon O. Jacobson, M.D., director of the Franklin McLean Memorial Institute; Saunders MacLane, Max Mason Distinguished Service Professor, Mathematics; Paul

B. Moore and Thomas Schopf, professors in the department of geophysical Sciences; Karl F. Morrison, chair of the department of history; Arnold W. Ravin, professor of biology and microbiology and chair of the Committee on the Conceptual Foundations of Science; Robert Rosenthal, curator of the Department of Special Collections of the Regenstein Library; Edward Shils, Distinguished Service Professor of Sociology and Social Thought; Alvin B. Tarlov, chair of the department of medicine; Stephen Toulmin, professor in the Committee on Social Thought and the department of philosophy; and Robert B. Uretz, Deputy Dean for the Basic Sciences, Division of Biological Sciences.

The graduate degree program continued to develop through the department of history, where flexible rules permitted students to take their course requirements in history, in the sciences, in philosophy, or in any other department that offered courses of value to an individual's program. Although we never had a student who sought a joint M.D./Ph.D. with a specialty in the history of medicine, we anticipated this need by developing such a program with the department of medicine.

We actively sought to encourage an interest in the field on campus as much for the benefit of our students as for the University community as a whole. Chief among these activities was the Fishbein Lecture series. Generally, three such lectures were given each year, and the speakers represented some of the most prominent American and European scholars in the field. As I review the list of those who spoke prior to 1978, I note such major figures as Michael Hoskin (Cambridge), William McNeill (Chicago), Cyril Stanley Smith (M.I.T.), Morris Fishbein (who spoke on the changes in medicine he had witnessed in the sixty years since he had graduated from Rush Medical College in 1912), Hugh Trevor-Roper (Oxford), Richard S. Westfall (Indiana), William Coleman (The John Hopkins), Roger Hahn (University of California at Berkeley), Cesare Vasoli (Florence), and Stephen Jay Gould (Harvard).

We also succeeded in launching a series on the history of science and medicine with the University of Chicago Press for which I acted as Editor. I was particularly pleased that the first volume in this series was a revised version of a dissertation prepared by one of our students, William B. Provine.[12]

Our program also benefited from the proximity of so many major libraries. The holdings of the Regenstein Library at the University of Chicago have great strength in our field, and these were occasionally highlighted through special displays. A major exhibit of books from the Fishbein Collection in the history of medicine took place in 1972. Five years later, a similar display of works from the Schaffner Collection in the history of science was mounted. Moreover, by the mid-seventies preliminary discussions were already underway regarding the possible move of the John Crerar Library to the University of Chicago. However, in addition, our students profited from the early printed works of the Newberry Library on the North Side of Chicago, and even, on occasion, from the extensive collections in the history of science housed at the Memorial Library at the University of Wisconsin.

It was evident that there would always be a problem in obtaining satisfactory coverage for fields as vast as the history of science and medicine. Because of this, we sought to broaden our contacts with other historians of science and medicine in the Chicago area. To this end, we initiated annual meetings of local scholars and graduate students at different universities and libraries. On a number of occasions, we drew on this wider pool of specialists for the guidance of our students.

We also tried to bring to the attention of our group the regular meetings of the Society of Medical History of Chicago. I even served as its president for several years in the seventies. And, in concert with George Daniels at Northwestern University, we organized a history of science program for the Big Ten schools through the Committee on Institutional Cooperation. We began a fellowship program for our students, although there never seemed to be enough money to fund it properly, and we held an annual open house at our home in Deerfield for students and faculty alike. Morris and Anna Fishbein always attended these gatherings, which he mentioned in his *Season's Greetings.* After one of the first (24 October 24 1970), he noted, "in the afternoon to the home of Allen Debus in Deerfield where came all the graduate students in medical history with their wives and many members of the medical faculty of the University of Chicago." I can recall Morris Fishbein in front of the house greeting everyone who attended. However, in fact, many of those present represented fields other than medicine or medical history.

In spite of the Center's many successes, as I look back on those six years, I also recall some failures. The Fishbein funds yielded only about $30,000 a year, making it very difficult either to carry on an extensive program or to expand. We did offer two modest Fishbein fellowships, but we needed to fund more since every year we had more well-qualified applicants than we could support. Of far greater importance we needed to add to our faculty in the fields of recent science and medicine. We argued for this every year. In the early seventies, Chicago may well have had the finest faculty anywhere for the study of science through the eighteenth century, but we were keenly aware of our deficiencies in the period after 1800. We saw an essential need for specialists in the history of both the recent physical and biological sciences as well as in the history of recent medicine. Every year, we organized search committees and nominated specific candidates to the department of history. My files contain numerous folders related to these search committees. But even though we obtained departmental approval regularly for our nominations, little came of our efforts. With Bylebyl's departure and the freeing of his salary, we were able to appoint Robert J. Richards. In him, we secured a specialist on Darwin and post-Darwinian evolution whose subsequent work has more than justified our early confidence. At the same time we obtained Thomas Jobe, a psychiatrist, who was working toward a doctorate in our program. Jobe's research paper in my seminar had just received the Schumann Prize of the History of Science Society, and we hoped that his joint appointment with the department of psychiatry would lead to an active joint program in the history of medicine with the Medical School. However, Jobe left after only a year and eventually became the head of clinical psychiatry at the University of Illinois at the Chicago Circle Campus. If he had remained, I believe that our program in the history of medicine might have made more headway at the time.

By 1976–1977, however, there seemed little hope of making an appointment in the recent physical sciences. The 1970s had witnessed declining revenues at academic institutions and the end of the sixties period of expansion. In addition, the departmental chairs who followed McNeill and Franklin clearly had less interest in the development of the program in the history of science and medicine. The pressures of the debates with the philosophers of science now seemed a thing of the past, and the Center had its own—albeit small—budget, something few other fields in history could boast. Thus, although the Fishbein Center was already well-known, and although much had been accomplished, there seemed little hope for continued expansion at this point.

My second term as director ended 30 June 1977, and Arnold W. Ravin was chosen to succeed me. Since he had to postpone this assignment for a year, William McNeill served as the Center's acting director for 1977–1978. I was pleased to see McNeill take this position because of his many efforts on behalf of the history of science over the years, and I did all I could to help him during his interim directorship. Also in 1978 I was appointed the Morris Fishbein Professor of the History of Science and Medicine, and Elizabeth Bitoy took over the secretarial post left vacant by Helen Little's departure for the department of Far Eastern Languages and Civilizations. We have been fortunate that Mrs. Bitoy has remained in this position since that time.

Arnold Ravin, as a former chair of the Committee on the Conceptual Foundations of Science, worked for a closer union of the center and the committee. However, not long after he became the director he fell seriously ill. Before he died in 1981, I spoke to him about the administration of the Center and called a meeting to take care of the normal business that could not otherwise be handled in his absence.

George W. Stocking, Jr., succeeded Ravin as Director in 1981, and he remained in that position until 1993, when he was succeeded by Bob Richards. During his successful regime, I have remained on the center's Executive Committee, but I have consciously avoided any major role in its administration. The number of graduate students in the programs of both the department of history and the Committee on the Conceptual Foundations of Science has continued to grow as has the actual membership of the center. The most recent published brochure lists twenty-seven members not only from the departments of history and philosophy, but also from anthropology, radiology, statistics, sociology, mathematics, public policy, chemistry, social thought, astronomy, the university library, and the committee on art and design. There have been an increasing number of lectures by guest speakers, perhaps too many for students who should be devoting most of their time to their studies and their own research. The center is also somewhat richer today, and this has resulted in an increased number of fellowships and prizes for students.

In retrospect, I am generally pleased with the decisions I made from 1961 to 1977. The department of history provided truly fertile ground for the growth of new programs under William McNeill. He encouraged us to develop the field of the history of science. The department's very flexible rules then permitted students to cross departmental and divisional lines to draw on the riches of the faculty in all areas of the University. Above all, the field committee structure of the department allowed specialists to make decisions regarding students, faculty, and all areas of the program itself. This proved particularly important for the history of science which had often developed as an independent program at other institutions. I remain convinced that extreme flexibility is required for any program in the history of science or the history of medicine. Although these fields have moved more toward the social relationships of science and medicine than toward technical and internalists studies, any successful program must be able to accommodate both approaches. We recognized this in the sixties, and it was one of the main points of contention between the historians and the philosophers of science. At the same time, I believe that one area in which the historians of science must receive training is history.

They should be well-grounded in the intellectual, social, and political history of the period in which they plan their ultimate research. Above all, they should be students of the historiography of their field and of historical method in general.

It is, however, one thing to say what one believes and quite another to apply it to a given situation. All departmental chairs are not like William McNeill. At Chicago, in spite of a reluctance to his plan on the part of some members of the department, McNeill directed the expansion of the scope of the studies in history. I am well aware that elsewhere historians sometimes are hesitant to introduce new fields. Indeed, in many places support for the history of science and medicine remains in the individual science departments and medical schools. Unfortunately, I thought the excellence of the departmental structure in the sixties was so evident that it would be permanent. But an attack on this structure in the early 1980s abolished the field system, which had proved so advantageous to the history of science program. Today, departmental committees drawn from all areas make decisions in regard to students and programs alike. Thus, students working in the history of science or applying for admission in the field must deal with committees composed of faculty members who may have little knowledge of either the history of science or the history of medicine.

If the decisions of a quarter century ago were made today, I would just as firmly oppose the exclusiveness of the proposed program in the history and philosophy of science, but I would perhaps insist less on maintaining the history of science within the department of history. It is important to keep in mind that students applying for admission to a program in the history of science or medicine may have widely divergent backgrounds. Some may come from the sciences, but many do not. Others may never have had a course in history in their undergraduate training. Admission and fellowship decisions often differ significantly from those in more traditional areas of history. There is no doubt that with the field committee organization of the department in the 1960s and 1970s, historians of science had far more control over the admission of students and over their progress within the program than they do today. Furthermore, the field committee program maintained considerable flexibility and allowed students to take fields in other departments and divisions of the university. Nevertheless, the final degree was in history, a real advantage because the availability of jobs in history will always be greater than in more specialized areas. This semi-autonomous arrangement gave the faculty authority where it was most necessary, while it preserved the necessary connection with history, and ensured that our Ph.D. students would be historians, not just historians of science. Since the history of science had developed originally outside of traditional history programs, this arrangement also provided our program with a certain needed identity. We were not simply submerged within a large and traditional department.

With the termination of the field committee system, the history of science has in a sense lost this identity. Today, I believe that to regain the position we had as recently as ten years ago, it would be necessary to establish a separate department or degree-granting committee of the history of science and medicine—something I have always opposed in the past. In today's climate, this arrangement would have the clear advantage of giving the history of science and medicine the autonomy they need. Although students would be required to take one or more fields in history, they would lose something by not obtaining a degree in history. And at the same time, the faculty would lose by having to take on otherwise unnecessary administrative duties.

Teaching and Research

In his notice of 10 June 1977 to the faculty of the Morris Fishbein Center for the Study of the History of Science and Medicine, Dean William Kruskal stated that "Professor Allen G. Debus will return to full time research and teaching on 1 July 1977." In fact, my research and teaching have always been my chief interests. Surely, the opportunity to work with the talented young men and women who have come to the University of Chicago has been a great privilege for me. It was particularly gratifying to see so many of my former students at the conference that has resulted in the present book.

Unfortunately, teaching is not a skill that is taught, rather it is learned of necessity. I believe that satisfactory teaching really depends on personal research. I have found too often that relying on historical accounts of others can be dangerous—especially if one has a different approach to the sources. I recall once telling Marshall Clagett that the more I read in the sources the less confident I became of the truth of secondary accounts. He agreed and added that this lack of confidence continues to grow the more one reads.

And yet, what is a young Ph.D. to do? After five to eight years of studying in a major research institution to become an authority on a very specific subject, one takes a job and is suddenly faced with the need to teach general courses. A graduate in European history who may be a specialist in some aspect of nineteenth-century central European history will find it necessary to teach a course covering everything from ancient Egypt to the present. The historian of science may have to survey the development all the sciences from antiquity to the present. This is indeed what George Sarton had done in a two-year course at Harvard. When I was a student, Cohen offered a condensed–but similarly spirited–one-year version. Cohen once said that graduate students should take detailed notes of the lectures of their teachers since they would need them when they began teaching. He was quite right. For the beginner, the pressure of teaching almost forces the use of notes from graduate courses and of secondary accounts, even if one would prefer not to. And if one is opening a new field as I was at Chicago, it is important to offer a broad course. When I joined the Chicago faculty in 1961, Cyril Stanley Smith suggested that I offer a course on sixteenth- and seventeenth-century chemical analysis because he had heard me speak on this subject in New York, but I felt that to establish the field I had to give a survey course. Over the course of my first few years at Chicago, I did develop a three-quarter course beginning with ancient science and continuing to the beginning of the twentieth century. Only later did I begin to offer more intensive seminars on sixteenth- and seventeenth-century science and medicine.

My primary interest has always centered on the period of the Scientific Revolution, which had traditionally been presented in terms of the development of astronomy and the physics of motion and of the triumph of the new experimental philosophy of the seventeenth century. My own research–strongly influenced by the work of Walter Pagel–has presented the chemical philosophy as an early modern alternative to the ancient-modern dichotomy normally presented in the textbooks. This led me, in 1971, to introduce a course on the historiography of the history of science and medicine that indicated the development of the various interpretations of scientific and medical history from the Renaissance to the present. It also led me to see the need for primary sources for students. The major works of Galileo and Newton were readily available along with Harvey's work on the circulation of the blood and Boyle's *Sceptical Chymist.* There was even a translation of some of the shorter works of Copernicus. The work of Paracelsus and the chemical authors of the period were not easily accessible, however. As a result, I have always fostered the publication of a variety of primary texts in this field for use in the classroom and elsewhere. (See below.)

I was also convinced of the importance of beginning courses with readings in these difficult and often mystical texts. If the students could understand the goals of the chemical philosophers from the beginning, I knew that they would have no trouble dealing later with the more modern thought of Galileo or Newton. In all of this, the discussion approach employed at Chicago is essential. A student from the sciences must be able to question why such "non-scientific" material is being used in a course or a seminar on the history of science. And future historians can only learn the significance of early modern concepts of nature through discussion.

Above all, seminars are important for they give students the opportunity to define research topics of interest to them. No less important, there they will find out whether they have acquired the necessary skills to pursue those topics. In my own case, I eventually published all of the seminar papers I prepared at Harvard. Although there are those who would deny it, the academic world is truly a "publish or perish" profession. Because of the importance of publications, I have always tried to impress on my students that the research that they carry out in a seminar should not be wasted. They should consider this work as a preliminary study related to their dissertation or as a potential paper for publication. In fact, many papers from my seminars have eventually appeared as journal articles, and some have led to published books.

In short, the teaching of graduate students only partially involves the imparting of information. Students should be taught to apply their own interests and skills to their reading of the primary sources, and they should realize that their instructors have interpreted the subject they teach based on their own personal backgrounds, just as the students themselves will in the future.

When I began to focus on the history of science, the field reflected the positivistic legacy of George Sarton. The Scientific Revolution of the sixteenth and seventeenth centuries dominated the literature, and there was a special emphasis on the period leading from Copernicus to Newton. Most historians of science had been trained in mathematics or physics and relatively little attention was paid to the sciences other than astronomy and the physics of motion. William Harvey and the circulation of the blood seemed to be the major exception to this rule. As mentioned above, my initial contact with the history of science at Indiana University resulted in a paper on Robert Boyle's *Sceptical Chymist* (1949–1). I remember how pleased I was when John F. Fulton, the compiler of *A Bibliography of the Honourable Robert Boyle* (1932), included a reference to it in the second edition of his book (1961).

During the five years I spent at Abbott Laboratories prior to entering Harvard's graduate program in the history of science, my research necessarily shifted to chemistry (see 1958–1, 1959–1, 1959–2, 1960–2, and 1963–3), but I also began to write a column on collectors' recordings. The latter appeared in *Hobbies* regularly from 1954 to 1956 and then sporadically down to 1967. I have over the years continued to speak and publish actively on American popular entertainment from the 1890s through the early 1930s. I was particularly gratified when the American Musical Theater group of the Smithsonian Institution asked me to prepare a three-record (LP) boxed set of music of Victor Herbert (1979–2). This contained original cast and other contemporary recordings from my collection ranging from 1898 to 1913. A long acquaintance with the discographer, Brian Rust, dating from the fifties resulted in our *Complete Entertainment Discography* (1973–1), which has become a standard guide for collectors of personality recordings and which has recently been revised and expanded (1989–1). My activities as a collector in this field have been detailed in a published interview by Michael G. Corenthal (1986–6).

Harvard was like a new world after spending five years at Abbott Laboratories. There I pursued my interest in the followers of Paracelsus first in Jordan's seminar on Tudor England. This resulted in a paper on "The Paracelsian Compromise in Elizabethan England" which won the Bowdoin Prize in the Natural Sciences for 1957 and was later published in *Ambix* (1960–1). From the beginning of my graduate work, it seemed to me that the standard works on sixteenth- and seventeenth-century science paid far too little attention to the debates centered on chemistry. The Paracelsians spoke of their "chemical philosophy" which was opposed to the natural philosophy and medicine of the ancients no less than to the much-vaunted mechanical philosophy. I planned then to prepare a work on the chemical philosophy and hoped to make this the subject of my thesis. I should add that I never thought of this work in terms of a rejection of the traditional history of science. Rather, I felt then as now that a shift from a positivistic to a more contextual approach would add a new dimension to our understanding of the period.

At first, I was unaware of the volume of sixteenth- and seventeenth-century literature on chemistry and chemical medicine. Part of the problem stemmed from the fact that this literature had gone practically untouched by scholars. However, I continued to examine specific aspects of the material. Another seminar paper, "Robert Fludd and the Circulation of the Blood," won a second Bowdoin Prize in 1958 (1961–1). This paper and the earlier one on the Paracelsian Compromise pointed me toward a more thorough examination of the English Paracelsians. This came to define the subject of my dissertation, "The English Paracelsians: A Study of Iatrochemistry in England to 1640," and several short papers based on material that had not formed part of my major work (see 1961–2 through 1962–3).

After my appointment at the University of Chicago in 1961, I had little time for research other than for the preparation of short papers. Ilza Veith suggested that I submit my dissertation on the English Paracelsians to the University of Chicago Press. I did, and it was returned with a number of suggestions that I had no time to attend to then. In 1962, Michael Hoskin approached me about the possibility of publishing my text in a history of science series he had recently initiated at the Oldbourne Press in London. My initial reaction was negative because I remained convinced that there was relatively little interest in the Paracelsian chemists among historians of science. Walter Pagel once told me to write books, not articles, because individual papers quickly become lost in the vast periodical literature. This advice, in addition to Hoskin's persistence finally prompted me to send him the manuscript. *The English Paracelsians* (1965–1) became one of the early books in the Oldbourne History of Science Library in 1965 and appeared one year later in an American edition by Franklin Watts. Oddly enough, the Franklin Watts stock of the Oldbourne series was later taken over by the University of Chicago Press.

As alluded to above, my early years at Chicago made me keenly aware of the need for more primary sources for students. Accordingly, I began to publish editions of sixteenth- and seventeenth-century texts. The first of these, the history of chemistry and chemical medicine written by R. Bostock in 1585 (1964–2) underscores the importance of the Paracelsians' belief in the antiquity of their medicine. Three years later, I wrote an introduction to a reprint (1967–1) of Elias Ashmole's *Theatrum Chemicum Britannicum* (1652), the most important of the seventeenth-century English alchemical collections. At this time, my interest was also growing in the 1654 debate between the chemist, John Webster, and the mechanists, Seth Ward and John Wilkins, over the reform of the English universities. Several papers on this exchange preceded a reprint of the tracts themselves in *Science and Education in the Seventeenth Century: The Webster-Ward Debate* (1970–1).[13] I still use the tracts of Webster and Ward in my seminars to illustrate the confrontation between chemists and mechanists in the mid-seventeenth century.

Several years later, I began to edit a series entitled "Primary Sources from the Scientific Revolution" for Neale Watson's Science History Publications. Although a number of texts were planned for the series, only two actually appeared. The first of these reprinted John Dee's "Mathematicall Praeface" to the 1570 English translation of Euclid (1975–1), and the second was the first publication of Robert Fludd's *Philosophicall Key* (1979–1). The latter manuscript, which dates from about 1618, presents Fludd's defense of his system of the macrocosm and the microcosm as well as for his description of the experiment on wheat in which he tried to isolate the vital spirit.

My interest in Fludd began as a graduate student. He is important not only because he sought a new understanding of the world based on hermetic thought and Scriptural authority, but also because he extensively used observational and experimental evidence. Fludd's views ran counter to those of other scientists of his day, and he engaged in lengthy debates with Kepler, Mersenne, and Gassendi. I had included a chapter on Fludd in *The English Paracelsians,* and I continued to examine aspects of his work for many years. One of my first publications had concerned his connection with William Harvey and his views on the circulation of the blood (1961–1), a theme I returned to nearly ten years later in a paper on "Harvey and Fludd: The Irrational Factor in the Rational Science of the Seventeenth Century" (1970–5). I examined Fludd's application of magnetism to his cosmology and his medicine in "Robert Fludd and the Use of Gilbert's *De Magnete* in the Weapon-Salve Controversy" (1964–5), and I explored his views on the importance of the sun and its place in the universe in a paper presented in Brussels in 1963 (1965–2). More general accounts, which attempted to present an overview of his thought, came out in several publications (1966–1, 1967–2, 1967–3, 1972–2) and a comparison of the reception of the works of Fludd and van Helmont was presented at a conference on Capri in 1973 (1975–2). A study of Fludd's application of the barometer/thermometer appeared as the "Key to Two Worlds: Robert Fludd's Weather-Glass" (1982–3).

The introduction of figures such as Fludd and other more mystical Paracelsians to the history of science clearly involved a shift from a positivistic approach to the field to a more contextual one. My interest in historiography dates from my first years at Chicago when Louis Gottschalk asked me to contribute a paper to the *Cahiers d'Histoire mondiale*, and I wrote on "The Significance of Early Chemistry" (1965–3). A series of historiographical papers followed. I discussed the relation of alchemy to the history of science in (1967–5) and contributed an article on "The Chemical Philosophy of the Renaissance" to George Basalla's *The Rise of Modern Science: Internal or External Factors* (1968–3). More recently, I have analyzed the varying interpretations of Paracelsus from the Renaissance to the present (1984–2) and have even commented on some of the problems related to the study of Islamic science (1989–3).

While spending a year at the Institute for Research in the Humanities at Madison in 1981–1982, I received a letter from Professor A. J.

A. Gouveia, a former rector and chair of the department of chemistry at the University of Coimbra. He wrote of his interest in the history of chemistry and of his plans to introduce the history of science in Portugal. A pleasant correspondence followed that resulted in an invitation to give a series of lectures at Coimbra in April 1983. For these lectures, I drew on the historiographical material I had been teaching. The four lectures were published by the University of Coimbra Press the following year as (1984–1). Since then I have returned to Portugal a number of times and have been elected a Foreign Associate of the Academia das Ciências de Lisboa (1987).

Aside from my work on the Scientific Revolution, I occasionally turned to other projects. In the early sixties, the A. N. Marquis Company, publishers of *Who's Who in America* asked me to edit a "Who's Who" in science (1968–2). I enlisted the help of a large number of Chicago students to search for information on historical figures in the sciences. My colleague, Edward Collins, and our graduate student, Ronald Calinger, served as assistant editors on this project, which in the end included about 25,000 scientists, roughly half of whom were living, and the remaining covered the sciences from antiquity to the twentieth century.

The meeting of the International Congress of the History of Science in Paris in 1968 led to a series of discussions with scholars who had been deeply influenced by the work of Walter Pagel. The idea of a *Festschrift* in his honor met with real enthusiasm, and I acted as editor of the two-volume set which appeared four years later (1972–1). This work was to be the first publication of Neale Watson's Press, Science History Publications. I undertook a similar task in 1971. UCLA's C. Donald O'Malley had, over the years, organized a series of successful symposia on the history of medicine and was working on one on seventeenth-century English medicine before his death. I participated in this conference with a paper on Noah Biggs and medical reform in the mid-seventeenth century and subsequently served, with the assistance of Diana Menkes, as editor of the resulting volume, *Medicine in Seventeenth Century England* (1974–1).

In spite of this time spent on other projects, my primary goal remained the one I had set for myself as a graduate student, a study of the sixteenth- and seventeenth-century chemical philosophy as a whole. Most of the papers I wrote for conferences in the sixties and seventies were related directly to this end. Thus, in 1967 I published on the role of fire analysis in the determination of the elements and showed the dependence of Robert Boyle's *Sceptical Chymist* on the earlier work of van Helmont (1967–4). The first year we spent at Churchill College required a public lecture from the Overseas Fellows, and I presented an overall view of the Chemical Philosophy in "The Chemical Dream of the Renaissance" (1968–1). Through Walter Pagel's arguments for the influence of Aristotle and Galen on Paracelsus and his followers, I became particularly interested in the actual differences between the Paracelsians and their competitors both ancient and modern. Accordingly, I studied the Paracelsians' use of mathematics (1968–5) and motion (1973–7), since the mathematization of the sciences and the new views of local motion are the subjects we normally associate with the mechanical philosophy of the seventeenth century.

However, the Paracelsians differed not only with the mechanists but also with the Galenic medical establishment. I discussed "The Chemical Compromise in Early Modern Medicine" in my paper for the Pagel *Festschrift* (1972–1) and then distinguished between the Paracelsian theorists and the practice of chemistry in a paper for *Clio Medica* (1972–3). A second article for *Clio Medica* investigated "The Pharmaceutical Revolution of the Renaissance" (1976–1), and a piece in *History of Science* (1974–1) gave an overview of the theory and practice of chemical medicine from Paracelsus to van Helmont.

An invitation to speak at a conference on the history of geology provided the opportunity to explore the manner in which sixteenth-century chemical theory had been applied to the growth of metals (1969–1), while the chance purchase of a seventeenth-century book by Hugh Plat led me to look into the application of chemical theory to agricultural reform in the early modern period (1968–6). I also contributed a long article on alchemy to the *Dictionary of the History of Ideas* (1973–3) and even looked at the strange chemical concept of palingenesis (1973–8).

We spent the year 1972–1973 at the Institute for Advanced Study at Princeton, and during that year I was finally able to turn my attention completely to the long-delayed work on the chemical philosophy. Perhaps the delay was beneficial since during the intervening years I had prepared a number of research papers on significant points that I had planned to develop in the book. As my writing proceeded, I agreed to let Neale Watson publish the volume. The manuscript was essentially complete by the end of the year at Princeton, but a series of new developments held up its publication for four years. Covering a chronological spectrum from the chemical background of Paracelsus to the late seventeenth century, this is the work for which I would most like to be remembered (1977–1).

Very early in my research, I had become convinced of the necessity of detailing the relationship of the Paracelsians and the chemical physicians to mainstream history of science. For this reason, I had readily accepted an invitation to contribute a volume on Renaissance science to the planned history of science series being edited for Wiley by my former classmates at Harvard, William Coleman and George Basalla. I signed my contract for this volume in the early 1960s and discussed the book in depth with Coleman in Poland in 1965 at the time of the International Congress of the History of Science. Due to the ongoing debates over the future of the history of science at Chicago—as well as to my feeling that it was necessary to complete my work on the chemical philosophy first—I was unable to begin work on this book until some ten years after signing the contract. A fellowship from the National Endowment for the Humanities taken at the Newberry Library in 1975–1976 finally gave me the time to write *Man and Nature in the Renaissance* (1978–1). On its completion, I sent the manuscript to Wiley and was surprised to hear that they had dropped the series. No one had told me this—nor had Coleman or Basalla been informed. They immediately began to look for a new publisher and finally chose the Cambridge University Press. My book was the first in the series to be published by Cambridge. As I write, this book has gone through eleven printings in English and has been translated into Italian, Spanish, Japanese and Chinese. Plans are currently underway for translations into Greek and Portuguese.

I had published three books in 1977, 1978, and 1979; *The Chemical Philosophy: Paracelsian Science and Medicine in the Sixteenth and Seventeenth Centuries* (1977–1), *Man and Nature in the Renaissance* (1978–1) and *Robert Fludd and His Philosophicall Key* (1979–1). *The Chemical Philosophy* received the Pfizer Award of the History of Science Society and the Edward Kremers Award of the American Institute of the History of Pharmacy in 1978.

After the completion of *The Chemical Philosophy* I was not sure at first what major project to pursue. I did know that I wanted to work on the late seventeenth and eighteenth centuries rather than return to the Renaissance. My papers from this period reflect some of this indecision. I was extremely interested in the *Philosophical Essay* (1672) of Thomas Sherley (1980–1) which sought to found a new science on the work of van Helmont and Robert Boyle—an unlikely union from today's viewpoint, but seemingly quite reasonable for Sherley. Meetings in Aleppo and Istanbul sparked an interest in the spread of western chemical medicine to Islam in the seventeenth century (1988–3 and 1989–3). I returned to the ideas of Robert Fludd (1982–3) and touched on the late eighteenth-century medico-astrological work of Ebenezer Sibley (1982–2). Plans for a Festschrift in honor of I. Bernard Cohen finally led me to collect my notes on eighteenth-century chemical medicine into a paper on "The Paracelsians in Eighteenth Century France: A Renaissance Tradition in the Age of the Enlightenment" (1981–1). I was able to discuss the effect of the Enlightenment on Paracelsian thought again in my inaugural address (1984–2) to the International Congress for the History of Pharmacy in Washington, D.C. in 1983.

By 1980, I had decided that my next book would focus on the development of Paracelsian thought and chemical medicine in France. I had already touched on the French chemico-medical debates of the seventeenth century in *The Chemical Philosophy,* but I hoped to center this new study on the rich and largely ignored literature of the eighteenth-century literature. In 1981–1982, I had a National Endowment for the Humanities fellowship to study at the Institute for Research in the Humanities at Madison, and with the resources of the Duveen Collection at the Memorial Library, I thought that I would successfully complete this book by the end of the academic year. However, this was not to be. Although I made considerable progress, the American Musical Theater group at the Smithsonian pressured me to complete another collection for them—this time on the black musical comedy star, Bert Williams. Then, a fall on the ice in November resulted in a broken arm and brought a temporary halt to my writing. An invitation to present a paper at the International Conference on "Hermeticism in the Renaissance" in March permitted me to present some of my then current research in a paper on "Alchemy and Paracelsianism in Early Eighteenth Century France." This meeting also resulted in an invitation to edit the conference volume with Ingrid Merkel of the Catholic University of America (1988–2).

An international meeting on the historical concept of "Spiritus" in Rome (1983) gave me the opportunity to gather together two decades of work on the vital spirit and the aerial niter (1984–3). And at a symposium on "Representation and Value: Literature, Philosophy and Science" in Atlanta (1986), I explored the use of myth and allegory in the alchemical literature of the Renaissance (1987–2). In 1985 the Catholic University of Louvain celebrated the three hundredth year of the establishment of its first chair in chemistry. I was awarded an honorary doctorate, and this visit to Belgium was the occasion for the presentation of several papers, including a study of the introduction of chemistry to the European universities in the seventeenth century (1986–5) and a more general discussion of "The Role of Chemistry in the Scientific Revolution" (1986–2).

In 1987 I was given the Dexter Award of the Division of the History of Chemistry of the American Chemical Society, and in my acceptance address, I turned to the relation of chemical quantification and medical motivation in the early modern period (1989–2). The same year a collection of thirteen of my earlier papers was published by Variorum as *Chemistry, Alchemy and the New Philosophy, 1550–1700* (1987–1). Covering the period from 1961 to 1983, these papers included studies on key aspects of the chemical philosophy as well as a number of papers on the work of Robert Fludd.

I was fortunate to spend the first six months of 1987 as a National Endowment for the Humanities Fellow at the Folger Shakespeare Library in Washington. While there, I hoped finally to write my long-planned book on French medical chemistry in the eighteenth century. My earlier work in *Ambix*, the Cohen *Festschrift,* and *The Chemical Philosophy* had given me a head start on this project. However, as I began to work on an introductory chapter on the sixteenth- and seventeenth-century background, I came to realize how little was really known of this early period. My introductory chapter soon evolved into a plan for a new book on French chemistry from the early sixteenth through the early eighteenth centuries. Although this work was for the most part completed in Washington in 1987, it did not appear as *The French Paracelsians* (1991–1) until 1991. Another short book (in Italian) is to come from a series of lectures given at the Istituto italiano per gli Studi filosofici in Naples at the end of March 1990. There, I gave an overview of my research covering the period from Paracelsus through the later seventeenth century.

Over the years, I have become ever more convinced that we must include the development of chemistry as a major factor in our understanding of the rise of modern science. To do so will involve a change in the viewpoint of historians of science. The Paracelsians openly rejected the texts of Aristotle, Galen, and their commentators, while praising Plato and the mystical authors of antiquity in both medicine and natural philosophy. It is thus impossible to separate the history of medicine and the history of science when discussing their work. At the same time, the Paracelsians lauded their own reliance on new observations, but found themselves at odds with those who sought a new mathematically and mechanically based science. The chemical debates with both the so-called "ancients" and "moderns" forms an essential background to the chemical revolution that took place in stages from Paracelsus down to Lavoisier. To ignore the sixteenth-, seventeenth-, and early eighteenth-century developments in this story will perpetuate the relegation of chemistry to a lesser role in the Scientific Revolution proper. This is the message I presented in "The Iatrochemical Background to the Chemical Revolution" (1990–1) at the international conference on alchemy held at the University of Groningen in April 1989 (1990–1), reiterated in Brazil the following year, and developed further in an address to the Italian historians of chemistry at a meeting in Venice in November 1991. In Venice, I discussed early modern chemistry as an essential component for a proper interpretation of the Scientific Revolution as a whole.

Postscript

As I look back on the past thirty years, I realize that although the debates we had regarding the relationship of the history of science to the philosophy of science were at times painful, they served to define these fields not only at the University of Chicago, but perhaps elsewhere as well. Earlier assumptions that the two fields form an essential unity may have received their impetus from the rapid development of the philosophy of science and the enormous influence of Alexandre Koyré. The fact that the department of history at Chicago was so receptive to this field in the 1960s resulted in the rapid growth of a new history of science program, but almost inevitably resulted in the confrontation with the philosophers of science. In the past few decades, the importance of an historical basis for the field has become more evident. In any case, the experience of participating in the development of a new academic program was interesting and rewarding, if at times frustrating.

The history of science has undergone a series of transformations in recent decades. There is no longer the overwhelming emphasis on the sixteenth- and seventeenth-century Scientific Revolution that characterized my time as a student. Since then there has been a turn toward nineteenth- and twentieth-century research—first in regard to Darwin and the development of evolutionary thought, and now more frequently toward the recent physical sciences. Above all, there is an increasing interest in the relation of the sciences to society. Here I regret only the frequent uncompromising attitude of each new wave of scholars. In fact, there is so much material still to read and digest that we need the contributions of everyone. We are still far from the point where we can say that there is a single correct history of this field. Indeed, we may never reach such a state.

As for myself, I plan to continue my work on the chemical debates of the early modern period and to carry this story well into the eighteenth century. I would like to establish these developments as essential for an understanding not only of the background to Lavoisier but also of the Scientific Revolution as a whole. I think that there is already a greater appreciation of the importance of the role of chemistry in the sixteenth and the seventeenth centuries, and I am pleased to have played some part in this.

1 March 1992
Deerfield, Illinois

[*Editors' Note:* In 1994, Allen G. Debus received the Sarton Medal of the History of Science Society for his lifetime contributions to the field.]

Notes

1. On both sides the family were people who had emigrated to the United States and Chicago in the middle decades of the last century. My father's grandfather had practiced his trade as a shoemaker in Chicago, and my mother's father, Hans Schwenneke, had worked in a Chicago kite factory as a boy. There were some adventurous members of the family. My mother's maternal grandfather, Theodore Susan, had enlisted in the Illinois Infantry Volunteers in 1862 when he was over forty and fought to the end of the Civil War. My father's maternal grandfather, August Studt, had left his family and his lumber yard in 1898 to hunt for gold in the Klondike. My father, George Debus, was born while he was away.

2. My father's father, William H. Debus, sold imported Irish lace and linen and my mother's father worked for a box manufacturer before opening his own set-up box factory. My father learned the box business from him and then established his own company, Modern Boxes, Inc., in 1935.

3. I had been exposed to very little science before entering high school in 1940. There I took a number of courses, but none was more interesting to me than the survey of inorganic chemistry taught by Mr. Anderson. I am sure that his teaching influenced my future far more than I would have guessed at that time. In history a series of courses on American and World History taught by Emma Rogers and a senior-level course on modern European history by Mr. Ulrey gave substance and organization to my fascination with history that had been developing in a rather haphazard fashion earlier. This was also a time when the

study of Latin was still considered an important subject for all students who planned to continue on to college.

4. This reference, as well as similar ones below, refers to the bibliography found below in this volume.

5. The most interesting and valuable of these patents is U.S. patent 2,935,525 (1960–2), "Esterification of Organic Acids," which presented a new method for the preparation of the local anesthetics, novocaine (named procaine in its Abbott form). It was the first major improvement in this synthesis in some forty years.

6. A few lectures were given by Everett Mendelsohn, who did not yet have his degree. Edward Grant, John Murdoch, and Erwin Hiebert, recent graduates of the program at the University of Wisconsin who had single year appointments at Harvard, also taught in the program. Even then I was called on to give a lecture on early systems of element theory for the survey course in the history of science during my final year as a student.

7. It was not as common for American graduate students to work in England as it is today. I thought then it might be our only time to visit Europe. For this reason, we traded our old Buick for an English Ford Consul to be delivered in London. With it, we drove all over England and Scotland during our stay, and shortly before our return home in the summer of 1960, we motored through France, Switzerland, Austria, Germany and Belgium. Of course, since then we have returned to England and the Continent many times, but we have never regretted the effort we made to travel during our first trip abroad.

8. In addition to the preparation of so many papers, we also had the usual problems of day-to-day life: organizing shopping trips to the grocery stores, taking the children to different schools, and then picking them up in the afternoon. More serious was the fact that at the end of October my mother died, and we all returned home for the funeral. This came as a shock to us all, even though it was not entirely unexpected. We were concerned about leaving my father at home alone, but he agreed to visit us in England at Christmas. When he did come, we took a short vacation to Spain. Upon our arrival there I broke my arm playing with the boys. In short, by July we looked back on a very busy year and decided to fly home rather than booking passage on a ship. This decision was dictated in part by the fact that I had already agreed to attend a conference on the history of geology in New Hampshire early in September.

9. Derek J. de Solla Price, "A Guide to Graduate Study and Research in the History of Science and Medicine," *Isis* 58 (1968): 393–403.

10. With the decision made to remain at Chicago, we decided to move from Park Forest where our house was now much too small for our family. After briefly considering houses in Hyde Park, we decided to move to Deerfield which involved a long commute to the University but made it possible to be closer to my father in Northbrook.

11. Collins had been a brilliant and popular teacher. He had finally completed his dissertation on the science of Descartes toward the end of his fifth year as an Instructor at Chicago. Although the department voted overwhelmingly for his promotion, this recommendation was not approved by the administration. His departure represented a real loss to our program.

12. William B. Provine, *The Origins of Theoretical Population Genetics* (Chicago: University of Chicago Press, 1971).

13. In addition to the tracts by Webster and Ward, I had included a short work by Thomas Hall presenting the Aristotelian view toward the proposed reform. Hall's comments had served as an appendix to a longer work. I had discussed the make-up of the book with the publishers in London, but when they actually printed the book, they did not choose the correct section by Hall. An attempt was made to recall all of the copies, but this was only partially successful. And although the book was reprinted by them correctly, today both versions turn up for sale by booksellers.

SELECT BIBLIOGRAPHY

Allen G. Debus

This is a selected list of publications. Book reviews—other than essay reviews—are not included, nor are most publications related to American and British entertainment and popular music.

1949–1. "Robert Boyle and His Sceptical Chymist," *The Indiana Journal for Bookmen* 5 (1949): 39–47.

1954–1. "Current Collectors' Recordings." A series of articles that appeared in *Hobbies* magazine beginning in February, 1954. The series appeared regularly through 1956 and then sporadically to 1967.

1958–1. Patent: U.S. 2,858,329. "Preparation of Bis-diloweralkylaminoalkyl Loweralkyl Dicarboxylates," 2 pages, 28 October 1958 (with Willard C. Braaten and Gregory A. Bauer).

1959–1. Patent: U.S. 2,874,180. "Production of Procaine Dihydrogen Phosphate," 1 page, 17 February 1959.

1959–2. Patent: British 815,144. "Process for the Preparation of Esters of Tertiary Amino Alcohols," 4 pages, 17 June 1959. Based on U.S. 2,935,525.

1960–1. "The Paracelsian Compromise in Elizabethan England," *Ambix* 8 (1960): 71–97.

1960–2. Patent: U.S. 2,935,525. "Esterification of Organic Acids," 3 pages, 3 May 1960.

1961–1. "Robert Fludd and the Circulation of the Blood," *Journal of the History of Medicine and Allied Sciences* 16 (1961): 374–-393.

1961–2. "Gabriel Plattes and His Chemical Theory of the Formation of the Earth's Crust," *Ambix* 9 (1961): 162–165.

1962–1. "The Paracelsian Influence on English Medicine," *The Proceedings of the Institute of Medicine of Chicago* 24 (1962): 53.

1962–2. "Sir Thomas Browne and the Study of Colour Indicators," *Ambix* 10 (1962): 29–36.

1962–3. "Solution Analyses prior to Robert Boyle," *Chymia* 8 (1962): 41–61.

1962–4. "John Woodall—Paracelsian Surgeon," *Ambix* 10 (1962): 108- 118.

1963–1. "Paracelsian Doctrine in English Medicine," *Chemistry in the Service of Medicine,* ed. F. N. L. Poynter (London: Pitman, 1963), pp. 5–26.

1963–2. "Pierre Gassendi and His 'Scientific Expedition' of 1640," *Archives internationales d'Histoire des Sciences* 6 (1963): 129–142.

1963–3. Patent: Canadian 659,596. "Esterification of Organic Acids," 12 pages, 19 March 1963. Based on U.S. 2,935,525.

1964–1. "The Aerial Niter in the 16th and Early 17th Centuries," *Actes du dixième Congrès international d'Histoire des Sciences* 2 vols., (Paris: Hermann, 1964), 2:835–839.

1964–2. "An Elizabethan History of Medical Chemistry," *Annals of Science* 18 (1962, published 1964): 1–29.

1964–3. "A Forgotten Chapter in the Introduction of the New Chemistry in Italy," *Ambix* 11 (1963, published 1964): 153–-157.

1964–4. "The Paracelsian Aerial Niter," *Isis* 55 (1964): 43–61.

1964–5. "Robert Fludd and the Use of Gilbert's *De Magnete* in the Weapon-Salve Controversy," *Journal of the History of Medicine and Allied Sciences* 19 (1964): 389–417.

1965–1. *The English Paracelsians* (London: Oldbourne Press, History of Science Library, 1965). American edition, New York: Franklin Watts, Inc., 1966.

1965–2. "The Sun in the Universe of Robert Fludd," *Le Soleil à la*

Renaissance—Sciences et Mythes, Travaux de l'Institut pour l'Étude de la Renaissance et de l'Humanisme (Brussels: P.U.B./ P.U.F., 1965), 2:257–278.

1965–3. "The Significance of the History of Early Chemistry," *Cahiers d'Histoire mondiale* 9 (1965): 39–58.

1966–1. "Renaissance Chemistry and the Work of Robert Fludd" in *Alchemy and Chemistry in the Seventeenth Century,* ed. Allen G. Debus and Robert P. Multhauf (Los Angeles: William Andrews Clark Memorial Library, University of California, 1966).

1966–2. "The Chemical Philosophy of the Renaissance," *The Proceedings of the Institute of Medicine of Chicago* 26 (1966): 79.

1967–1. Elias Ashmole, *Theatrum Chemicum Britannicum . . . , A Reprint of the London Edition 1652, With a New Introduction by Allen G. Debus,* Sources of Science No. 39 (New York/-London: Johnson Reprint Corporation, 1967).

1967–2. "Renaissance Chemistry and the Work of Robert Fludd," *Ambix* 14 (1967): 42–59.

1967–3. "Robert Fludd and the Chemical Philosophy of the Renaissance," *Organon* (L'Institut d'Histoire de la Science et de l'Académie polonaise des Sciences) 4 (1967): 119–126.

1967–4. "Fire Analysis and the Elements in the Sixteenth and the Seventeenth Centuries," *Annals of Science* 23 (1967): 127–-147.

1967–5. "Alchemy and the Historian of Science," *History of Science* 6 (1967): 128–138. Essay review of C. H. Josten's *Elias Ashmole.*

1968–1. *The Chemical Dream of the Renaissance,* Churchill College, Cambridge, Overseas Lecture Series—No. 3 (Cambridge: Heffer, 1968). Reprinted Indianapolis: Bobbs-Merrill, 1972, as HS-16 in "The Bobbs-Merrill Reprint Series in the History of Science."

1968–2. *World Who's Who in Science: A Biographical Dictionary of Notable Scientists from Antiquity to the Present,* ed. Allen G. Debus (Chicago: The A. N. Marquis, Co., 1968).

1968–3. "The Chemical Philosophy of the Renaissance," in *The Rise of Modern Science,* ed. George Basalla (Lexington, Mass: D. C. Heath, 1968), pp. 82–88.

1968–4. "Philosophical Chemistry and the Scientific Revolution," *Actes du XIe Congrès international d'Histoire des Sciences,* 5 vols. (Warsaw: Académie polonaise des Sciences, 1968), 4:26–30.

1968–5. "Mathematics and Nature in the Chemical Texts of the Renaissance," *Ambix* 15 (1968): 1–28, 211.

1968–6. "Palissy, Plat, and English Agricultural Chemistry in the 16th and 17th Centuries," *Archives internationales d'Histoire des Sciences* 21 (1968): 67–88.

1969–1. "Edward Jorden and the Fermentation of the Metals: An Iatro-chemical Study of Terrestrial Phenomena," in *Toward a History of Geology: Proceedings of the New Hampshire Inter-Disciplinary Conference on the History of Geology, September 7–12, 1967,* ed. Cecil J. Schneer (Cambridge, Mass: M.I.T. Press, 1969), pp. 100–121.

1969–2. "'The Devil upon Dun,' A Seventeenth-Century Attack on the Quack Chemist," in *Die ganze Welt ein Apotheken, Festschrift für Otto Zekert,* ed. Sepp Domandl, Salzburger Beiträge zur Paracelsusforschung, vol. 8 (Vienna: Verlag Notring der wissenschaftlichen Verbände Österreichs, 1969), pp. 47–62.

1969–3. "The Webster-Ward Debate of 1654: The New Philosophy and the Problem of Educational Reform," *The Proceedings of the Institute of Medicine of Chicago* 27 (1969): 248–249.

1970–1. *Science and Education in the Seventeenth Century: The Webster-Ward Debate* (London: Macdonald and Co., Ltd./-New York: American Elsevier, 1970).

1970–2. "Becher, Johann Joachim," *Dictionary of Scientific Biography,* ed. Charles C. Gillispie (New York: Charles Scribner's Sons, 1970), 1:548–551.

1970–3. "The Webster-Ward Debate of 1654: The New Philosophy and the Problem of Educational Reform," *L'Univers à la Renaissance: Microcosme et Macrocosme,* Travaux de l'Institut pour l'Étude de la Renaissance et de l'Humanisme (Brussels: P.U.B./ P.U.F., 1970), 4:33–51.

1970–4. "John Webster and the Educational Dilemma of the Seventeenth Century," *Actes du XIIe Congrès international d'Histoire des Sciences,* 12 vols. (Paris: A. Blanchard, 1970), 3B:15–23.

1970–5. "Harvey and Fludd: The Irrational Factor in the Rational Science of the Seventeenth Century," *Journal of the History of Biology* 3 (1970): 81–105.

1971–1. "Chemistry and the Scientific Revolution," *Teaching the History of Chemistry: A Symposium, San Francisco, California, U.S.A., April 1968* (Budapest: Akadémiai Kiadó, 1971), pp. 101–111.

1971–2. "Duchesne, Joseph," *Dictionary of Scientific Biography,* ed. Charles C. Gillispie (New York: Charles Scribner's Sons, 1970), 4:208– 210.

1971–3. "The History of Chemistry and the History of Science," *Ambix* 18 (1971): 169–177.

1971–4. "The Relationship of Science-History and the History of Science," *The Journal of Chemical Education* 48 (1971): 804–805. Published in Chinese translation in the *Journal of Central China University Researchers* (1990–1991), pp. 153–-155.

1972–1. *Science, Medicine and Society in the Renaissance: Essays to Honor Walter Pagel,* ed. Allen G. Debus, 2 vols. (New York: Science History Publications/London: Heinemann, 1972). Includes an introduction by Allen Debus on the life and work of Walter Pagel, 1:1–9, and a paper by Allen Debus,

"Guintherius, Libavius and Sennert: The Chemical Compromise in Early Modern Medicine," 1:151–165.

1972–2. "Fludd, Robert," *Dictionary of Scientific Biography,* ed. Charles C. Gillispie (New York: Charles Scribner's Sons, 1970), 5:47–49.

1972–3. "The Paracelsians and the Chemists: The Chemical Dilemma in Renaissance Medicine," *Clio Medica* 7 (1972): 185–199.

1972–4. "Some Comments on the Contemporary Helmontian Renaissance," *Ambix* 19 (1972): 145–150. Essay review of the reprint of J. B. van Helmont's *Aufgang der Artzney-Kunst* (1683:1971).

1973–1. Brian Rust with Allen G. Debus, *The Complete Entertainment Discography: From the Mid-1890s to 1942* (New Rochelle, N.Y.: Arlington House Publishers, 1973).

1973–2. "Paracelso y los Paracelsistas" (with Walter Pagel), in *Historia universal de la Medicina,* ed. Pedro Laín Entralgo (Barcelona/ Madrid: Salvat Editores, S.A., 1973), 4:107–129. Divided into (a) "Paracelso" (Pagel): pp. 107–118; (b) "El Mundo de los Paracelsistas" (Debus): pp. 118–126; and (c) "Bibliografía" (Pagel and Debus): pp. 127–129.

1973–3. "Alchemy," *Dictionary of the History of Ideas,* ed. Philip P. Wiener, 5 vols. (New York: Charles Scribner's Sons, 1973–-1974), 1:27–34.

1973–4. "The Medico-Chemical World of the Paracelsians," *Changing Perspectives in the History of Science: Essays in Honour of Joseph Needham,* ed. Mikulas Teich and Robert Young (London: Heinemann, 1972), pp. 85–99.

1973–5. "La Philosophie chimique de la Renaissance et ces Relations avec la Chimie de la Fin du XVIIe siècle," *Sciences de la Renaissance: VIIIe Congrès international de Tours* (1965), ed. Jacques Roger (Paris: J. Vrin, 1973), pp. 273–283.

1973–6. "The Sciences and the Public in the Twentieth Century," in

The First Fifty-Year History of the International Association for Dental Research, ed. Frank J. Orland (Chicago: University of Chicago Printing Department, 1973), pp. 1–8.

1973–7. "Motion in the Chemical Texts of the Renaissance," *Isis* 64 (1973): 4–17.

1973–8. "A Further Note on Palingenesis: The Account of Ebenezer Sibley in *The Illustration of Astrology* (1792)," *Isis* 64 (1973): 226–230.

1973–9. "Elsie Janis, the First World War, and the Introduction of Jazz to England," *Popular Music and Society* 2 (1973): 334–-339.

1974–1. *Medicine in Seventeenth Century England: The C. D. O'Malley International Symposium,* ed. Allen G. Debus (Berkeley, Los Angeles, and London: The University of California Press, 1974). Includes paper by Allen G. Debus, "Paracelsian Medicine: Noah Biggs and the Problem of Medical Reform," pp. 33–48.

1974–2. "The Chemical Philosophers: Chemical Medicine from Paracelsus to van Helmont," *History of Science* 12 (1974): 235–259.

1974–3. "Motion in the Chemical Texts of the Renaissance," in *Actes du XIIIe Congrès international d'Histoire des Sciences: Moscou, 18–24 Aout, 1971,* 12 vols. (Moscow: Editions "Nauka," 1974), 7:196–204.

1975–1. John Dee, *The Mathematicall Praeface to the Elements of Geometrie of Euclid of Megara (1570),* introduction by Allen G. Debus, Primary Sources from the Scientific Revolution (New York: Science History Publications, 1975).

1975–2. "The Chemical Debates of the Seventeenth Century: The Reaction to Robert Fludd and Jean Baptiste van Helmont" in *Reason, Experiment, and Mysticism in the Scientific Revolution,* ed. M. L. Righini Bonelli and William R. Shea (New York: Science History Publications, 1975), pp. 18–47, 291–-298. Included in this volume are commentaries on my paper by William R. Shea, "Trends in the Interpretation of Seventeenth Century Science," pp. 1–17, and by Cesare Vasoli, "Alchemy in the Seventeenth Century: The European and Italian Scene," pp. 48–58.

1975–3. "Van Helmont and Newton's Third Law," in *Paracelsus, Werk und Wirking, Festgabe für Kurt Goldammer zum 60. Geburtstag,* ed. Sepp Domandl, Salzburger Beiträge zur Paracelsusforschung 13 (Vienna: Verband der wissenschaftlichen Gesellschaften Österreichs Verlag, 1975), pp. 45–52.

1975–4. "Severinus, Petrus," in *Dictionary of Scientific Biography,* ed. Charles C. Gillispie (New York: Charles Scribner's Sons, 1970), 12:334– 336.

1976–1. "The Pharmaceutical Revolution of the Renaissance," *Clio Medica* 11 (1976): 307–317.

1976–2. "Valentine, Basil," in *Dictionary of Scientific Biography,* ed. Charles C. Gillispie (New York: Charles Scribner's Sons, 1970), 13:558–560.

1976–3. "Webster, John," in *Dictionary of Scientific Biography,* ed. Charles C. Gillispie (New York: Charles Scribner's Sons, 1970), 14:209–210.

1976–4. "Medical Themes in American Vaudeville," *Newsletter of the Society of Medical History of Chicago,* no. 4 (June 1976), p. 3.

1976–5. "Morris Fishbein, Historian," *Newsletter of the Society of Medical History of Chicago,* no. 6 (November 1976), pp. 2–3.

1976–6. "Paracelsians and Pragmatists" in *The Times Literary Supplement* (18 June 1976), p. 746. Review of Owen Hannaway, *The Chemists and the Word: The Didactic Origins of Chemistry* (Baltimore: The John Hopkins University Press, 1975).

1977–1. *The Chemical Philosophy: Paracelsian Science and Medicine in the Sixteenth and Seventeenth Centuries,* 2 vols. (New York: Science History Publications, 1977).

1977–2. "A Tribute to Morris Fishbein," *Bulletin of the History of Medicine* 51 (1977): 153–154.

1977–3. "'The Drug Store Cabaret': Pharmacy and Vaudeville in 1920," *Pharmacy in History* 19 (1977): 39–42.

1977–4. "Chimica: Storia," in *Enciclopedia Europea* (Milan: Aldo Garzanti Editore, 1977), 3:295–298.

1978–1. *Man and Nature in the Renaissance,* Cambridge History of Science Series (Cambridge: Cambridge University Press, 1978). Subsequently reprinted in English in 1980, 1982, 1984, 1985, 1986, 1987, 1988, 1990, 1991.

1978–2. "Chemistry, Pharmacy and Cosmology: A Renaissance Union," *Pharmacy in History* 20 (1978): 125–137. The Edward Kremers Award Address, American Institute of the History of Pharmacy, Montreal Meeting, 16 May 1978.

1978–3. "The Pseudo-Sciences and the History of Science," *The University of Chicago Library Society Bulletin* 3 (1978): 3–20.

1978–4. "The Arabic Tradition in the Medical Chemistry of the Scientific Revolution," *Proceedings of the First International Symposium for the History of Arabic Science, April 5–12, 1976,* ed. Ahmad Y. Al-Hassan, Ghada Karmi and Nizar Namnum (Aleppo: Institute for the History of Arabic Science, 1978), 2:275– 290. Reprinted in *Essays on Science: Felicitation Volume in Honour of Dr. S. Mahdihassan,* ed. Hakim Mohammed Said (Karachi: Hamdard Foundation Press, 1987), pp. 33–50.

1978–5. "Victoria's World: The Sound of the English Music Hall," *The Hillandale News,* no. 105 (December 1978), pp. 149–-154.

1979–1. *Robert Fludd and His Philosophicall Key: Being a Transcription of the Manuscript at Trinity College, Cambridge, with an Introduction by Allen G. Debus,* Primary Sources from the Scientific Revolution (New York: Science History Publications, 1979).

1979–2. *Music of Victor Herbert* [Three record boxed set in the Smithsonian American Musical Theater Series: The Smithsonian Collection DPM 30366 R 017A Instrumental Selections; R 017B *The Fortune Teller*; R 017C *Babes in Toyland* and *Mlle. Modiste*; R 017D *The Red Mill* and *Naughty Marietta*; R 017E *Natoma* and *Madeleine*; 017F *Sweethearts*]. "Programming and annotation by Allen G. Debus, Original recordings from the collection of Allen G. Debus." Includes 16-page booklet: Allen G. Debus, *The Early Victor Herbert: From the Gay Nineties to the First World War* (Washington, D.C.: The Smithsonian Collection, 1979).

1979–3. "Science vs. Pseudo-Science: The Persistent Debate," The Morris Fishbein Center for the Study of the History of Science and Medicine, Publication No. 1 (1980), vi + 18 pp. Inaugural Lecture as Morris Fishbein Professor of the History of Science and Medicine.

1979–4. "Mysticism and the Rise of Modern Science," *Proceedings of the International Congress of the History and Philosophy of Science, Islamabad 8–13 December 1979,* ed. Hakim Mohammed Said (Karachi: Hamdard Foundation Press, n.d. [1983]), pp. 46–61. Reprinted in the *Journal of Central Asia* 3 (1980): 46–61; and in *Studies in History of Medicine* 4 (1980): 199–213.

1980–1. "Thomas Sherley's *Philosophical Essay* (1672): Helmontian Mechanism as the Basis of a New Philosophy," *Ambix* 27 (1980): 124–135.

1980–2. Comment on "Pathological Science: Towards a Proper Diagnosis and Remedy" by Ray Hyman in the *Zetetic Scholar,* no. 6 (July 1980), pp. 50–51. Reprinted in Ray Hyman, *The Elusive Quarry: A Scientific Appraisal of Psychical Research* (Buffalo: Prometheus Books, 1989), pp. 263–264.

1981–1. "The Paracelsians in Eighteenth Century France: A Renaissance Tradition in the Age of the Enlightenment," *Ambix* 28 (1981): 36–54. Modified version printed in *Transformation and Tradition in the Sciences: Essays in Honor of I. Bernard Cohen,* ed. Everett Mendelsohn (Cambridge: Cambridge University Press, 1984), pp. 193–214.

1982–1. *L'Uomo e la Natura nel Rinascimento,* trans. Amilcare Mantegazza (Milan: Editoriale Jaca Book, 1981). Italian translation of *Man and Nature in the Renaissance* (see 1978–1).

1982–2. "Scientific Truth and Occult Tradition: The Medical World of Ebenezer Sibley (1751–1799)," *Medical History* 26 (1982): 259- 278.

1982–3. "Key to Two Worlds: Robert Fludd's Weather-Glass," *Annali dell'Istituto e Museo di Storia della Scienza di Firenze* 8 (1982): 109–144. Invited article for the Festschrift issue honoring Maria Righini Bonelli.

1983–1. "The History of Science Today," *Memórias da Academia das Ciências de Lisboa: Classe de Ciências* 25 (1983): 89–111.

1983–2. "Walter Pagel (1898–1983)," *Bulletin of the History of Medicine* 57 (1983): 611–615.

1984–1. *Science and History: A Chemist's Appraisal* (Coimbra: Edição do Serviço de Documentação e Publicações da Universidade de Coimbra Subsidiado pela Fundação Calouste Gulbenkian, 1984). The fourth lecture of this volume, "The Scientific Revolution: A Chemist's Reappraisal," was translated into Chinese and published in *Zirankexue Zhexuewenti* [*Problems in Natural Science and Philosophy*] 1 (1987): 57–63.

1984–2. "History with a Purpose: The Fate of Paracelsus," *Pharmacy in History* 26 (1984): 83–96. Inaugural Lecture at the International Congress for the History of Pharmacy, 22 September 1982, Washington, D.C.

1984–3. "Chemistry and the Quest for a Material Spirit of Life in the Seventeenth Century," *Spiritus: IV° Colloquio internazionale del Lessico intellettuale Europeo, Roma, 7–9 gennaio 1983,* ed. M. Fattori and M. Bianchi (Rome: Edizioni dell'Ateneo, 1984), pp. 245–263.

1984–4. "Walter Pagel, M.D.—12 November 1898–25 March 1983,"

The British Journal for the History of Science 17 (1984): 83–84.

1984–5. "Walter Pagel, 1898–1983," *Archives internationales d'Histoire des Sciences* 33 (1983, published 1984): 333–334.

1984–6. "John Webster (1610–1682)," *Biographical Dictionary of British Radicals in the Seventeenth Century,* ed. Richard L. Greaves and Robert Zaller, 3 vols. (Brighton: Harvester Press, Ltd., 1981–1984), 3:297–299.

1985–1. *El Hombre y la Naturaleza en el Renacimiento,* trans. Sergio Lugo Rendón (Mexico City: Fondo de Cultura Económica, 1985). Spanish translation of *Man and Nature in the Renaissance* (see 1978–1). Reprinted Mexico City, 1986.

1985–2. "The Significance of Chemical History," *Ambix* 32 (1985): 1–14.

1986–1. *Runesansu no shizenkan: Riseishugi to shinpishugi no soukoku* [*The View of Nature in the Renaissance: Conflict Between Rationalism and Mysticism*], trans. Shuntaro Ito and Mariko Hashimoto (Tokyo: Saiensu-Sha Co., Ltd., 1986). Japanese translation of *Man and Nature in the Renaissance* (see 1978–1).

1986–2. "The Role of Chemistry in the Scientific Revolution," *Lias: Sources and Documents Relating to the Early Modern History of Ideas* 13 (1986): 139–150. Paper presented at the honorary degree ceremony at the Catholic University of Louvain, 14 November 1985.

1986–3. "Chemistry and Iatrochemistry in Early Eighteenth-Century Portugal: A Spanish Connection," in *Historia e Desenvolvimento da Ciência em Portugal: I Colóquio—até ao Sécolo XX. Lisboa, 15 a 19 Abril de 1985,* 2 vols. (Lisboa: Academia das Ciências de Lisboa, 1986), 2:1245–1262.

1986–4. "The History of Science Today," *Technologia* 9 (1986): 133–146. Slightly altered version of second Coimbra lecture (see 1984–1) presented in Brussels 12 November 1985. Includes abstracts in English, French, and Dutch.

1986–5. "Chemistry and the Universities in the Seventeenth Century," *Academiae Analecta: Klasse der Wetenschappen* 48 (1986): 13–33. Paper presented at the Catholic University of Louvain, 13 November 1985. Reprinted in *Estudos Avançadas* [Brazil] 4 (1990): 173–196.

1986–6. "Dr. Allen G. Debus: Collector of Pioneer Recording Artists." Interview by Michael G. Corenthal published in his book, *The Iconography of Recorded Sound, 1886–1986* (Milwaukee: Yesterday's Memories, 1986), pp. 110–121.

1987–1. *Chemistry, Alchemy and the New Philosophy, 1550–1700* (London: Variorum Reprints, 1987). Collection of previously published papers (1961-1, 1962-3, 1964-4, 1964-5, 1967-4, 1968-6, 1972-3, 1973-7, 1974-2, 1979-3, 1981-1, 1982-3, 1984-3).

1987–2. "Myth, Allegory and Scientific Truth: An Alchemical Tradition in the Period of the Scientific Revolution," *Nouvelles de la Republique des Lettres,* Part 1 (1987): 13–35.

1987–3. "The Paracelsian Chemical Philosophy," in *Science and Culture in the Western Tradition: Sources and Interpretations*, ed. John G. Burke (Scottsdale, Arizona: Gorsuch Scarisbrick, Publishers, 1987), pp. 55–57. Extract from *Man and Nature in the Renaissance* (see 1978–1).

1988–1. *Wenyi fuxing shiqi di ren yu ziran,* trans. Liu Yuan and Wu Zhong (Hangzhou Provincial Press, 1988). Chinese translation of *Man and Nature in the Renaissance* (see 1978–1).

1988–2. *Hermeticism and the Renaissance: Intellectual History and the Occult in Early Modern Europe,* ed. Ingrid Merkel and Allen G. Debus (Washington, D.C.: The Folger Shakespeare Library/London and Toronto: Associated University Presses, 1988). Includes "Introduction" by the editors, pp. 7–13, and Allen G. Debus, "Alchemy in an Age of Reason: The Chemical Philosophers in Early Eighteenth Century France," pp. 231–250.

1988–3. "The 'Geber' Tradition in Western Alchemy and Chemistry," *Journal of the History of Arabic Science* 8 (1984; printed 1988): 3–29.

1988–4. "The Chemical Philosophy and the Scientific Revolution," in *Revolutions in Science: Their Meaning and Relevance,* ed. William R. Shea (Canton, Mass.: Science History Publications, 1988), pp. 27–48.

1988–5. "The 1987 Dexter Address: Quantification and Medical Motivation: Factors in the Interpretation of Early Modern Chemistry," *Bulletin for the History of Chemistry: Division of the History of Chemistry of the American Chemical Society* 1 (1988): 4–5. Condensed version of Address presented 1 September 1987.

1989–1. Brian Rust and Allen G. Debus, *The Complete Entertainment Discography From 1897 to 1942,* 2d ed. (New York: Da Capo Press, 1989). A greatly expanded revision of 1973–1.

1989–2. "Quantification and Medical Motivation: Factors in the Interpretation of Early Modern Chemistry," *Pharmacy in History* 31 (1989): 3–11. Complete text of the Dexter Award Address presented at the ACS meeting in New Orleans 1 September 1987 (for an earlier condensed version of the address, see above 1988–5). Reprinted in the *Journal of Central China Normal University* (Natural Sciences) 23 (1989): 137–148 (with summary in Chinese).

1989–3. "Islamic Scientific History: Where to Next?" *Bulletin of Islamic Medicine, 3, Proceedings of the Third International Conference on Islamic Medicine, Istanbul, 18 September to 2 October 1984,* ed. Ali Al-Saif, Ahmed Ragai El-Gindy, and Hakeem Mohammed Zahooral Hasan (Kuwait: n.p, 1989), pp. 51–57.

1990–1. "Iatrochemistry and the Chemical Revolution," in *Alchemy Revisited: Proceedings of the International Conference on the History of Alchemy at the University of Groningen 17–19 April 1989,* ed. Z. R. W. M. van Martels (Leiden/New York/-Copenhagen/Cologne: E. J. Brill, 1990), pp. 51–66.

1990–2. "The English Music Hall Performer and American Vaudeville," *The Hillandale News,* no. 176 (October 1990), pp. 99–111.

1991–1. *The French Paracelsians: The Chemical Challenge to Medical and Scientific Tradition in Early Modern France* (Cambridge: Cambridge University Press, 1991).

1991–2. "Drugstore Cabaret: Reflections of Pharmacy in Popular Music"—descriptive notes by Allen G. Debus (Madison: American Institute of the History of Pharmacy, 1991). Thirty-two page booklet printed to accompany cassette tape by Max Morath introduced at the annual meeting of the American Pharmaceutical Association, New Orleans, 9–12 March 1991.

1991–3. "A Cienca e as Humanidades: A Função renovadora da Indação histórica," *Revista da Sociedade brasileira de História da Ciencia,* No. 5 (Janeiro-Junho 1991): 3–13.

1991–4. "Allen G. Debus: Devolta à Revolução quimica," *Ciência Hoje* 12 (1991): 60–66. Entrevista concendida a Carlos A. L. Figueiras e Roberto Barros de Carvalho.

1991–5. "A Longa Revolução qímica," *Ciência Hoje* 13 (1991): 34–43. Given at the University of Campinas, Brazil, 28 August 1990, based on 1990–1.

1991–6. "Afterword: Alchemy and the History of Science," *Cauda Pavonis,* n.s. 10 (1) (1991): 13–14. Commentary presented at session on Renaissance Hermeticism at Modern Language Association meeting, Chicago, December 1990.

1992–1. "La Chimica nell prima Etá moderna," in *Storia della Scienza*, vol. 3, *Natura e Vita,* ed. Ferdinando Abbri and Renato Mazzolini (Milan: Elemond Editori Associati, 1992; reprinted Turin: Giulio Einaudi Editori, 1993), pp. 428–477.

1992–2. "Alchemy and Iatrochemistry: Persistent Traditions in the Seventeenth and Eighteenth Centuries," *Química nova* 15 (1992): 262–268. Paper presented at Ribeirão Preto, Brazil, 22 August 1991.

1992–3. "History of Chemistry: Key to Modern Science," *Atti del IV Convegno nazionali di Storia e Fondamenti della Chimica: Venezia, 7–9 novembre 1991,* estratto dal "Rendiconti della Accademia nazionali dell Scienze detta dei XL," *Memorie di Scienze fisiche e naturali,* ser. V, 16/2 (1992): 117–131.

1992–4. "The Chemical World View in the Twentieth Century," *História e Desenvolvimento da Ciência em Portugal no Séc. XX: Lisboa, 13 a 17 de novembro de 1989,* 3 vols. (Lisboa: Publicações do II Centenário da Academia das Ciêcias de Lisboa, 1993), 3:1393–1402.

1992–5. "Science and History: The Birth of a New Field" and "The Scientific Revolution: A Chemist's Reappraisal" [Lectures 1 and 4 from *Science and History: A Chemist's Appraisal* (1984-1)], in *Science, Pseudo-Science, and Utopianism in Early Modern Thought,* ed. Stephen McKnight (Columbia, Mo., and London: University of Missouri Press, 1992), pp. 1–36, 37–54. Texts slightly altered from first edition.

1993–1. "Paracelsus and the Medical Revolution of the Renaissance: A Five-Hundredth Anniversary Celebration," in *Paracelsus, Five Hundred Years: Three American Exhibits* (Bethesda, Md.: Friends of the National Library of Medicine, Inc., for the Hahnemann University Library, the National Library of Medicine, and the Washington University School of Medicine, 1993). Descriptive booklet and paper (16 pp.) prepared for these exhibits.

Currently I have in press a Japanese translation of *The Chemical Philosophy* (1977–1) to be published by Heibonsha Ltd. (Tokyo) in 1996. *Man and Nature in the Renaissance* (1978–1) is being tranlated in to Greek by the Crete University Press. Four lectures on "Paracelsus and the Paracelsian Tradition" that were given at the Istituto italiano per gli Studi Filosofici in Naples in 1990 are currently being translated into Italian for publication in 1996 as a monograph. "Chemical Medicine in Early Modern Europe" has been prepared for a three-volume history of medicine being edited by Mirko D. Grmek and Bernardino Fantini. A number of papers dealing with Paracelsus and the chemical philosophy, delivered to audiences in Leiden, Glasgow, Ouro Preto and São Paulo (Brazil), and Coibra and Lisbon (Portugal), are currently in press. I have begun a major project tentatively titled "Chemistry and Medical Debate: 1650–1750."

NOTES ON CONTRIBUTORS

MARTHA BALDWIN is currently post-doctoral fellow and teaching assistant in the Department of the History of Science, Harvard University. Her published articles on early modern medicine have appeared in *Isis, Bulletin for the History of Medicine,* and *Ambix.*

I. BERNARD COHEN is the Victor S. Thomas Professor (emeritus) of the History of Science, Harvard University, and the author of numerous books, articles, and editions of works on the Scientific Revolution. His most recent book is *Science and the Founding Fathers,* a study of the role of science in the political thought of Jefferson, Franklin, Adams, and Madison.

ALLEN G. DEBUS is Morris Fishbein Professor Emeritus of the History of Science and Medicine at the University of Chicago. He has worked extensively on chemical aspects of the Scientific Revolution and has recently published *The French Paracelsians.* He is currently working on a major project entitled, "Chemistry and Medical Debate: 1650–1750." In 1994 he was the recipient of the Sarton Medal of the History of Science Society.

TERENCE D. MURPHY teaches European history at The American University of Paris. His research focuses on the French medical profession under the Old Regime. In a series of studies on the social foundations of medical knowledge, he has investigated the transformation of traditional medical culture within the context of the development and decline of the absolute monarchy.

KAREN HUNGER PARSHALL is associate professor of history and of mathematics at the University of Virginia, Charlottesville. She has completed a book with David E. Rowe, entitled *The Emergence of the American Mathematical Research Community, 1876–1900: J. J. Sylvester, Felix Klein, and E. H. Moore* (1994), and is preparing a volume of selected correspondence with historical and mathematical commentary of the British mathematician, James Joseph Sylvester.

ROBERT J. RICHARDS is professor of history, philosophy, biology, and of the conceptual foundations of science at the University of Chicago. He has recently published *The Meaning of Evolution: The Morphological Construction and Ideological Reconstruction of Darwin's Theory.*

PAUL THEERMAN is associate archivist, Smithsonian Institution Archives. He studies the nineteenth-century popularization of science, especially in Britain, and has recently co-edited, with Adele Seeff, *Action and Reaction,* a series of articles assessing the ideas and influence of Isaac Newton.

MICHAEL T. WALTON received his Ph.D. in history from the University of Chicago. He has published and presented papers on medicine in fifteenth-century London, the influence of the rediscovery of Hebrew on sixteenth-century science, chemistry in the seventeenth century, and Christian-Jewish relations in the sixteenth century. He is presently an independent scholar and president of Walton Marine in Salt Lake City, Utah.

After doing graduate work in the history of science, PHYLLIS J. WALTON received a law degree from the University of Utah. She has, with her husband, worked in the areas of Christian-Jewish relations in the sixteenth century, legal history in the fifteenth and sixteenth centuries, and the influence of Hebrew study on the development of science. She is active in the practice of law and in business.

KATHLEEN WELLMAN is associate professor in the department of history at Southern Methodist University. Her book, *La Mettrie: Medicine, Philosophy, and Enlightenment,* was published by Duke University Press in 1992.

Index

The University of Western Ontario Series in Philosophy of Science

1. J. Leach, R. Butts and G. Pearce (eds.): *Science, Decision and Value.* 1973
ISBN 90-277-0239-X; Pb 90-277-0327-2
2. C.A. Hooker (ed.): *Contemporary Research in the Foundations and Philosophy of Quantum Theory.* 1973
ISBN 90-277-0271-3; Pb 90-277-0338-8
3. J. Bub: *The Interpretation of Quantum Mechanics.* 1974
ISBN 90-277-0465-1; Pb 90-277-0466-X
4. D. Hockney, W. Harper and B. Freed (eds.): *Contemporary Research in Philosophical Logic and Linguistic Semantics.* 1975
ISBN 90-277-0511-9; Pb 90-277-0512-7
5. C.A. Hooker (ed.): *The Logico-algebraic Approach to Quantum Mechanics.*
Vol. I: Historical Evolution. 1975 ISBN 90-277-0567-4; Pb 90-277-0613-1
Vol. II: Contemporary Consolidation. 1979
ISBN 90-277-0707-3; Pb 90-277-0709-X
6. W.L. Harper and C.A. Hooker (eds.): *Foundations of Probability Theory, Statistical Inference, and Statistical Theories of Science.*
Vol. I: Foundations and Philosophy of Epistemic Applications of Probability Theory. 1976 ISBN 90-277-0616-6; Pb 90-277-0617-4
Vol. II: Foundations and Philosophy of Statistical Inference. 1976
ISBN 90-277-0618-2; Pb 90-277-0619-0
Vol. III: Foundations and Philosophy of Statistical Theories in the Physical Sciences. 1976 ISBN 90-277-0620-4; Pb 90-277-0621-2
7. C.A. Hooker (ed.): *Physical Theory as Logico-operational Structure.* 1979
ISBN 90-277-0711-1
8. J.M. Nicholas (ed.): *Images, Perception, and Knowledge.* 1977
ISBN 90-277-0782-0
9. R.E. Butts and J. Hintikka (eds.): *Logic, Foundations of Mathematics, and Computability Theory.* Part One: Logic, Foundations of Mathematics, and Computability Theory. 1977 ISBN 90-277-0708-1
10. R.E. Butts and J. Hintikka (eds.): *Logic, Foundations of Mathematics, and Computating Theory.* Part Two: Foundational Problems in the Special Sciences. 1977 ISBN 90-277-0710-3
11. R.E. Butts and J. Hintikka (eds.): *Logic, Foundations of Mathematics, and Computability Theory.* Part Three: Basic Problems in Methodology and Linguistics. 1977 ISBN 90-277-0829-0
12. R.E. Butts and J. Hintikka (eds.): *Logic, Foundations of Mathematics, and Computability Theory.* Part Four: Historical and Philosophical Dimensions of Logic, Methodology and Philosophy of Science. 1977 ISBN 90-277-0831-2
Set (9-12) ISBN 90-277-0706-5
13. C.A. Hooker J.J. Leach and E.F. McClennen (eds.): *Foundations and Applications of Decision Theory.*
Vol. I: Theoretical Foundations. 1978 ISBN 90-277-0842-8
Vol. II: Epistemic and Social Applications. 1978 ISBN 90-277-0844-4
14. R.E. Butts and J.C. Pitt (eds.): *New Perspectives on Galileo.* 1978
ISBN 90-277-0859-2; Pb 90-277-0891-6

The University of Western Ontario Series in Philosophy of Science

15. W.L. Harper, R. Stalnaker and G. Pearce (eds.): *Ifs. Conditionals, Belief, Decision, Chance, and Time.* 1981 ISBN 90-277-1184-4; Pb 90-277-1220-4
16. J.C. Pitt (ed.): *Philosophy in Economics.* 1981 ISBN 90-277-1210-7; Pb 90-277-1242-5
17. M. Ruse: *Is Science Sexist?* And Other Problems in the Biomedical Sciences. 1981 ISBN 90-277-1249-2; Pb 90-277-1250-6
18. N. Rescher: *Leibniz's Metaphysics of Nature.* A Group of Essays. 1981 ISBN 90-277-1252-2; Pb 277-1253-0
19. L. Laudan: *Science and Hypothesis.* Historical Essays on Scientific Methodology. 1981 ISBN 90-277-1315-4
20. W.R. Shea (ed.): *Nature Mathematized.* Historical and Philosophical Case Studies in Classical Modern Natural Philosophy. 1983 ISBN 90-277-1402-9
21. M. Ruse (ed.): *Nature Animated.* Historical and Philosophical Case Studies in Greek Medicine, 19th-Century and Recent Biology, Psychiatry, and Psychoanalysis. 1983 ISBN 90-277-1403-7
22. W.R. Shea (ed.): *Otto Hahn and the Rise of Nuclear Physics.* 1983 ISBN 90-277-1584-X
23. H.F. Cohen: *Quantifying Music.* The Science of Music at the First Stage of the Scientific Revolution (1580-1650). 1984 ISBN 90-277-1637-4
24. R.E. Butts: *Kant and the Double Government Methodology.* Supersensibility and Method in Kant's Philosophy of Science. 1984 ISBN 90-277-1760-5; Pb 90-277-2384-2
25. J.R. Brown (ed.): *Scientific Rationality.* The Sociological Turn. 1984 ISBN 90-277-1812-1
26. F. Wilson: *Explanation, Causation and Deduction.* 1985 ISBN 90-277-1856-3
27. J.C. Pitt (ed.): *Change and Progress in Modern Science.* 1985 ISBN 90-277-1898-9
28. H.B. Hollinger and M.J. Zenzen: *The Nature of Irreversibility.* A Study of Its Dynamics and Physical Origins. 1985 ISBN 90-277-2080-0
29. K. Okruhlik and J.R. Brown (eds.): *The Natural Philosophy of Leibniz.* 1985 ISBN 90-277-2145-9
30. G. Oddie: *Likeness to Truth.* 1986 ISBN 90-277-2238-2
31. F. Wilson: *Laws and Other Worlds.* A Human Account of Laws and Counterfactuals. 1986 ISBN 277-2232-3
32. J. Earman: *A Primer on Determinism.* 1986 ISBN 90-277-2240-4; Pb 90-277-2241-2
33. R.E. Butts (ed.): *Kant's Philosophy of Physical Science.* Metaphysische Anfangsgründe der Naturwissenschaft (1786-1986). 1986 ISBN 90-277-2309-5; Pb 90-277-2310-9
34. I.B. MacNeill and G.J. Umphrey (eds.): *Advances in the Statistical Sciences.* Festschrift in Honor of Professor V.M. Joshi's 70th Birthday (6 volumes).
Vol. I: Applied Probability, Stochastic Processes, and Sampling Theory. 1987 ISBN 90-277-2393-1
35. I.B. MacNeill and G.J. Umphrey (eds.): *Advances in the Statistical Sciences.* Vol. II: Foundations of Statistical Inference. 1987 ISBN 90-277-2394-X

The University of Western Ontario Series in Philosophy of Science

36. I.B. MacNeill and G.J. Umphrey (eds.): *Advances in the Statistical Sciences.* Vol. III: Time Series and Econometric Modelling. 1987 ISBN 90-277-2395-8
37. I.B. MacNeill and G.J. Umphrey (eds.): *Advances in the Statistical Sciences.* Vol. IV: Stochastic Hydrology. 1987 ISBN 90-277-2396-6
38. I.B. MacNeill and G.J. Umphrey (eds.): *Advances in the Statistical Sciences.* Vol. V: Biostatistics. 1987 ISBN 90-277-2397-4
39. I.B. MacNeill and G.J. Umphrey (eds.): *Advances in the Statistical Sciences.* Vol. VI: Actuarial Science. 1987 ISBN 90-277-2398-2
    Set ISBN (Vols 34-39) 90-277-2399-0
40. N. Rescher: *Scientific Realism.* A Critical Reappraisal. 1987
    ISBN 90-277-2442-3; Pb 90-277-2528-4
41. B. Skyrms and W.L. Harper (eds.): *Causation, Chance, and Credence.* 1988
    ISBN 90-277-2633-7
42. W.L. Harper and B. Skyrms (eds.): *Causation in Decision, Belief Change and Statistics.* 1988 ISBN 90-277-2634-5
43. R.S. Woolhouse (ed.): *Metaphysics and Philosophy of Science in the 17th and 18th Centuries.* Essays in Honor of Gerd Buchdahl. 1988
    ISBN 90-277-2743-0
44. R.E. Butts and J.R. Brown (eds.): *Constructivism and Science.* Essays in Recent German Philosophy. 1989 ISBN 0-7923-0251-6
45. A.D. Irvine (ed.): *Physicalism in Mathematics.* 1989 ISBN 0-7923-0513-2
46. J. van Cleve and R.E. Frederick (eds.): *The Philosophy of Right and Left.* Incongruent Counterparts and the Nature of Space. 1991 ISBN 0-7923-0844-1
47. F. Wilson: *Empiricism and Darwin's Science.* 1991 ISBN 0-7923-1019-5
48. G.G. Brittan, Jr. (ed.): *Causality, Method and Modality.* Essays in Honor of Jules Vuillemin. 1991 ISBN 0-7923-1045-4
49. W. Spohn, B.C. van Fraassen and B. Skyrms (eds.): *Existence and Explanation.* Essays in Honor of Karel Lambert. 1991 ISBN 0-7923-1252-X
50. J.C. Pitt: *Galileo, Human Knowledge, and the Book of Nature.* Method Replaces Metaphysics. 1992 ISBN 0-7923-1510-3
51. V. Coelho (ed.): *Music and Science in the Age of Galileo.* 1992
    ISBN 0-7923-2028-X
52. P. Janich: *Euclid's Heritage: Is Space Three-Dimensional?* 1992
    ISBN 0-7923-2025-5
53. M. Carrier: *The Completeness of Scientific Theories.* On the Derivation of Empirical Indicators within a Theoretical Framework: The Case of Physical Geometry. 1994 ISBN 0-7923-2475-7
54. P. Parrini (ed.): *Kant and Contemporary Epistemology.* 1994
    ISBN 0-7923-2681-4
55. J. Leplin (ed.): *The Creation of Ideas in Physics.* Studies for a Methodology of Theory Construction. 1995 ISBN 0-7923-3461-2
56. J.E. McGuire: *Tradition and Innovation.* Newton's Metaphysics of Nature. 1995 ISBN 0-7923-3617-8

The University of Western Ontario Series in Philosophy of Science

57. R. Clifton (ed.): *Perspectives on Quantum Reality.* Non-Relativistic, Relativistic, and Field-Theoretic. 1996 ISBN 0-7923-3812-X
58. P.H. Theerman and K. Hunger Parshall (eds.): *Experiencing Nature.* Proceedings of a Conference in Honor of Allen G. Debus. 1997 ISBN 0-7923-4477-4

KLUWER ACADEMIC PUBLISHERS – DORDRECHT / BOSTON / LONDON